For Chantal

Tourist Destinations: Structure and Synthesis

Douglas G. Pearce

Emeritus Professor of Tourism Management, Victoria University of Wellington

CABI

CABI is a trading name of CAB International

CABI
Nosworthy Way
Wallingford
Oxfordshire OX10 8DE
UK

Tel: +44 (0)1491 832111
Fax: +44 (0)1491 833508
E-mail: info@cabi.org
Website: www.cabi.org

CABI
WeWork
One Lincoln St
24th Floor
Boston, MA 02111
USA

Tel +1 (617)682-9015
E-mail: cabi-nao@cabi.org

A catalogue record for this book is available from the British Library, London, UK.

References to Internet websites (URLs) were accurate at the time of writing.

Library of Congress Cataloging-in-Publication Data

Names: Pearce, Douglas G., 1949- author.
Title: Tourist destinations : structure and synthesis / Douglas G Pearce.
Description: Oxfordshire ; Boston, MA : CABI, [2021] | Includes bibliographical references and index. | Summary: "Destination research today has become increasingly fragmented as studies become more specialized. This book provides a systematic examination and combines spatial and organizational structure across destinations from the national to the local scale to produce a comprehensive understanding of destinations and a complete integrative framework of them"-- Provided by publisher.
Identifiers: LCCN 2020023970 (print) | LCCN 2020023971 (ebook) | ISBN 9781789245837 (hardback) | ISBN 9781789245844 (ebook) | ISBN 9781789245851 (epub)
Subjects: LCSH: Tourism--Planning. | Tourism--Marketing. | Place marketing.
Classification: LCC G155.A1 P3598 2021 (print) | LCC G155.A1 (ebook) | DDC 910.68/8--dc23
LC record available at https://lccn.loc.gov/2020023970
LC ebook record available at https://lccn.loc.gov/2020023971

ISBN-13: 9781789245837 (hardback)
9781789245844 (ePDF)
9781789245851 (ePub)

Commissioning Editor: Claire Parfitt
Editorial Assistant: Emma McCann
Production Editor: Shankari Wilford

Typeset by SPi, Pondocherry, India
Printed and bound in the USA by Integrated Books International, Dulles, Virginia

Contents

Figures

Tables

1 Introduction: the Spatial and Organizational Structure of Destinations

Destinations have long been a key feature of tourism and the focus of much tourism research. Most tourism research prior to the Second World War concerned places that were attracting visitors on holiday and that were becoming recognized as different types of places and forms of land use. The post-war growth in tourist travel saw such places multiply and the amount of tourism research increase. These two trends gave rise to more and more research on destinations from an ever-widening range of disciplines as their importance was acknowledged, not just as places visited by and increasingly developed for tourists but as places experiencing often significant changes brought about by the impacts occasioned by those visits and that development. These characteristics have led to destinations being studied in different ways and from a variety of perspectives with the result that destination research has become increasingly fragmented as studies have become progressively more specialized. No broad consensus of a definition of destinations exists, myriad concepts are used and multiple approaches to destination research are taken. Multiple definitions, concepts, perspectives and approaches can be confusing but they can also shed new light on a common topic by examining it from different angles and in different ways. What is needed now is a more integrative approach, one which systematically draws together different research threads to provide a more comprehensive and coherent picture and fuller understanding of destinations, their structure and how they function. The goal of this book is to provide such a synthesis.

Synthesizing a large and disparate body of research on destinations is, however, an ambitious and challenging task. Focus and selectivity are needed to keep it manageable and to achieve an appropriate balance between the levels of generalization and detail. To that end this book focuses on the structure of destinations and draws together two approaches; one dealing with the spatial structure of destinations, the other with their organizational structure. Each of these approaches provides the scope to consider a broad range of destination features. The spatial structure of destinations concerns the physical location, distribution, configuration and interconnectedness of products, services and actors and the factors which underlie the resultant patterns of these. Organizational structure, as used here, refers to the make-up or composition of destinations and the various configurations and ways in which and the extent to which multiple actors, collectively and individually, behave and interact to produce the experiences sought by tourists. Examining the organizational structure of destinations thus provides an opportunity to synthesize the multiplicity of actors involved, not just thus firms or public- and third-sector organizations, conventionally the subject of organization research, but also individuals, including the tourists themselves. In general, the streams of research on the spatial and organizational structure of destinations are largely non-convergent and only occasionally intersect explicitly. Drawing the two approaches and their associated findings together thus provides a means of developing a more comprehensive picture of the overall structure of destinations and new insights into how they function.

Developing a sound understanding of the structure of destinations is important for research, for destination management and marketing and for tourism planning and policy making. How we conceive the structure of destinations influences the problems we address, shapes our research design and methodology, and conditions the interpretation of our findings. In order to know where and how to act or intervene to achieve specific management and policy objectives or to address particular problems, at a firm or destination level, we need to understand a destination's make-up and how it functions. A better understanding of destination structure will enable firms to identify their place within that structure more clearly and thereby to operate and compete more effectively. Understanding destinations, what they offer and how they are structured is also important for travel intermediaries in the markets so that they can serve their customers well.

Within this focus on destination structure different levels and forms of synthesis are required. Systematic juxtaposition and discussion of concepts, approaches and findings can more readily enable commonalities to be identified and gaps in knowledge to be established. Underlying conceptual and theoretical frameworks need to be compared along with the different methodologies employed. Most destination research is framed in terms of a single concept. Where other frameworks are referenced it is often their shortcomings which are emphasized. There is, however, much to be gained by being less dismissive of what is different and putting more emphasis on common or complementary features. Considerable scope exists to integrate conceptual and empirical research: conceptualizations of destinations are frequently not tested by empirical applications; much empirical work is not underpinned by any explicit theorization. Issues of supply and demand are often treated separately and need to be drawn together more closely in order to obtain a more comprehensive picture. Each of these also requires more complete coverage and greater integration. Many demand studies deal with either domestic tourists or international tourists but not both. Specialized studies of accommodation and attractions need to be set alongside each other so as to identify the impact each may have on the other. Research on hotels needs to be extended to other forms of accommodation and that on airlines to other modes of transport. Much destination research continues to focus on particular types of destination; for example, urban or rural, coastal or alpine, with little consideration given to any common features they may share or discussion of what it is that distinguishes one type from another. Likewise, research on local destinations is generally not connected to what is happening at regional and national levels even though these higher level destinations are composed of sets of the former and local patterns and processes may be influenced by the latter. Consequently, the structures of national, regional and local destinations must be systematically considered and the relationships between them examined.

A particular challenge in synthesizing destination research is that most of the literature consists of case studies of individual destinations, and then of only specific aspects of them. These case studies generally provide the scope to examine in some detail local, regional and national characteristics and the influence of context on structure. Some, such as those dealing with resort morphologies, attempt a fairly comprehensive approach but many focus on specific aspects of destination structure, such as studies of hotel location (Baum and Haveman, 1997; Tsang and Yip, 2009) or tourist movements (Bauder and Freitag, 2015; de Cantis *et al.*, 2016). Comparative case studies which take a common approach and use comparable data to examine multiple destinations are especially helpful in identifying and accounting for differences and similarities and developing generalizations about destination structure (Bocquet, 2008; García-Palomares *et al.*, 2015; Favre-Bonté *et al.*, 2016; Pulido-Fernández and Merinero-Rodríguez, 2018).

Systematic reviews of case study material can provide many insights into the structure and functioning of destinations but the search for generalities in patterns and the factors underlying them is inevitably constrained by issues of equivalence, whether in terms of definitions, data sources or means of analysis (Pearce, 1993a). These issues and the impact of contextual factors need to be weighed up and taken into account. Assessment of the extent to which the experience of particular destinations reflects broader structures is enhanced when case studies, as well as comparative research, are explicitly framed with regard to a particular concept or theory, such as clusters (Varisco, 2004; Perles-Ribes *et al*., 2015). In comparative research and attempts to take a more generalized view of the structure of destinations from case studies it is also helpful to use generic terms rather than referring solely to the named features of specific destinations as is the case with many ideographic studies. In terms of morphological studies, for example, it is useful to think of specific physical elements not just by their proper names (e.g. the Eiffel Tower, the Hilton Hotel or Malaga airport) but as a type of element and the role it plays in a destination (respectively an attraction, an upmarket hotel, a transport node). Likewise, in organizational research it is helpful to identify actors in such generic terms as destination management organizations (DMOs), non-governmental organizations (NGOs), local government, hotel chains and local accommodation providers in addition to their proper names.

These considerations have shaped the structure of this book. To set the subsequent examination of the spatial and organizational structure of destinations in context, Chapter 2 outlines different perspectives on what constitutes a destination: the evolving use of the term; formally designated destinations; destinations as functional nodes; and the tourists' perspective. Chapter 3 then provides a fuller conceptual foundation by reviewing a set of the more commonly used frameworks in destination research. Given the diversity of approaches and sources of data used by destination researchers, Chapter 4 systematically addresses core methodological issues which arise in analysing the structure of destinations so as to allow a more fluid discussion of patterns, processes and explanation in succeeding chapters and also provide a toolkit which might be used in future research on the topic. More specific techniques and methodological issues, however, are dealt with as they occur in those chapters. Although some common threads run though research undertaken at different scales, national, regional and local destinations tend to be treated as sub-fields in the literature. That approach is initially retained here but with a view to subsequently bringing scale matters together in Chapter 10. Chapter 5 and Chapter 6 deal respectively with the structure of destinations at a national and a regional scale. Local level destinations are much more numerous and consequently considerably more research has been carried out at this scale. Urban destinations are examined in Chapter 7, coastal resorts in Chapter 8 and ski resorts and rural and natural area destinations in Chapter 9. In these five chapters, discussion of issues related to the spatial structure of supply and demand are followed by those pertaining to organizational structures and finally to any links between them. This common format, together with the preceding discussion of perspectives, concepts and methodological issues, more readily enables a systematic overview of the material presented and general points to be identified and integrated in Chapter 10.

Broad though this coverage is it is not exhaustive and the limitations of the literature reviewed and the examples used need to be taken into account in the synthesis provided in Chapter 10, which is based on the material presented in the preceding chapters. Although a wide range of international examples are cited there is a greater emphasis on developed Western destinations than those in developing countries. Examples of island destinations with their distinctive contextual characteristics are included in Chapters 5 and 6 but not treated separately in depth. It is also recognized that the boundaries between types of destination are not always hard and fast. At some point, for example, coastal resorts may evolve into cities and other cities may have elements of a coastal destination. The approach taken is not to get bogged down in semantics but rather to concentrate on the dominant characteristics of the examples being used. The frequent use of French and Spanish examples reflects the fact that France and Spain are among the leading destinations worldwide in terms of international visitor arrivals and thus a considerable amount of research has been undertaken there, including some of the cutting-edge work on destinations. It is also a function of the author's familiarity with these destinations and languages (all translations are those of the author), factors which also have enabled the inclusion of a number of Latin American examples. In other instances, such as the rapidly expanding body of work from China, the material reviewed is that which appears in English. The literatures reviewed and the discussion and interpretation of them will also have been inevitably shaped by the author's disciplinary background, first as a tourism geographer, later as a professor of tourism management.

The remainder of this chapter introduces the core themes of spatial and organizational structure.

Spatial Structure

Spatial structure has been defined as 'the manner in which space is organized by the cumulative locations of infrastructure, economic activities and their relations' (Rodrigue *et al.*, 2006, p. 272) and has come to refer to 'any more or less ordered spatial arrangement, assemblage or system: the orderings through which space is implicated in the operation and outcome of social and biophysical processes' (Sheppard and Gregory, 2009, p. 715). Those orderings function in two ways: spatial processes lead to arrangements in space and conversely spatial structure shapes socio-economic processes (Rogers *et al.*, 2013). The concern here is with how different tourism phenomena are arranged in space, how and why those arrangements have come about and how knowledge of that arrangement helps us to understand what destinations are and how they function. Three fundamental and interrelated attributes of spatial structure are distribution, location and spatial interaction.

Distribution refers to the spread of phenomena throughout an area, region or country. A prime concern is with establishing whether the phenomena in question are distributed uniformly or not and in identifying the factors or processes which give rise to the patterns identified. Patterns of distribution can be characterized in such terms as concentration, dispersion and clustering.

Location is used here as a relative notion. In a destination context the location of accommodation can be considered in terms of accessibility to points of interest, transport convenience and the surrounding environment (Bull, 1994; Yang *et al.*, 2018). Bull (1994, p. 11) observed:

> The location of any lodging property includes more than one component. At the very least, two factors are involved:
>
> distance from, or access to, one or more specific places, such as a beach, city centre, airport or highway;
> neighbourhood amenity or quality, such as quietness, views from rooms, nature of the property's surroundings.

As a result trade-offs may have to be made: '[T]here may be specific advantages which one location has in access distance from, say, an attraction or city centre; trade-offs in access distances between two or more attractions, and inherent advantages from specific neighbourhood characteristics' (Bull, 1994, p. 10). Similarly, the location of an attraction might be considered in terms of proximity to other attractions, accommodation and ease of access.

Spatial interaction concerns the ways in which places or particular objects within a place are interconnected by tangible or non-tangible flows such as those involving tourists, money and information. In general, such flows are subject to the friction of distance and a distance decay effect occurs. The volume of tourists, for example, normally decreases with distance away from the origin as costs in terms of time, money and effort increase and awareness of what places have to offer decreases (Greer and Wall, 1979; Pearce, 1995a). The effect of distance on the tourist flows is moderated by the attributes of the places between which interactions occur, such as population size or the range and quality of attractions on offer. Moreover, as Greer and Wall (1979, p. 230) note, the potential supply of recreational and vacational opportunities will increase geometrically with distance from the origin 'as each successive unit of distance give access to increasingly larger areas of land.' However, '[R]ather than simply being about movement per se, spatial interaction seeks to place such movement in context, and to recognize the interdependence of interactions; that is, flow has consequences at origin, destination, and along the route' (Rogers *et al.*, 2013).

With tourist flows two major and distinct forms of interaction occur: origin–destination interactions and circuit travel which involves interaction between two or more destinations along the route as well as with the origin (Pearce, 1995a). These two types of flows are closely related to different motivations for travel. Gray (1970) identified two basic reasons for pleasure travel: wanderlust and sunlust. He saw wanderlust as that trait which causes people to seek out and temporarily exchange the ordinary for the exotic while sunlust depends on the existence elsewhere of different or better amenities or resources for a particular purpose, eponymously a sunny climate but also other features such as better skiing conditions. While this is to take a rather stereotyped view and later research provides a much more nuanced analysis of tourist motivations, Gray's basic typology does draw attention to the need to consider why and how people are travelling. Sunlust travel often involves travel to a single destination whereas much circuit travel, especially that involving longer tours, reflects forms of wanderlust. Places which constitute nodes on a circuit may have different functions, often more than one. Some, for example, may act as a gateway and primary destination; others may have a role as a secondary destination or as a hub from which side-trips are made.

Scale

The scale of analysis influences the ways in which both the spatial and organizational structure of destinations is examined. Scales are constructs, 'not fixed, separate levels of the social world' (Paasi, 2004, p. 542). As such, scale is a rather fluid concept that can be viewed in different ways and from different perspectives: in terms of areal extent (size), place in a hierarchy (level) or as relation (Howitt, 1998). Scale as size and level is commonly used as an ordering device, a way to structure and make sense of the world. Scale as relation is particularly important from a systems perspective. Two types of relational approaches structured around scale might be identified. In the first, the object of analysis and/or the way it is studied is seen to vary from one scale to another in a nested or hierarchical fashion. When the focus is on local level destinations such as coastal or ski resorts the emphasis is on their internal structure. When the focus shifts to a regional or national level the resorts become components of the larger destination and the resorts are analysed in more general terms. In the second relational approach the emphasis is on explicitly investigating and explaining the linkages between phenomena at different scales, with a focus on the strength, direction, nature and outcomes of relationships (Pearce, 1995a). More generally, Molloy *et al.* (2011, p. 582) argue: 'Any level within a hierarchical level cannot be understood in isolation because it is shaped by – and in turn changes – other system levels . . . Therefore it is imperative to attend to whole–part relationships.' This is especially the case when analysing nodal functions. More generally, the analysis of hierarchical relationships, as well as horizontal ones, requires decisions about what sorts of relationships are to be examined and how they are to be measured.

Boundaries

Questions of scale in turn give rise to the vexed issue of destination boundaries. Whether the concern is with delimiting the size and extent of a particular destination or differentiating between destinations horizontally or vertically, boundaries need to be drawn. What boundaries delimit the destination and how are these determined? Where, for example, does the local end and the regional begin? Most commonly, the scale of destinations is depicted according to administrative or institutional limits. Between the lower local/municipal/district/county level and the nation state a number of intermediate divisions exist, such as provinces, regions, states and autonomous communities. Analysing destinations in this framework may be very appropriate from a planning and administrative perspective as the relevant agencies and organizations are frequently, but not necessarily, structured in this way (Pearce, 1992, 2015a; Pearce *et al.*, 2017). Moreover, the most readily available destination data are generally disseminated in such units. This applies not only to tourism data but also to data on other sectors which may be needed to characterize and contextualize the destination. Tourists, however, do not necessarily see and experience destinations in terms of conventional administrative and political boundaries. Destinations as viewed by tourists might cut across administrative boundaries (Blasco *et al.*, 2014; Saarinen, 2014); differ from planners' concepts (Ashworth and de Haan, 1987); or vary according to the context (Pearce and Schänzel, 2013). Consideration therefore also needs to be given to how destinations might best be delimited from the perspective of different stakeholders.

Hernández-Martín *et al.* (2016, p. 777) suggest that 'Tourism territories ... very rarely present clear boundaries due to their progressive evolution. It is normal to find a juxtaposition or superposition of land uses.' Homogeneity of tourist typologies and tourism supply is one of six criteria that they used to operationalize micro-destinations in the Canary Islands. More often, tourism is concentrated in particular locations within administrative units at a range of scales. Barrado Timón (2004) sees such sectoral concentrations as destinations within larger territories. Saarinen (2014, p. 48) points out that 'Generalization of ... "tourist hotspots" to a wider administrative regional unit may be misleading in development work and strategies.' On the other hand, dispersed tourist activities and facilities might be linked in such a way that they collectively function as a destination. The creation of the Mont Blanc 'ultra-trail', an extreme sports event, crosses 19 communes in three countries (France, Italy and Switzerland) and is contributing to the development of a new regional identity (Bessy, 2016).

Regions

Formal regions are characterized in terms of definable criteria which are distributed in a uniform fashion. They are 'areas of likeness or similarity' (Symanski and Newman, 1973, p. 350). However, as Dauphiné (1979, p. 39) cautions: '[T]he concept of spatial homogeneity is not absolute but relative: spatial homogeneity depends on the scale of observation.' Sub-regional differences, for example, may be revealed in what appears to be a uniform region when a finer lens of analysis is applied (Merinero-Rodríguez *et al.*, 2014). Nodal regions are based around a node or series of nodes and are defined in terms of the interactions and linkages between these. Both types of region may be examined in terms of pattern and process. 'Pattern is what we describe and identify and process is what we search out to make pattern more meaningful' (Symanski and Newman, 1973, p. 351). Formal regions are commonly studied by mapping and analysing the distribution of tourism-related phenomena such as bednights or hotel rooms (Pearce, 1995a) whereas the analysis of internodal linkages revealed, for example, by tourist flow patterns is the focus of research on nodal regions (Liu *et al.*, 2012; Asero *et al.*, 2016).

Dauphiné (1979) asserted that a region could be decomposed into sub-systems and if all of these were independent then a region did not exist; it was simply an aggregate of the sub-systems. Conversely, if no differences between the systems could be discerned then a state of total regional unity or uniformity existed, something he considered an extreme case. In reality, he contended, interdependence is strong but not complete. He cites the Côte d'Azur as an intermediate case, one where each hinterland area is linked to the coast but not to each other. Dauphiné also drew attention to another critical consideration, that of the dominant element that imposes its mark on other regional characteristics and determines how it is designated. In the case of the Côte d'Azur, Dauphiné (1979, p. 40) argued that 'tourism conditions all the other activities, agriculture, industry, commerce and the spatial structure without, however, being able to speak of a homogenous region.' The question here is does tourism have to dominate before we can speak of a tourism region? This depends on the focus of the research. If our concern is with studying tourism in the context of the broader economy, then the notion of tourism regions being defined as those regions where the sector is dominant is appropriate. However, if the focus is directed specifically at tourism then the characteristics of tourism within a given region might be studied regardless of whether or not the sector is the dominant regional activity, as measured, for example, in terms of jobs or share of regional gross domestic product (GDP). Now tourism regions, or the sub-regions of which they are composed, might be classified in terms of some tourism specific element, such as whether they are predominantly a domestic or international tourism region or whether they are a cultural or a coastal tourism region. This question also applies to local level destinations. We can examine the structure of urban destinations regardless of whether or not tourism is the dominant activity in the city. The extent to which a destination, regional or local, depends on tourism, however, is likely to influence its structure.

Morphology as spatial structure

A significant body of research on the physical form of local-scale destinations emerged in the 1970s and 1980s (Pearce, 1995a; Liu and Wall, 2009). Much of this focused on the morphology of coastal resorts and, to a lesser extent, ski resorts whose distinctive form and growing importance attracted the attention of geographers, especially in Europe. Morphology 'is concerned mainly with the forms and functions of places, the relationships between them, and how they change over time' (Liu and Wall, 2009, p. 34). Morphological studies of urban destinations started to flourish in the 1980s and 1990s added a new dimension, that of destinations in which tourism is a significant activity but not the dominating or defining one of the places labelled resorts (Gordon and Goodall, 2000; Liu and Wall, 2009).

Part of the early interest in the physical form or built environment of destinations might be attributed to the visible and tangible nature of what was being mapped and studied: types of accommodation, attractions, tourist-oriented shops ... But beyond this is the belief that form reflects function; that studying the physical arrangement of land uses reveals the particular character of destinations and the activities that occur there (Wall, 1975; Pearce, 1978). Moreover, there was a well-established, broader tradition of morphological studies in geography, especially relating to urban landscapes (Pearce, 1978; Whitehand, 1992; Liu and Wall, 2009). Many of the early destination studies were not explicitly linked to this broader tradition but the ideas associated with the latter contribute to developing a better understanding of the spatial structures of destinations and how these evolve. Whitehand (1992) attributed differences between landscapes in an urban area to differences inherent in the functions they perform, in the ways in which developments occur and in the roles and functions of the agents of development. To date these different components have not been brought together in a very integrated fashion in the destination literature. Morphological studies tend to be fairly descriptive and to relate either to specific destinations or particular types of destinations, predominantly coastal resorts or cities. There has been little attempt to derive more generalized patterns across destination types (Pearce, 1978, 1995a).

In his essay on urban spatial structure Bourne (1982) provides a useful set of terms and concepts which might be applied systematically to destinations and to frame research on their spatial structure. Not only are many cities major destinations but the terms and concepts he uses recur frequently in research on destinations of many kinds, though usually not in such an explicit and integrated fashion. Bourne's (1982, p. 29) statement that 'interrelationships are undoubtedly the essence of the city and of urban spatial structure' might readily be paraphrased and applied to destinations. Bourne (1982, p. 30) portrays urban spatial structure in terms of three cumulative concepts:

- Form: 'the spatial pattern or arrangement of individual elements – such as buildings and land uses ... as well as social groups, economic activities and public institutions' – within a given area.

- Interaction: 'the underlying set of interrelationships, linkages and flows that acts to "integrate" the pattern and behaviour of individual land uses, groups and activities' into functional entities or sub-systems.
- Spatial structure: the combination of form and patterns of behaviour and interaction within sub-systems brought about by some organizing mechanism which links these together according to a set of 'organizational rules' which may be internal, external or some mix of the two.

In summary, a fairly comprehensive set of concepts and approaches exists to examine the spatial structure of destinations but these are more developed and have been more widely applied in terms of urban destination and resort morphologies than at other scales.

Organizational Structure

Different aspects of the organizational structure of destinations are emphasized depending on the theoretical stance taken. As destinations are generally characterized by the presence of multiple, interdependent organizations the appeal of some form of interorganizational analysis is evident (Selin and Beatson, 1991; Pearce, 1992; Mwesiumo and Halpern, 2019). With interorganizational analysis the focus is not on a single organization but rather 'a number of distinguishable organizations engaged in a significant amount of interaction with each other' (Benson, 1975, p. 230). Two main approaches have emerged, one based on exchange theory as presented in the seminal paper by Levine and White (1961), the second stressing a power dependency approach as expressed in Benson's (1975) paper. Levine and White emphasized voluntary activity, goal-oriented exchange in a broad sense, not just the transfer of material goods. Organizations perceive the chances of attaining their goals in some domains are greater by acting jointly rather than behaving independently. Benson (1975) laid greater stress on the importance of resource acquisition, particularly the acquisition and defence of money and authority. Forms of interorganizational analysis can be used to investigate relationships both within a destination and between destinations, particularly from a hierarchical perspective.

Figure 1.1 presents a framework applied to the interorganizational analysis of territorial tourism organizations at various levels. It is hierarchical in nature and depicts both horizontal and vertical (top-down, bottom-up) relationships between organizations, sets the network of organizations in the broader environment, a sub-environment of which is the tourism environment, and includes a temporal dimension (T_1, T_2, T_3) to acknowledge changes over time (Pearce, 1992). The emphasis is on a comprehensive interorganizational approach to analysing what tourism organizations actually do, a concern expressed by focusing on their goals and functions, and how the relationships between organizations influence and are influenced by these. The links between the destinations are thus represented by the relationships between the organizations responsible for aspects of their management and marketing. At each scale these will be affected by the degree of spatial homogeneity or diversity of each destination as reflected by the scale and economic and social significance of tourism; the composition of the tourism sector; the type of tourism; the pattern of growth; and the type of market. As Paddison (1983) stressed, power is distributed geographically, with the manner in which the function of governing is divided within a state between national and sub-national governments varying along a continuum from those with a markedly centralist system to others which are federal in structure. The structure of destination organizations is heavily influenced by these broader political structures. These organizations seek resources from their respective environments in return for which outputs are provided such as marketing campaigns and development plans.

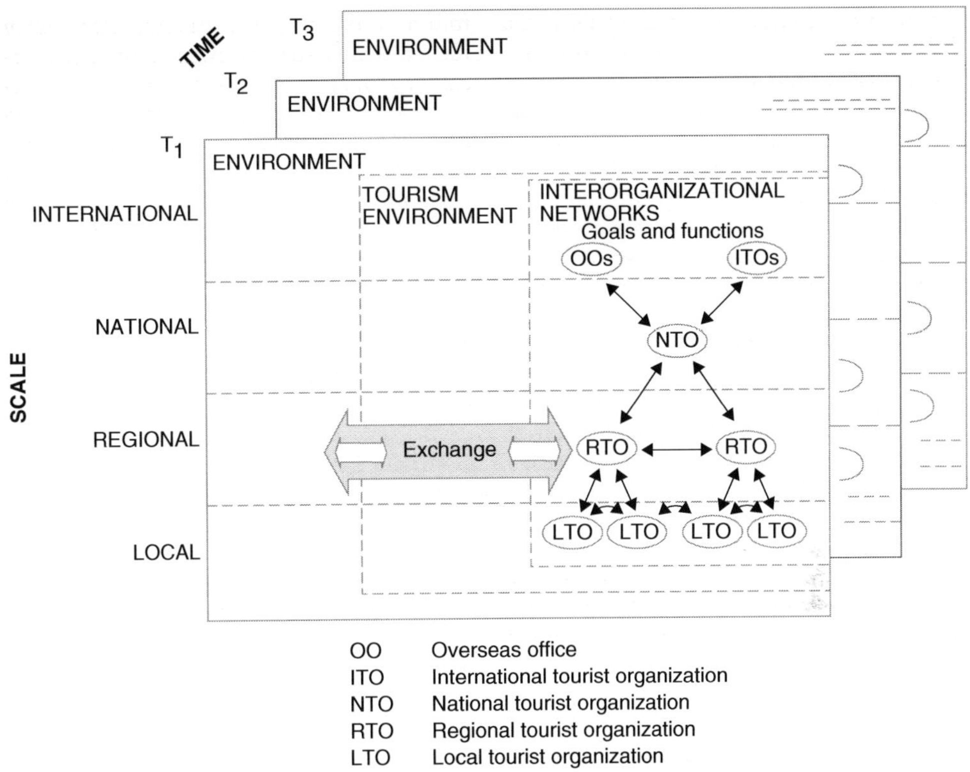

Fig. 1.1. A conceptual framework for the interorganizational analysis of tourist organizations. From Pearce (1992), Figure 1.1, with permission of Pearson Education Limited.

Destinations might also be seen in terms of market structure, defined as 'the number of firms in the market that provide similar products and services as well as the relationships of the firms in the market' (Ma *et al.*, 2015, p. 120). Different market structures – perfect competition, monopoly and oligopoly – are related to the competitive behaviour and performance of firms (Bull, 1995). The main elements of the market structure of an industry 'are usually taken to be the degree of seller concentration, conditions of entry, buyer concentration and product differentiation' (Akehurst, 1984, p. 25). To date, however, relatively little work on destinations has been framed explicitly in these terms or focused on structural characteristics which vary according to the different products that the destination has to offer (Bull, 1995). Papatheodorou (2004) considered conditions of entry into tourism and argued that ease of operations spreads uncertainty among customers, a situation which favours a small number of reputable firms which invest part of their increased profitability in service quality. This is said (p. 222) to result in a virtuous cycle of constant enlargement in which the 'promotion of standard products becomes necessarily associated with large operating scale to satisfy the increasing flows, while small size operations focus on specific market niches.'

Andergassen *et al.* (2013) developed an economic model for tourism destinations which draws on two key issues: their conceptualization of the tourism product as a bundle of complementary and substitutable goods and services produced and sold within a given territory which provide the overall holiday experience and their interpretation (p. 88) of a destination as a '(meta) economic agent taking important decisions from the supply-side at a level that is intermediate to the micro (firms and tourists) and macro-levels (the whole economic system, usually the country).' The bundle of products and services feature of the tourism product has two main implications. First, it is seen to be anti-common in nature; that is, ownership is fragmented among multiple providers. This in turn gives rise to the need for coordination and cooperation if tourists are to be provided with everything they need for their holiday. For the sake of simplicity they focus on price coordination and assume coordination in quantity and quality is already in place. Second, consideration must be given to what goods and services should be provided. In particular, what is the appropriate balance between developing natural and/or cultural resources and offering a more varied (sophisticated) product? Following Dixit and Stiglitz (1977), the assumption is made that tourists show 'a love for variety', that is, a wider availability of local goods and services (e.g. restaurants, recreational facilities) beyond the enjoyment of the main resource will increase their satisfaction and result in increased demand.

Their economic modelling produced two theorems.

Theorem 1 The Coordination Theorem. Given the anti-common property of the tourism product, coordination among firms in the destination, which can either be provided by a DMO or by a tour operator, increases profits from tourism. Price coordination is shown to result in more efficient tourism activity compared to when there is no coordination. As a corollary, local profits are less when coordination is provided by a foreign tour operator rather than by a DMO or other local authority.

Theorem 2 The Love for Variety Theorem. As long as the negative externalities on tourism quality are small, reorganization of the tourism destination toward increasing the variety of available goods and services raises tourists' welfare and their willingness to spend on tourism at the expense of non-tourism consumption, thereby stimulating the economic development of the destination.

Andergassen *et al.* recognized that due to variations in the local context and decisions taken, different real-world situations will be found. With regard to coordination they identify three sets of destinations:

- Individually based destinations where there is no coordination between local tourism firms.
- 'Community-managed destinations' where local firms are coordinated by a DMO.
- Corporate-based destinations where coordination is provided by a tour operator, local or foreign.

Three types of destination are identified in terms of tourism sophistication:

- Resource-based destinations which depend primarily on local natural, cultural or artificial resources and have a very limited variety of differentiated goods.
- 'Sophistication-based destinations', where local resources are very limited but the tourism product is based on a large variety of local goods and services
- Mixed-based destinations having a balance between local resources and a certain degree of sophistication of the tourism product.

The three classes in each of the two dimensions thus result in taxonomy of destinations with nine classes. Italian examples of some of these situations are given but the authors note the need for empirical testing of the model as well as attention to methodological issues such as how best to measure sophistication and price coordination.

However, the most common approach to studying the organizational structure of destinations has been to draw on a broader framework from outside the domain of tourism which gives particular weight to the way in which the relationships between multiple interdependent actors are configured and function. In many of these scope exists to incorporate a wider range of actors, including communities and tourists, not just organizations *per se* which are the focus of the approaches just outlined. Given the stress on the place dimension of destinations and the need to explain how multiple product and service elements come together, it is not surprising that researchers have turned to territorial-based industrial organization models such as industrial districts and clusters. Other frameworks have also been adopted, including networks, systems, social constructs and assemblages. The nature of these frameworks and their application to destinations are discussed in detail in Chapter 3.

Spatial and organizational structure

Interdependence and interconnectedness are common themes in studies of both spatial and organizational structure; in the former in terms of the physical expression of the products and services and associated patterns of demand; in the latter, with regard to the interrelationships between multiple actors. The question this raises is in what ways and to what extent are these two structural forms interrelated? Early researchers analysing the development and morphology of ski resorts in the French Alps recognized that the different forms and processes which they were describing were the product of different organizational and institutional structures (Cumin, 1970; Préau, 1970). Co-location of firms in a designated geographic area is a feature of the territorial-based industrial organization frameworks of industrial districts and clusters (Chapter 3).

Terhorst and Erkuş-Öztürk (2011) took a systems perspective and argued that a tourism production system is a system due to the technical complementarity between the goods and services consumed by the tourists, noting differences may occur between places in the degree of this complementarity, as well as the institutional complementarity which influences the ways in which these goods and services are produced, distributed and regulated under different modes of economic governance. They contend (p. 72) that:

> A tourism production system is in itself a spaceless concept ... but ... it is clear that it is impossible to analyze a tourism production system without simultaneously spatializing it. There is only a technical complementarity between sets of specific tourism goods and services when they are produced in a specific place. And it is hard to speak of institutional complementarity without referring to space either. Tourism places are made up of a specific mix of modes of economic governance at the local level that complement each other while they often simultaneously show many characteristics of a broader national social system of production in which they are embedded.

Jansen-Verbeke and Ashworth (1990, p. 619) argued 'tourism development depends upon concentration rather than on dispersal, functional combination rather than segregation, and multi-functional environments rather than mono-functional ones.' Demand associations, they suggested, can be considered in terms of the space–time budgets and paths of visitors, on their action space, on what they do and where they go during the course of the day or their visit. On the supply-side they identified four sets of relationships between spatial clustering and functional associations, only one of which involves some form of association between spatially proximate businesses (Table 1.1). These include linkages through organizational structure or cooperative activities and functional linkages resulting from spatial clusters serving similar customer groups or 'benefitting from environmental "shadow effects" conferred by other proximate users.' Jansen-Verbeke and Ashworth do little more than outline these ideas but these do provide some basis for exploring the nature of destination structure.

Later, French scholars caution against the assumption that co-location necessarily implies cooperation or interaction and take

Table 1.1. Relationships between spatial clustering and functional associations. From Jansen-Verbeke and Ashworth (1990).

Category	Association	Proximity	Example
Non-spatial associations	+	0	Organization chains
Spatial coincidence	0	+	Premises links
Spatial association	+	+	Immobile joint supply
Non-spatial non association	0	0	Isolated supply

a more nuanced approach to proximity, distinguishing between geographic proximity, organized proximity and cognitive proximity (Bocquet, 2008; Marcelpoil and François, 2008; Clergeau and Violier, 2013). The work of Torre and Rallett (2005), which two of these studies draw upon, suggests relationships between geographical and organized proximity merit greater consideration. Geographical proximity expresses the physical or kilometric distance between two units. It might also be measured in terms of cost or time distance as well subjectively. Organized proximity is relational; it refers to the strength of relationship between actors, whether geographically close or distant, and to 'the ability of an organization to make its members interact' (Torre and Rallet, 2005, p. 49). Belonging to an organization, they argue, facilitates interaction through a logic of belonging and a logic of similarity. The intersection of these two types of proximity form a matrix resulting in four different types of interaction (Table 1.2). Quadrant 1 depicts a situation of agglomeration in which there is a co-location of activities but no direct relations between the actors or organizations. Quadrant 2 is characterized by strong interactions between co-localized actors resulting from organized proximity as reflected in 'the intensity of the client–supplier relationships, the exchange of knowhow or the existence of an "atmosphere"' (Torre and Rallet, 2008, p. 50), conditions that are attributed to industrial districts, localized systems of production and clusters. Quadrant 3 reflects the situation where both forms of proximity occur but only temporarily, for example where business trips enable long-distance coordination. In the case of supra-local organizations (Quadrant 4), coordination between units in different places is maintained by organizational rules and norms and information and communication technologies with little face-to-face interaction.

Tables 1.1 and 1.2 suggest that a wide range of potential situations exist when spatial and organizational proximity are considered together and that as a consequence much diversity in destination structure might be expected. Bocquet's (2008) application of Torre and Rallet's framework to French ski resorts, presented in Chapter 9, is distinctive on two levels: its use in relation to destinations and as a framework to explore different types of destination structure. Other research tends to focus on aspects of only one of the quadrants of Table 1.2 or one of the types of relationships in Table 1.1 and to be concerned with a particular sector rather than the destination as a whole. Attention has been given to organization chains, the non-spatial association example given in Table 1.1, in the form of research on hotel chains (Ivanova and Ivanov, 2015; Le *et al.*, 2018). These studies do not go beyond the sector, tend to focus on business aspects and generally do not examine the spatial dimension in much detail. However, as is shown in Chapter 5, at the national level hotel chains have distinctive geographical characteristics. The analysis of spatial clusters at a national level (Chapter 5) incorporates multiple tourism variables (Capone and Boix,

Table 1.2. Intersection of geographical and organized proximity. From Torre and Rallet (2005).

	Geographical proximity	Organized proximity
Geographical proximity	1) Agglomeration without interactions: I have a lot of neighbours but I do not know them (no direct coordination) Agglomeration due to infrastructures, for example	2) Local systems of innovation, production, clusters: Geographical proximity activated by organized proximity
Organized proximity	3) Non-permanent co-localization: Long-distance coordination implies temporary face-to-face (sales representatives, researchers)	4) Supra-local organizations: Coordination between sedentary people located in different places, supported by organizational rules and information and communication technologies

2008; Majewska, 2015) but research on agglomeration at the local level (Chapter 7) has focused almost exclusively on hotels from a sectoral perspective (Baum and Haveman, 1997; Chung and Kalnins, 2001; Canina *et al.*, 2005; Tsang and Yip, 2009). There is a concern in both cases with analysing the externalities arising from spatial concentrations but little attempt to identify other forms of interaction which might exist or to consider any links between spatial clusters within countries and those which occur within cities. Relationships between co-located actors – spatial associations in Table 1.1; Quadrant 2 in Table 1.2 – have attracted an increasing amount of attention (Chapter 3) in destination research but as later chapters will show there is relatively little attempt to examine the spatial dimension of such relationships beyond the actors being clustered together in a particular place.

Much scope exists then to draw together these still largely non-convergent streams of research on the spatial and organizational structure of destinations. The following chapters provide a way forward by synthesizing work on each theme and illustrating how these might be integrated more effectively.

2 Perspectives on Destinations

It took some time for the term destination to emerge in the literature and to be widely used. Moreover, as references to destinations became more frequent, differences developed in meaning and use. This trend is summarized in the next section, which identifies commonly recurring characteristics attributed to destinations, including a sub-section on use of the term in the French tourism literature. These essentially academic definitions and concepts are then complemented by consideration of formally designated destinations, illustrated by examples from France and Chile. Subsequent sections deal with destinations as functional nodes and with tourists' views on what destinations are. The final section draws these different threads together.

Evolving Use of the Term 'Destination'

Some of the earliest research in this field drew attention to the way in which the development of tourism and the emergence of this new form of land use was starting to change the landscape of places visited by tourists and to give rise to new and distinctive urban forms (Jones, 1933; Brown, 1935; Selke, 1936; Carlson, 1938; Gilbert, 1939; Eiselen, 1945; Deasy, 1949; Wolfe, 1951, 1952). Although some attempts were also made to distinguish commonalities in the forms examined (Gilbert, 1939), much of this research took the form of ideographic local and regional case studies not related to wider issues. In these and later studies the places frequented by tourists were yet to be commonly referred to as destinations. Rather, some other term was used, often qualified by an adjective reflecting the nature of the place: tourist towns (Jones, 1933), hill stations (Senftleben, 1973) and coastal, ski or health resorts (Gilbert, 1939; Stansfield, 1969, 1971; Standley, 1971). Similarly, French scholars were writing about *stations* (resorts) of various types: *stations balnéaires, stations de sports d'hiver* and *stations thermales* (Balseinte, 1958; Defert, 1960; Burnet, 1963; Barbier, 1968; Préau, 1968; Barbaza, 1970). The interest here was on places where tourism had become a dominant feature, places typically referred to as resorts or *stations*. In contrast, urban tourism had yet to garner much interest from researchers: although major or historic cities may have drawn large numbers of visitors, tourism was but one component of the urban economy. However, the expansion of tourism research in the 1970s brought with it a growing realization that tourism occurred in places other than just traditional resorts and that it could be conceived of at a range of scales, not just the local, thereby encouraging the use of a more general term. In their influential text, Burkart and Medlik (1974, p. 46) wrote:

> The geographical unit visited by a tourist may be a self-contained centre, a village or a town or a city, a district or a region, an island, a country or a continent. *This geographical unit may be described as the tourist destination* ... The tourist destination, however defined geographically, provides a convenient focus for the examination of the tourist movement and of its manifold impact and significance.
>
> (Emphasis added)

Other strands of research had also emerged by this time which examined demand, tourist travel patterns and the way in which the places frequented by tourists were linked to their places of residence. These early studies again initially tended to use some term other than destination to refer to the places being visited: summer dormitory zone (Wolfe, 1951); vacation service region (Campbell, 1967); tourist receiving country (Guthrie, 1961). However, by 1974 Plog had published his oft-cited article on 'why destination areas rise and fall in popularity' and the use of destination was gaining some currency. Plog (1974) did not define destination, a term he uses interchangeably in places with resort. His model of the psychographic positions of destinations arose out of research on air travel and includes destinations at a range of scales from Africa to Coney Island. The notion of travel *to* a destination, as an end point, in subsequent travel pattern research may also have led to the proliferation in usage of the term. Lundgren's (1982) model of a hierarchy of tourist flows, for example, included metropolitan destinations, peripheral urban destinations, peripheral rural destinations and natural environmental destinations. Nevertheless, other terms continued to be used, such as in Butler's (1980) *tourist area* life cycle (emphasis added).

The concept of tourist destinations continued to evolve as tourism research expanded in the 1980s and 1990s. Laws (1995, p. 25) suggested the traditional notion of a destination as 'a place where people spend their holidays' was being superseded by more recent concepts of it, namely as 'an area where people choose to spend their holidays, and the effects of their activities', with a concomitant need for 'managing the demand for tourism, and managing its effects on the destination.' By the turn of the 21st century destination was such a commonly used term that its meaning was often assumed and many authors offered no definition. In other cases it was explicitly recognized that destinations had multiple dimensions and that they were being defined and conceptualized in different ways. Researchers would review these in varying depth before opting for one or elaborating their own definition.

The proliferation of definitions and concepts in turn resulted in more comprehensive reviews of destinations which systematically sought to identify common threads and themes (Framke, 2002; Kadri *et al.*, 2011; Saraniemi and Kylänen, 2011; Miller *et al.*, 2013; Pearce and Schänzel, 2013; Fayos-Solà *et al.*, 2014; Flores and Mendes, 2014; Pearce, 2014; Viken, 2014). The specifics of the multiple definitions of destinations found in the literature vary and weight is given to different features depending on the disciplinary background of the researcher, the purpose of the study and the approach taken in studying particular problems. Framke (2002), for instance, compared conventional business oriented and sociological understandings of the term. Kadri *et al.* (2011) identified geographic, marketing and systematic perspectives.

Elaborating on Pearce and Schänzel (2013), three commonly recurring characteristics of destinations in the literature can nevertheless be identified. First, destinations are made up of multiple elements. They are variously said to consist of an amalgam, bundle, cluster, composite, package, portfolio, set or the sum of activities, attractions, attributes, experiences, places, products or services (Hu and Ritchie, 1993; Buhalis, 2000a; Murphy *et al.*, 2000; Framke, 2002; Presenza *et al.*, 2005; Vanhove, 2005; Capone, 2006; Hanlan *et al.*, 2006; Ekinci *et al.*, 2007; López Bonilla and López Bonilla, 2007; Cracolici and Nijkamp, 2008; Rusko *et al.*, 2009). These features are interdependent and together provide or constitute an integrated, holistic, final or total experience, product, brand or 'natural whole' (Buhalis, 2000a; Murphy *et al.*, 2000; Fuchs and Weiermair, 2004; Ryglová, 2008).

Second, destinations also have a significant geographical dimension. The features cited above are associated with particular locations, localities, areas, regions or spaces at a range of scales from the local to the macro-regional (Davidson and Maitland, 1997; Pechlaner and Fuchs, 2002; Ejarque, 2005; Presenza *et al.*, 2005; Vanhove, 2005; Capone, 2006; Nicolau and Más, 2006; López Bonilla and López Bonilla, 2007; Rygová, 2008; Tinsley and Lynch, 2008; Botti *et al.*, 2009; Bornhorst *et al.*, 2010; Barros *et al.*, 2011; Beritelli *et al.*, 2014). Debate occurs over whether the boundaries of destinations are fixed or fluid, administrative or functional. Beritelli *et al.* (2014) are particularly critical of the use of conventional administrative limits to bound destinations and argue it is necessary to deframe the destination construct to take into account the variable use of products and services of multiple visitor segments. Instead they propose (p. 415) the concept of variable geometry based on 'multiple different, partly overlapping spaces, consisting of dynamic flows' and the identification of strategic business areas which serve particular groups of visitors. In some cases the geographical areas are qualified further. Ejarque (2005) suggests a destination must have a sufficient influx of visitors that tourism becomes one of the area's economic bases. This also implies that other activities occur in those geographical areas but these are frequently ignored in studies focusing wholly on tourism. For Frisk (1999, p. 9, cited by Tinsley and Lynch, 2008) a destination is 'a geographic place or region where some sort of "touristic society" inside of the "ordinary" community exists.' Flores Ruiz (2014, p. 23) states the area should be 'a homogenous space with common socio-economic and environmental characteristics that enable unified management.'

Third, destinations are commonly defined from either a demand or supply perspective. In these the two sets of characteristics above are seen either in terms of the tourists' needs, perceptions and experiences (Bieger, 1998; Buhalis, 2000a; Fuchs and Weiermair, 2004; Zehrer *et al.*, 2005; Pechlaner *et al.*, 2008; Botti *et al.*, 2009; Bornhorst *et al.*, 2010) or with regard to the multiple private-sector providers and relevant public-sector organizations who offer a range of products and services (Pavlovich, 2003; Candela and Figini, 2012). According to Pechlaner *et al.* (2008, pp. 109–110) 'the guest decides what appeals to him/her and thus what a destination is' whereas the economists Candela and Figini (2012, p. 10) consider the tourism destination as 'a systemic mix of firms.'

Although Candela and Figini (2012) contend that a destination is not a firm but a mix of firms, other researchers have drawn parallels between a firm and a destination (Flagestad and Hope, 2001; Melian-Gonzalez and Garcia-Falcon, 2003; Westermann *et al.*, 2005). Drawing on resource-based theory and various conceptualizations of the firm they suggest destinations too can be seen as a 'bundle of resources' (Penrose, 1959) and as a 'collection of interrelated economic activities' (Porter, 1985, 1991); as possessing a set of goals and objectives; and as having to adapt to their environment if they are to survive. At the same time, differences occur, particularly in respect to a lack of central management control over the multiplicity of firms and other stakeholders. Barros *et al.* (2011, p. 143), however, make the assumption that:

> a destination is a geographical area in which a tourist can live at least one tourism experience and which, from a Destination Management Organization (DMO) point of view, can be managed i.e. can be organized and developed to attract tourists. In this case, a destination is seen as a firm with inputs (its tourism attractions) and outputs (bed-nights, employments) and the DMO objective is to achieve maximum efficiency through a proper use of its inputs.

This is to ascribe a larger degree of control to the DMO than generally occurs.

Occasionally a more comprehensive approach is adopted, one which combines demand and supply perspectives and a geographical dimension. One of the distinguishing characteristics of destinations in this regard is that they are places of both production and consumption. According to Swisscontact (2014, p. 9) a destination is:

> a specific geographic area with particular characteristics and features. It has:
> attractions and services which motivate and facilitate the tourist's stay;
> adequate price-value means of access;
> support of the local population which is involved in the development of different aspects of the visitor experience;
> a brand to market the area based on its all-round character.

Murphy *et al.* (2000) conceptualize the destination product in terms of a three-layered model consisting of a set of destination environment (place) factors and service infrastructure elements which together influence the tourist's destination experience. A similar 'destination cake model' is offered by Ivanova (2017) in which she differentiates between the service providers and the products that they offer which create customer value. Underpinning these layers of the cake is the territory, characterized by the local community, tourist resources and the general infrastructure. Each of the layers contributes to an overall set of destination attributes.

A French perspective

The title of a collection of volume of francophone studies – *Destinations et territoires: co-présence à l'oeuvre* (Destinations and Territories: Co-Presence at Work) – captures the essence of various contributions set mainly in France and Canada (Lemasson and Violier, 2009). According to Violier (2009, p. 1), *destination* is a term that has been mainly used in marketing to refer to the places to which tourists go while geographers have taken a more territorially based perspective using concepts such as *espace touristique* or *lieu touristique* to refer to 'spatial objects transformed by tourism.' Hazebroucq (2009) notes that the term destination in French has a double meaning: *lieu où on se dirige, où l'on est dirigé* (a place one makes one's way to or is sent to) and *finalité* or *usage* (purpose). He then goes on to discuss the various approaches taken by French researchers to studying destinations, emphasizing spatial and economic aspects, processes of change and the role of different stakeholders.

Similarly, Kadri *et al.* (2011) distinguish between destination as a *lieu où se rendre* (a place one goes to) and *emploi d'un lieu* (how a place is used, the end to which it is put) in their semantic analysis in which they contextualized use of the term in the French literature with related concepts. They locate *lieu* and *emploi* as two semantic poles of a spectrum but argue that it is impossible to dissociate the two meanings without destroying the sense and terminological value of word; that is, a destination is both a place one goes to and the end to which that place is put. In practice, however, usage of the word by individual researchers tends to gravitate towards one pole of the spectrum with five variations being identified in terms of a range of criteria. A simplified version of their table is given in Table 2.1. The travel component is divided into two classes according to (I) whether the travel is independent, the place one goes to for leisure purposes, or (II) the travel is organized by an intermediary, that is, the place to which a tourist or group of tourists is sent. The use component, which is the most common, is divided into places developed for tourism (IIIA) or used by tourists (IIIB) and a range of place-based 'projects' (development, management, urban, economic or anthropological) or conscious attempts to 'construct' a destination (IV). The fifth variety (V) refers to places 'destined' to become destinations associated with certain levels of quality and prestige.

The variety of meanings identified in the French literature is attributed by Kadri *et al.* to the complexity of urban tourism, the focus of their study. Its value lies in the identification and juxtaposition of the related notions of a destination as a place one goes to and a place in which tourism has developed. Moreover, the variations in Table 2.1 also usefully differentiate independent travel from that which involves intermediaries.

Table 2.1. Variations of the term 'destination' used in the French literature. From Kadri *et al.* (2011), Tableau 5.

	I Direction of travel (independent)	II Direction of travel (organized)	IIIA Developed for tourism IIIB Used for tourism	IV Place-based projects	V Places destined to become destinations
Dominant semantic pole	Place one travels to ←	Place to be sent to	Use of the place	Use of the place →	

Formally Designated Destinations

Destinations are not just defined by academic researchers. In some countries destinations and related areas may have a legal status or be formally designated in plans and strategies. This designation is generally associated with policies aimed at supporting the development of tourism in specific areas and with regional and national marketing strategies. It may vary according to the context in which it is being used.

France

France has a long tradition of formal designation of *stations* (resorts). A 1919 law provided local authorities so designated with additional means of supporting the influx of visitors: tourist taxes, increased central government funding and casino licences. Five types of *stations* were recognized: *stations balnéaires; stations de tourisme; stations de sports d'hiver et d'alpinisme; stations hydrominérales (thermales)*; and *stations climatiques*. Administrative reforms in 2006 resulted in a single category of *stations classées de tourisme* (classified tourist resorts). Communes seeking to be officially classified as resorts must fulfil certain criteria: develop facilities and resources to foster tourism throughout the year; have a range of accommodation, shops and services catering to visitors; provide visitor information; facilitate access; have free wifi, have a land-use plan and appropriate infrastructure; and take steps to preserve the environment and cultural sites and monuments (Direction Générale des Entreprises, 2019). *Commune touristique* is a less stringent classification for communes that meet three criteria: have a tourist office; organize events and activities for tourists; and have accommodation for a non-resident population. The *communes touristiques* are also referred to as tourist destinations while the *stations classées* are considered tourist destinations of excellence.

As of February 2020, 438 communes or intercommunal groupings had been classified as *stations classées*. The majority were located on the coast. Others included winter sports resorts in the Alps, thermal resorts in the Pyrenees, Massif Central, Pre-Alpes and the Vosges. Marseille and Lyon are *stations classées* but other major cities such as Paris and Toulouse are not even though they receive large numbers of visitors each year. However, heavily visited places close to the capital such as Fontainebleau and Versailles are so designated. Of almost 35,000 communes, more than 1300 were classified as *communes touristiques*.

In 2014 a new national marketing strategy introduced the concept and practice of 'destination contracts' aimed at revitalizing French tourism through collective action by public and private actors 'constituting the destination.' The approach was territorially based but focused on the imagination of the visitors rather than being defined by administrative limits. By 2016, 22 destination contracts had been signed between the national marketing agency, Atout France, and the destination partners. In contrast to the local-level *stations classées*, the destinations in this marketing strategy were larger in scale and based around a thematic regional brand, for example:

Pyrenees: winter sports; outdoor activities, well-being
Côte d'Azur: major events
Normandy and Île-de-France: Impressionism
Provence: culture-Les Arts de Vivre en Provence
Paris: 'La Ville Augmentée.'

The contract for the latter was designed to promote a different image of Paris, one aimed at young European travellers and which emphasized the undiscovered and unusual, whether in terms of accommodation, culture or night-time activities.

Another formal designation was introduced in 2015 when *zones touristiques internationales* (ZTIs) were created to facilitate economic development through liberalizing shop trading hours on Sundays and in the evening to meet the 'reality' of tourist demand in particular geographic zones. By 2019, 21 ZTIs had been created, primarily in and around Paris, on the Cote d'Azur and in the coastal resorts of Deauville and La Baule.

Chile

In Chile 83 local-level *destinos* (destinations) were formally designated in the 2014 National Plan for Sustainable Tourism Development in Chile so as to provide a territorial focus for the strategies developed at a national level. The rationale for this approach was:

> Given that tourism is an activity that develops in a geographic area, in which the variables of nature, infrastructure, economic base, cultural life and history define the level of development, the action to be taken will promote areas that fulfil the condition of a tourism destination, with the aim of generating tourism with a local character.
> (Subsecretaria de Turismo, 2015, p. 22)

The level of tourism development for each of the country's 346 communes was measured by an index of tourist activity established by factor analysis of 27 supply and demand variables. Three levels of development were identified on this basis: mature; emerging; and potential. The plan aimed to strengthen consolidated destinations and drive the development of emerging ones and those with identified potential through steps that varied according to the level of development.

The *destino* boundaries, however, did not generally coincide with those of the communes. Rather, the *destinos* consisted of smaller areas within, or occasionally crossing, communal limits. Their boundaries were determined by discussion and negotiations between the national Sub-secretariat of Tourism and national and regional representatives of Sernatur (the agency responsible of implementing the sub-secretariat's policies).

Following a change in government, the list of *destinos* was revised and updated in 2018 using a similar procedure. This involved an analysis of 15 demand and supply variables including the number of arrivals and overnight stays, the floating population, the number of tourist attractions and tourist accommodation services and the identification and demarcation through discussion of areas deemed to fulfil the conditions of being a destination. The latter was said to correspond to a 2016 decree which defined a *destino turístico* as:

> A geographic area, delimited physically and administratively, according to a combination of natural and cultural attractions, tourist services, complementary facilities and infrastructure, accessibility, image, human resources and local identity which motivate tourist movements and the development of associated tourist activities.
> (Subsecretaria de Turismo, 2018, p. 1)

On this basis 89 tourist destinations, located in some 200 communes, were designated: 41 mature, 34 emerging and 14 potential. Moreover, they were classified according to type: coastal; urban; rural and natural alpine; lacustrine and fluvial; and insular. Some were characterized as combining features of two types: Iquique and Valparaiso, for example, were classified as both urban and coastal. Representative examples of the classification of the destinations by level of development and type are given in Table 2.2. The boundaries of the *destinos* as determined by staff in Senatur's planning section are irregular in form, an outcome primarily of the concentration of attractions and tourist services, a common local identity and tourist routes.

Destinations as Functional Nodes

The strand of literature dealing with tourist travel, that which corresponds in large part to Kadri *et al.*'s (2011) *lieu où se rendre*, recognizes that places perform different and often overlapping functions associated with different types of flows and that destination typologies might be derived on the basis of these functions and the destinations' position in a larger network or system (Chapters 5 and 6). Such an approach provides a more nuanced perspective, one which identifies different types of destinations or differentiates a destination from other types of nodes.

Table 2.2. Examples of formally designated tourist destinations in Chile, 2018–2019. From Subsecretaria de Turismo (2018).

2018–2019 tourist destinations	Level of tourism development	Type of tourist destination
Iquique	Mature	Urban and coastal
Colchane – Volcan Isluga N. Park	Potential	Rural and nature
San Pedro de Atacama	Mature	Rural and nature
Taltal	Emerging	Coastal
Valle de Acaoncagua	Emerging	Rural and natural
Portillo	Mature	Alpine
Valparaiso	Mature	Urban and coastal
Santiago	Mature	Urban
Valdivia-Corral	Mature	Lacustrine and fluvial
Torres de Paine National Park	Mature	Rural and nature
Tierra del Fuego	Potential	Rural and nature
Rapa Nui (Easter Island)	Mature	Insular
Robinson Crusoe	Emerging	Insular

Pearce's (1981) urban tourist flow model suggests that large cities can act not only as a source of tourists generating outbound flows but also can have several different and/or complementary functions as an international or domestic destination (Fig. 2.1). For international tourists the large city may constitute a gateway, a point of entry into or out of a national or regional territory. The city may also function as a regional staging post, a point of departure for travel onward to other centres or resorts. In the case of circuit tourism or multi-destination travel the city may be just one stop among several on a given tour. The city, of course, may also constitute a destination in its own right, in which case tourists will move around within it. The city may also provide a base from which day trips are made to surrounding areas. Similar functions are performed for domestic tourists from other areas. For certain tourists the city may play some or all of these roles: international tourists may arrive in the city, visit attractions there and make day trips from it before moving on to another destination. In other cases, the relative importance of these functions – as gateway, principal destination, staging post, stopover – may differ from one segment to another and also from city to city.

Lew and McKercher (2002) analysed the relative location of Hong Kong within the larger itineraries of its visitors, that is, its position and functions with regard to international travel to, in and beyond Asia. They proposed a fivefold classification of destination types: single destination, gateway (first) destination, egress (last) destination, touring (middle) destination and a hub (multiple visit) destination. In effect, the gateway and egress destinations both have gateway functions, one in terms of entry at the beginning of a multi-destination trip, the other, exit or egress at the end of it. They acknowledge that a hub may have a more precise meaning with regard to the notion of hubs and spokes in transportation but here use it simply to refer to any place that is visited more than once in a multiple destination itinerary. Lew and McKercher (2002, p. 609) also note that 'depending on their location within the overall trip itinerary, places can exhibit characteristics of one or more destination types.'

The ideas contained within these and other studies, including those from other fields such as transport geography, can be drawn together so that the concepts of the different nodal types and functions can be fleshed out. Two key interlinked characteristics of each node are position and function. Position here can be understood in terms of centrality, intermediacy and proximity (Burghardt, 1971; Fleming and Hayuth, 1994; O'Connor; 1995). Centrality, a concept developed by Christaller (1966) and others, is used by geographers to refer to 'the situation,

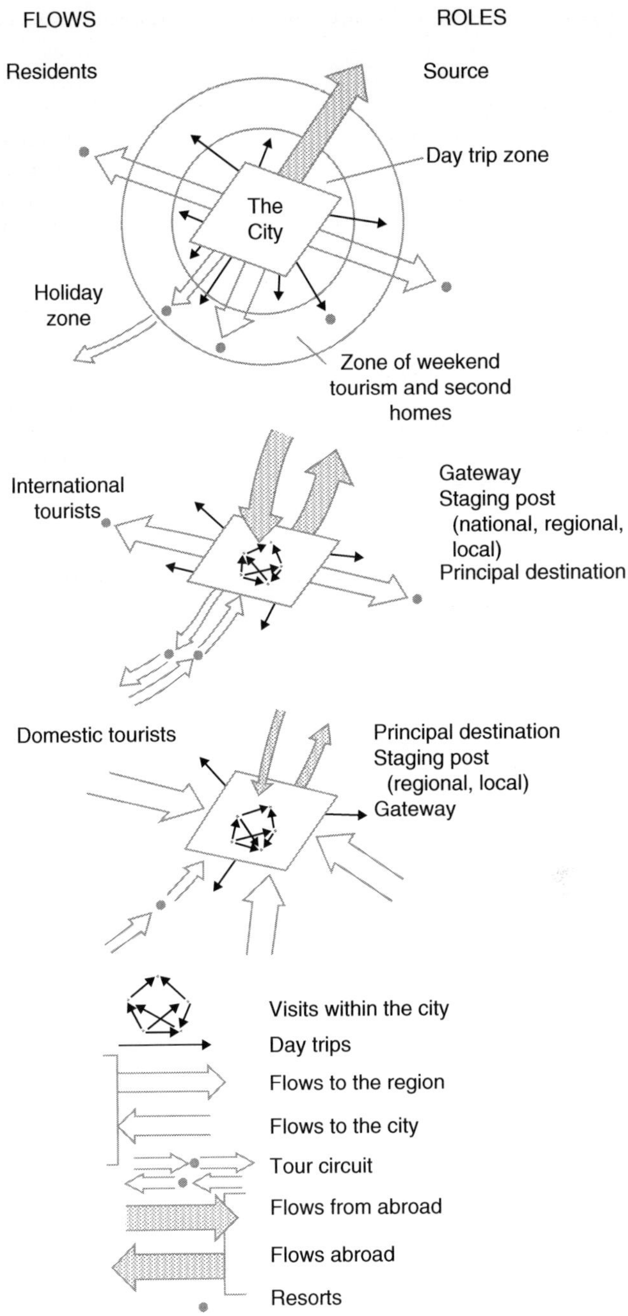

Fig. 2.1. Schematic representation of flows to and from major urban areas. Redrawn from Pearce (1981), Figure 1.

or relative location of the market centre and its external relationships, that is emphasized. A certain amount of productive land supports a market centre and 'the centre exists because essential services must be performed for the surrounding land' (Ullman, 1941, p. 203, cited by Fleming and Hayuth, 1994). Fleming and Hayuth (1994, p. 4) use the term intermediacy to describe 'locations between important origins and destinations – locations that are chosen as waystops, route junctions, break-in-bulk points, gateways, etc.' They continue: 'Central places, by definition, contain the generalizable locational attribute of centrality; en route places contain the generalizable attribute of intermediacy. Every large commercial place ... has, at some scale and to some degree, *both* of these locational attributes.' London's centrality, for example, is important at the regional and national scales; at a European and intercontinental scale the city is more peripheral and the condition of intermediacy comes to the fore. Proximity, the attribute of being close to a market or set of productive resources, is another locational factor included by Fleming and Hayuth, according to whom one of the clearest examples is gateway proximity to tourist attractions. They also note (p. 8) that 'proximity, in the sense of closeness to competitive locations, can have a negative impact on traffic levels.' O'Connor (1995) used the ideas of centrality, intermediacy and proximity to examine changes in airport development in Southeast Asia and, like Fleming and Hayuth, showed how the importance of these conditions can change over time with resultant impacts on the nodes concerned. In the context of destinations the relative weight of the conditions of centrality, intermediacy and proximity has a major bearing on the different types of nodal functions and issues of multi-functionality.

Gateways

Gateways in a general sense are major entry/exit points into or out of a national, regional or even local system. Gateways not only link other nodes within a national or regional network but also serve to link one network to another (Burghardt, 1971). Gateway cities are 'located towards one end of their tributary areas' and 'are in command of the connections between the tributary area and the outside world' (Burghardt, 1971, p. 269). Although local trade connections are important, 'the gateway is characterized best by long-distance trade connections' (Burghardt, 1971, p. 270). The converse applies to central cities. However, as they develop over time gateway cities may assume more central functions.

In addition to their role in transport systems of providing external links, gateways channel visitors into other parts of a country, region or locality. This may involve intermodal transfer connections. Packaged tours will generally begin at gateway nodes and many inbound tour operators will be located there. Information provision will be an important function for independent travellers who may still be finalizing their travel within a country or region. International gateways also have immigration and customs control responsibilities and tourists will be required to complete entry and exit procedures there. In the context of adventure tourism in Nepal, Zurick (1992, p. 620) wrote:

> The national gateways ... are often the country's capital, where travel contacts are made and adventure programs planned and provisioned. In Kathmandu one finds representatives of most major U.S. adventure firms, as well as numerous European and Nepalese companies. These firms handle the local tour logistics. Independent travellers may also join a group excursion in Kathmandu or arrange their own itinerary. In either case, Kathmandu remains the primary link in Nepal's adventure tourism spatial structure. Additional regional gateways that further channel the flow of tourists ... are located along roadways or other transportation links near the entrances to established adventure regions.

At a regional or local level the servicing functions of gateways are elaborated in terms of the roles of gateway or portal communities. Stange and Brown (2011, p. 137) define a gateway community as 'a geographically proximate community to an attraction that supplies numerous services to visitors.' While they include examples where the attraction is cultural, the term is most commonly used with reference to communities close to national parks or other protected areas and to the spatial separation of conservation and commercial functions (Ewert, 1996; Mules, 2005; Miller *et al.*, 2018).

Gateways may also have a symbolic function. International tourists have their first direct contact with a country as they arrive at a gateway and the first impressions they receive will often affect their enjoyment for the remainder of their stay (Pearce, 1984). According to Dredge (1999, pp. 787–788) 'gateways perform an important physical and psychological function, for example, to signify arrival in a destination region, to indicate a journey's end, to give a panoramic view of the destination, or to assist in the orientation of tourists. Gateways are statements about a destination region and accordingly should be given careful consideration in destination planning and design.' Likewise, Lew and McKercher (2002, p. 612) suggest that as the first place encountered the gateway is 'a liminal point of transition where the embodiment of the non-home experience becomes situated in place (as opposed to "in transit")' while on exit the gateway (egress destination in their terminology) has (p. 613) 'the opportunity to provide a sense of closure and preparation to re-enter the home place after a long tour itinerary.' In other instances, however, surface travel across regional and local administrative boundaries may go unnoticed by the tourist and the point of entry or gateway into a new regional or local destination may not be distinguished in any particular way.

Hubs

Although used traditionally to refer to places at a crossroads, in aviation studies the concept of a hub emphasizes an airport's gathering and transfer functions within a wider network (Dennis, 1994; Fleming and Hayuth, 1994; O'Kelly and Miller, 1994; O'Kelly, 1998; Meng *et al.*, 2016; Ivanova, 2017). According to O'Kelly (1998, p. 171), hubs are 'special nodes that are part of a network, located in such a way as to facilitate connectivity between interacting places.' This is achieved by 'the construction of a network where direct connections between all origin and destination pairs can be replaced with fewer indirect connections' (O'Kelly and Miller, 1994, p. 31). Airlines may develop a hub-and-spoke system whereby a central node – the hub – concentrates short- and medium-haul flights from regional or remote nodes for transfer of passengers on to longer-haul flights or redistribution to other regional nodes and enables a more efficient use of aircraft capacity (Ivanova, 2017). The gathering/distributing, transferring functions and management services of a tourism hub are depicted in Fig. 2.2. Meng *et al.* (2016) cite visa application as an example of services during the gathering phase, luggage transit at the transfer stage and additional itinerary services during the distributing phase. A comprehensive and coordinated approach to such services is especially important where transfer between terminals or transport modes is required. Meng *et al.* (2016) developed a set of indicators to assess the hub functions of Beijing and Shanghai. Their results indicated that the former had stronger functions in terms of gathering and distributing tourists, together with management and services, while the later had a superior transfer function. O'Kelly (1998) suggests

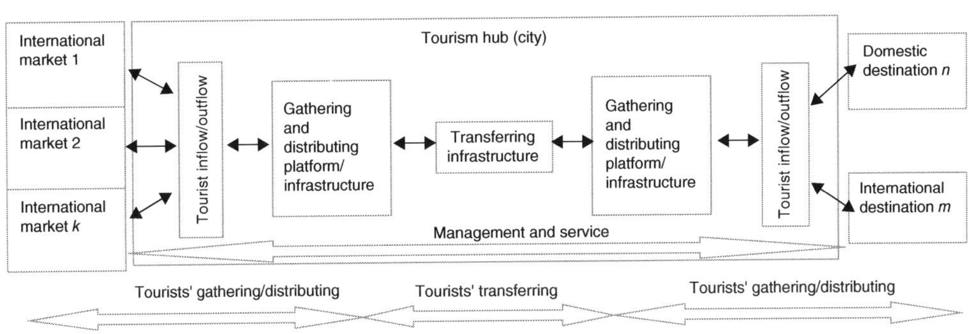

Fig. 2.2. Schematic diagram of tourism hub functions. From Meng *et al.* (2016), Figure 1.

that a hub increases a city's ability to attract conventions and business meetings.

The hub-and-spoke structure of airline networks has some parallels in the tourist trip literature. Effectively, the central nodes in Campbell's (1967) recreational vocational regional complex and in the base camp category of Lue *et al.* (1993) and Oppermann (1995) act as regional hubs. The difference with the airport hub is that the hub node in these other systems acts not so much as a gathering and redistributing centre but more as a central point from which a series of side-trips to secondary nodes is made. Moreover, such trips commonly involve overland travel rather than flights.

Stopovers and secondary centres

Stopover is a term used in the context of long-haul or circuit travel and implies a stop is being made to break a longer trip, whether international, domestic or intranational (Uysal and McDonald, 1989; Lohmann and Pearce, 2010; Pike and Kotsi, 2016). Often the term is not defined explicitly. Pike and Kotsi (2016) adopted the definition of the International Air Transport Association of a stopover as being at least a 24-hour stop at an intermediary port. Lohmann and Pearce (2010) distinguished the role of stopover from other nodal functions with reference to both time spent at a node and the reason for going there (see Chapter 4). Secondary centre is an allied term that is commonly used, but not frequently defined, to refer to places visited that are of lesser importance to some main destination on a larger circuit (Forer and Pearce, 1984).

Multiple functions and multi-functional nodes

The range of terms and concepts discussed here illustrates the multiple functions which places may have in a tourism system or network. How a particular place is viewed will depend on the perspective taken. In terms of travel patterns, a place for one segment of visitors may be their sole destination while for others it may function as a gateway, hub or a secondary destination on a larger itinerary. A gateway into a country or region may at the same time be a main or secondary destination. This multi-functionality may be expressed in hybrid terms: Lew and McKercher (2002) speak of gateway and hub destinations; Caves and Gosling (1999) refer to 'gateway hubs' in relation to gateways on the periphery of Europe which provide onward services to secondary centres. However, not all gateways and hubs are destinations, for example when transfer onward occurs without any overnight stay. These different nodal functions are often seen in terms of their relative position in a sequence of places visited but primary or secondary destinations may also be determined by levels of demand in a larger national or regional system (Forer and Pearce, 1984). As discussed in Chapters 5 and 6, national and regional destinations are made up of a set of different types of major and secondary local level destinations having different types of functions which are linked together in various ways.

The Tourists' Perspective

Many definitions of destinations are framed in terms of demand, of what products, services and experiences they seek, and much research based on visitor surveys has been undertaken on topics such as destination image and destination quality (Gallarza *et al.*, 2002; Pike, 2002; Lopez-Toro *et al.*, 2010). However, prior to Pearce and Schänzel's (2015) study there appears to have been no research which directly addressed the issue of what tourists consider a destination to be. Nor has there yet been research to explicitly explore how destinations are conceived by other stakeholder groups such as providers or residents though work has been done on related topics such as the determinants of success for destinations and DMOs (Bornhorst *et al.*, 2010).

Pearce and Schänzel (2015) held a series of focus groups with youth hostel guests in

New Zealand. Two broad themes emerged from the introductory, open-ended discussion of how the participants viewed and used the term 'destination': destinations were considered either with reference to travel patterns or to the characteristics of the places being visited. In each case multiple elements were mentioned.

A common response was to see a destination as an end point of a trip or part of a journey: 'Where you're aiming to get to sort of thing, end point of a particular section or the whole of the travels.' However, some saw it as a place you had reached and for others a destination was the place you were going to rather than where you were at. Participants also recognized that one might have multiple destinations on a single trip: 'Sometimes you can have one, an ultimate destination, but then on the way you just have mini destinations, where you kind of stop off on the way to your end goal.' Several also noted that destination was not a term that they would commonly use: 'Rather than saying "Oh, what's your destination tomorrow?" I'd say "Where are you headed?" or "Where are you going?"'

Destinations were viewed by others in terms of the characteristics of the places to be visited, viewing these especially as places with lots of activities and places of personal interest. They were also expressed as places to be planned for and purposively visited: 'It's a place you kind of plan for and read up about and you kind of prepare for and get excited about.' This is consistent with the mindful behaviour of holiday makers reported by Gyimóthy (1999). Differences between destinations were perceived with regard to the purpose of the trip, such as visiting family, going for work or to study. Several participants distinguished between destinations in functional terms: 'Some destinations are more to get stuff done, as in to catch a plane or to get your visa sorted, whereas other destinations are more cultural, or you want to do an activity.' However, in distinguishing between destinations, little or no reference was made to such widely used terms in the literature as urban, coastal, alpine, rural, domestic or international destinations.

As used by the participants in Pearce and Schänzel's (2015) study, the concept of destination was very fluid and readily encompassed layers of destinations as a journey unfolded. They could articulate clearly the changing scale of multiple destinations with respect to the nature of long circuit trips and to the context in which the term was being used – at what stage of the trip and with whom:

> For me, with my round the world trip, I'd have to think of countries. And it's a bit like a sort of a tree ... Your trunk is your country and then you start going off into branches which are your different places within that country. And then you might go off into ... leaves, let's put it as that, as saying what you're going to do in each one of those towns.
>
> I think it depends what you're talking about. Like if you're talking of your trip in general then your destination is probably the biggest main place, but if you're talking with someone who's travelling like you in New Zealand, 'Well what's your next destination?,' so it'll be the next city you're heading to.
>
> [I]f we were going away for like a ... two or four day city break... you might say 'I'm going to Spain' but you tend to say 'I'm going to Barcelona' ... because you're not going to travel around.

These findings lend support to the assertion of Mazanec *et al.* (2007, p. 88) that 'from a behavioural point of view, destination may be better conceived as a hierarchical and flexible concept that evolves during the trip decision process.' In this case it would appear to be not so much the evolution of the decision process but more the stage of the trip, the context in which it is being considered and with whom it is being discussed that influences the way in which destinations are conceived and the term is used.

When probed further as to whether there were differences between countries and cities being destinations, many participants agreed that both could be seen in these ways and for these reasons as destinations. Others though suggested that a clear distinction might be made between countries and cities depending on the purpose and type of trip, the type of places visited and the coverage of the country in question. Large cities received particular mention as sole focus destinations separate from their respective countries:

> [W]orld cities like London, Paris, New York ... They're very much destinations in themselves ... Cause I think countries like New Zealand or Australia ... there's not really any particular one focus, its more the country.
>
> [O]bviously the people ... we're going to meet in youth hostels and things are going to be backpackers and I guess they want to see the whole country, they just don't want to see a certain destination. Whereas, say if my parents were going to go on holiday to Spain they'd go to wherever, a beach, because they want to relax from their job, and that would be their only destination, so I guess they'd think of it as the city not the whole country.

Some lively discussion occurred in the final part of the focus groups when the participants were asked which of the following three definitions of a destination drawn from the literature they felt was the best:

> The location of a cluster of attractions and related tourist facilities which a tourist or tour group selects to visit or which providers choose to promote.
>
> (McIntyre, 1993)

> A geographical area within which the tourist enjoys various types of tourism experiences.
>
> (Botti *et al.*, 2009)

> A bundle of tangible and intangible attributes, and it can potentially be seen as a product and a brand.
>
> (Ekinci *et al.*, 2007)

Only two participants concurred with McIntyre's definition. Over half favoured the definition of Botti *et al.* (2009), particularly if all mention of tourist and tourism were dropped. There was also a view that it was sufficient for a place to offer a single experience rather than several. Most of the other participants felt that the definition of Ekinci *et al.* (2007) had something to offer: they liked the idea of a destination offering a bundle of tangible and intangible attributes but, with a few exceptions, were much less enthusiastic about expressing these as a product or a brand, features that smacked of commercialism. A few participants acknowledged that the definition to be used depended on the perspective adopted: 'For the people that are managing ... it's a product, you're selling it. For the people that are going, it's a place for doing stuff.' Taking account of these qualifications, several participants suggested combining key aspects of the second and third definitions so that a composite one might read: 'A geographical area with a bundle of tangible and intangible attributes that offer one or several types of experiences.'

Dolnicar and Grün (2017) too found consumers conceived of 'destination entities' at a range of scales: attractions, cities/towns, regions, countries and continents. The larger units (countries and continents) were considered by the majority of tourists for vacations far away from home but little difference occurred in the degree to which they thought about destinations as attractions, cities/towns and regions in terms of taking a vacation close to or far from home; levels of agreement were relatively high for each of these. Byrne and Skinner (2007) sought to identify the positioning and image of Dublin relative to Ireland as a prospective business tourism destination. Their interview respondents tended to think more in terms of city than country destinations while their focus group participants thought any such distinction 'very much depended upon the respective brand identities and levels of familiarity with the city and country' (Byrne and Skinner, 2007, p. 61). The European participants would consider Dublin before Ireland, whereas the Americans would consider Ireland before Dublin.

Discussion and Conclusions

As the preceding sections have shown, destinations have been studied from a variety of different perspectives for different purposes, they have been defined and conceptualized in a variety of ways and a commonly accepted and a widely used definition or concept of what a destination is has yet to emerge.

Nevertheless, certain recurring characteristics or features can be identified.

The most prominent of these is the geographic or territorial dimension: destinations are strongly associated with places, both as a place one goes to, an end point, and a place in which tourism has developed, that is, the end to which it is put. This usage and distinction were brought out most explicitly in the reviews of the French literature (Hazebroucq, 2009; Kadri *et al.*, 2011) and in the discussion of the tourists' perspective on what a destination is (Pearce and Schänzel, 2013). These two notions also tend to be dealt with separately in the English language research. The travel-related studies dealing with destinations as nodes largely sit within 'the place one travels to' domain.' Those discussed in the sections on the evolving use of the term and on the formally designated destinations are primarily framed in terms of places that are being developed for or by tourism. These two concepts are clearly closely related: places tourists travel to undergo tourism development and the development of tourism depends on and leads to growing numbers of visitors travelling to them. This point was recognized early on by Plog (1974, p. 57) who noted that increased visitation was associated with supply-side developments as a destination moved along the psychographic continuum of the type of tourist it attracts: to

> As a destination becomes more popular, the Mid-centric audience begins to pick it up. The continued increase of travellers leads to further development of resorts in terms of hotels, tourist shops, scheduled activities for tourists, and the usual services that are provided in a 'mature' resort area.

However, in the main the two basic concepts of destinations continue to be treated separately in studies which emphasize different characteristics or deal with them in different ways.

When destinations are defined or conceptualized as places developed by and for tourism the emphasis is generally on the interdependence of the mix of products and services that offer experiences for tourists, and the need for a full range of these to be available. In contrast, research that focuses on destinations as places tourists travel to puts greater stress on the different types of travel-related functions that particular nodes have and recognizes that some perform multiple functions. Some of these studies also draw attention to the products and services provided but the characterization of the nodes is primarily in terms of their relative position in different itineraries and the volume of various flows. These travel patterns, and the subsequent identification of different types of nodes, are generally demand-based whereas studies viewing destinations as places developed for tourism tend to cover a wider mix of demand and supply perspectives. When seen from a supply perspective account may also be taken of the suppliers involved, including the participation of the local community.

Geographical scale is a feature of both strands of research. The destinations as 'places of tourism development' strand recognizes that destinations can be considered at a range of scales but is divided over how destination boundaries should be drawn, whether they should be fixed or fluid, administrative or functional. While some argue that administrative boundaries do not reflect how visitors experience destinations, the participants in Pearce and Schänzel's (2013) study show tourists may be untroubled by the issue of boundaries as they appear quite fluid and flexible in how they conceive of destinations. The formally designated destinations reflect the need for boundaries to be set in order to implement national policies at a lower level while at the same time showing that such boundaries need not necessarily correspond to administrative divisions. The travel-related research provides another perspective on scale as nodal functions depend in part on the unit of analysis being considered. Analysis of regional or national destinations reveals they are made up of lower level nodes variously classified as gateways, hubs, primary or secondary destinations, each of which may in turn have its own structure when viewed at a smaller scale.

Clearly scope exists for drawing these different strands together more closely. What mix of elements, for example, is required where and for whom? Issues associated with destinations as an end point and as places in which tourism has developed need to be integrated further. Likewise, questions arising from a demand perspective need to be linked more explicitly and effectively to those seen from a supply standpoint. Multiple stakeholders will have different views on destinations and be involved in their development and functioning in different ways. Different conceptualizations of how these stakeholders interact are discussed in the next chapter.

3 Frameworks for Tourist Destinations

This chapter reviews the frameworks mentioned in Chapter 1 that have been taken from outside the tourism literature and applied to destinations, with a particular emphasis on the relationships between multiple inter dependent actors. It builds on and extends the author's earlier work (Pearce, 2014). Six sets of frameworks are discussed: industrial districts, clusters, networks, systems, social constructs and assemblages. The essence of each framework is outlined and the destination applications are reviewed. As the authors offer their own interpretation of each framework and do not consistently identify and apply the same elements or explore the same relationships it is necessary to go back to the foundation concepts rather than just concentrate on the tourism applications. Where there is less consistent reference to a specific author or paper the general characteristics of each concept are summarized from a range of papers before their tourism applications are considered. The boundaries between the concepts are not always clear cut and rigid: in some cases the terms are used loosely and even interchangeably; in others, a more rigorous terminology is adopted. These overlaps and variations in terminology are acknowledged but to avoid semantic paralysis the focus here is on the key features of each framework. Subsequent chapters address methodological issues relating to these frameworks and their application in empirical studies.

Industrial Districts

The comparison of tourism destinations to industrial districts is most evident amongst studies by European researchers (Hjalager, 2000; Gaido, 2002; Petrić and Mrnjavac, 2003; Capone, 2006; Maulet, 2006; Mottiar and Ryan; 2006; Sainaghi, 2006; Fernández Cuevas *et al.*, 2008; Martínez Pérez *et al.*, 2010; Cirer Costa, 2014a, 2014b). The underlying ideas about industrial districts derive from the work of the 19th-century British economist Alfred Marshall, who identified the external economies of concentrations of particular industries in particular localities in terms of the availability of skilled labour, the growth of supporting trades and the linkages between specialized enterprises. The modern conceptualization of the industrial district grew out of research by Becattini (1990) and others examining the success of industrial areas in Italy. Markusen (1996) offered a fourfold typology of industrial districts in which the Italianate Marshallian form illustrated by Becattini is complemented by three others: the hub-and-spoke district, the satellite industrial platform and the state-centred district. The industrial districts literature has continued to expand and evolve (Hervas-Oliver *et al.*, 2015) but tourism researchers have drawn essentially on the Italianate model.

Becattini (1990, p. 38) defined an industrial district as

> a socio-territorial entity which is characterized by the active presence of both a community of people and a population of firms in one naturally and historically bounded area. In the district, unlike in other environments ... community and firms tend to merge.

He argued (p. 37) that 'any phenomenon which exists through time is assumed to possess some internal logic' and devotes his paper to explaining that logic. In doing so he places particular emphasis on the way in which a specialized activity comprised mainly of small firms is embedded in the community and their behaviour is governed by local values and rules which depend on trust and personal relationships. Interdependence is a key feature: the firm is seen both as an entity in its own right and (p. 40) as 'a specific cog in a specific district.' Districts are said to be self-organizing. Becattini also outlines the dynamic nature of industrial districts, portraying them as 'adaptive systems' which evolve in response to market needs and changing conditions. Values are transmitted from one generation to another, human resources are constantly re-allocated and technological advances are readily accommodated as a gradual social process. In addition, Becattini underlines the importance of external connections; the surplus of production cannot be sold within the district and requires (p. 38) 'the development of a permanent network of links between the district and it suppliers and clients' and (p. 44) the need for 'an "image" of the district which is separated both from the single firms in it and from that of other districts.' The entrepreneur's 'special function lies in translating all the capabilities which are latent in the historical heritage of the district into products that can be sold in [the world market]' (p. 43).

Lazerson and Lorenzoni (1999, p. 362) emphasize the role of focal firms in expanding 'the district's horizons to incorporate new technologies, organizational skills, and markets.' Focal firms (p. 363) are those that 'occupy strategically central positions because of the greater number and intensity of relationships that they have with both customers and suppliers (Aldrich, 1979).' Such firms are often larger and have greater access to capital. Lazerson and Lorenzoni (1999, p. 363) assert that many small firms) in industrial districts have benefited directly from the presence of large firms, that the organizational structure of the firms in industrial districts is more likely to be heterogeneous than homogenous and that 'the importance of local municipalities, regions and various public agencies as agents of economic development within industrial districts has been overly exaggerated, minimizing the actual role of entrepreneurs.'

Tourism researchers suggest that similarities between industrial districts and destinations are readily identifiable in terms of the emphasis on the interdependence of generally small and medium-sized firms concentrated in a particular location serving a particular sector and the external economies that this may bring; strong inter-firm relations often characterized by informal cooperation; embeddedness in a social and professional milieu and supportive public policies (Hjalager, 2000; Petrić and Mrnjavac, 2003; Mottiar and Ryan, 2006). Hjalager (2000) notes that many of these features may be shared by tourism destinations but that destinations often have non-supportive governance structures, free-riding firms and a dependency on multinationals and concludes (p. 209) that 'it is debatable how far tourism research can go using the industrial districts approach.' Gaido (2002, p. 112) asserts that a district 'is "situated" and therefore able to be read' in terms of its geographic, social and cultural features. In economic terms it is also a 'mode of organization' involving the 'supple specialization' of small and medium-sized enterprises (SMEs), Maulet (2006, p. 46) focuses on structural issues, namely:

- the degree of systemic consciousness among firms, that is, 'the awareness of the firms that they are part of a transversal chain';
- the strategic means display which involves a bottom-up process to build a strategy at the system level and includes the role of local authorities and financial and technical assistance;
- the degree of cumulativeness resulting from how long the firms have been established, inter-firm relationships and the workers' competence; and
- the degree of spatial concentration reflected by the extent of firms' linkages to their local environment, the borders of the district and connections with external firms.

With the exception of Mottiar and Ryan (2006), these writers all draw parallels between industrial districts and destinations in conceptual terms. In examining the case of Killarney, Ireland, Mottiar and Ryan (2006) provide empirical evidence to support their argument that much of that destination's success can be interpreted in terms of the theory of industrial districts. In particular, they stress the interdependence of tourism firms and the importance of a strong social and professional milieu resulting from the early development of tourism in the area and the presence of businesses linked back through generations of local families. This has created an ambience in which much informal cooperation occurs between spatially proximate firms in a clearly defined geographic area and in which professional institutions have developed to foster collaborative marketing.

Clusters

A second territorial-based industrial organization model widely applied to destinations is the cluster, a concept popularized by Porter (1998, 2000). Porter sought to explain the changed role of location and how companies today continually create competitive advantage in a global economy through the benefits which proximity to interconnected firms bring. According to Porter (1998, p. 78), clusters are 'geographic concentrations of interconnected companies and institutions in a particular field. Clusters encompass an array of linked industries and other entities important to competition ... Clusters also extend downstream to channels and customers and laterally to manufacturers of complementary products.' The role of location is no longer the traditional one of giving places a comparative advantage; rather, it is the benefits of co-location that which makes them competitive. Clusters, Porter (1998, p. 79) argues, represent 'a kind of new spatial organizational form in between arm's length markets on the one hand and hierarchies, or vertical integration, on the other.' Such a form, he continues, is critical to improving competition and increasing productivity. 'A cluster allows each member to benefit *as if* it had greater scale or *as if* it had joined with others formally – without requiring it to sacrifice its flexibility.' Productivity is said to increase as clusters enable better access to employees and suppliers, to specialized information and to institutions and public goods; they provide better motivation and make measurement easier; and offer the benefits of complementarity, that is, the whole is greater than the sum of its parts. Porter (1998, p. 81) used tourism to illustrate this last aspect:

> In a typical tourism cluster ... the quality of a visitor's experience depends not only on the appeal of the primary attraction but also on the quality and efficiency of complementary businesses such as hotels, restaurants, shopping outlets, and transportation facilities. Because members of the cluster are mutually dependent, good performance by one can boost the success of the others.

Clusters are also said to drive the direction and pace of innovation by making opportunities more visible, by lowering the costs to experiment and by sheer competitive pressure. Similar factors also foster new business formation. Once formed, increased competition, better productivity, more innovation and the formation of new businesses lead clusters to grow in a 'self-reinforcing cycle.'

Porter (2000) underlines the influence which a location's business environment has on competitiveness and productivity and expresses this in terms of a 'diamond' composed of four interrelated sources of competitive advantage: factor inputs; context for firm strategy and rivalry; demand conditions; and related and supporting industries. 'A cluster', he asserts (p. 21), 'is a manifestation of the diamond at work.'

The merits of clusters and clustering advanced by Porter have no doubt contributed to the widespread adoption of the concept by researchers and policy makers. Martin and Sunley (2003), however, urge a more cautious approach, arguing the cluster is a somewhat chaotic concept which lacks clear boundaries, both geographic and industrial; that there are theoretical weaknesses relating to competitiveness; that difficulties arise with mapping and measuring clusters; and (p. 22) that the 'evidence of a positive association between clustering and innovation is not consistent.' None the less, an exponential growth in cluster studies has taken place since the late 1990s (Cruz and Teixeira, 2010; Trippl *et al.*, 2015). As the cluster literature has expanded diverse definitions have appeared. However, after reviewing these Cruz and Teixeira (2010, p. 1267) found a fair degree of consistency in the main features, concluding that: 'Spatial proximity, interrelatedness of capabilities/activities, interaction between agents, and institutional endowment are ... key elements of clusters.'

Tourism is often seen to share such characteristics. This, coupled with the claimed benefits for competitiveness and innovation, has led to a growing number of studies and policies throughout the world in which destinations are explicitly or more loosely framed in terms of clusters, often using the same rationale as Porter:

> A tourism cluster is a geographical space in which the entire tourist experience takes place... A cluster is an accumulation of tourist resources and attractions, infrastructures, equipments, services providers, other support sectors and administrative organisms whose integrated and coordinated activities provide customers with the experiences they expected from the destination they chose to visit.
>
> (Bordas Rubies, 2001, p. 39)

> The cluster concept's application to tourism destinations would appear to be particularly germane. Not only is tourism within a destination a conglomeration of activities and facilities creating a 'destination product' in Smith's words (1994) but increasingly these different components are brought together in destination associations and partnership marketing (Pearce, 1992).
>
> (Jackson and Murphy, 2002, p. 38)

> The concept of the cluster ... is applicable to tourism destinations. Tourism destinations are often regionally concentrated around certain key natural or man-made attractions. This is consistent with the external economies rationale behind clusters. Many different firms, specialising in different stages of the tourism product chain, contribute to the production of the destination's tourism product. They include accommodation, transport, and attraction operators. The interactions between these tourism business operators provide an example of a social dilemma of independence in the face of interdependence.
>
> (Huybers and Bennett, 2003, pp. 582–583)

> T and H [tourism and hospitality] clusters are essentially demand driven services nodes of interdependent, yet discrete firms operating within a single industry, which are heavily dependent on localised and destination-based resources to support services to tourists and residents.
>
> (Chhetri *et al.*, 2017, p. 402)

Other researchers have focused on adapting Porter's diamond to tourism. Although some limitations were noted and some modifications to the diamond were needed, Miller and Gibson (2006, p. 61) concluded that industrial cluster theory was one of the 'most promising professional paradigms' for examining tourism from an economic development perspective.

Clusters are associated both with cooperation between co-located firms and competition among them. According to Huybers and Bennett (2003, p. 582):

> Tourism operators in the region compete with each other on the basis of their individual competences. At the same time, however, they are mutually dependent and hence pursue collective benefits. In line with the concept of collective efficiency, the aim of cooperation between tourism businesses is to compete successfully as a destination with rival destinations on the basis of its collective distinctive features. The conceptual implication is that tourism businesses at a destination may be seen as an aggregate entity that operates at the higher level in a two-tiered market structure with both intra-regional and inter-regional dimensions of competition.

Moreover (p. 580), the repeated interactions between individuals brought about by geographic concentration are expected to foster cooperation and enable the 'credible threat of informal enforcement' to reduce free-riding.

Similarly, Michael (2003, p. 24) argued that a cluster is:

> a concept that implies a sense of co-operative exchange between both the firms and the community where they are in geographic co-location or close proximity to each other. The common location necessarily implies some level of interaction between firms, but in itself this may not translate to any recognizable benefit for those firms ... Firms with similar production processes or common markets can develop synergies in the supply of resources and infrastructure, marketing, information and the pooling of labour skills and distribution systems, which serve to lower costs and improve the competitive advantage of the clustered firms. A common location may also serve to pool the customer base, attracting buyers who know they will have a range of choices available, so reducing their search costs.

Fabry (2009) makes the point that in a tourism cluster the coming together of actors occurs around natural or created resources and properties and argues that the 'knot' binding the cluster together is its identity or major theme which acts as the mirror of the destination, a point she develops with reference to business travel to Paris. Fabry (2015, p. 297) later asserts categorically that 'a destination is not a cluster.' Rather, she contends, a tourism cluster is a mode of organization driven by information, innovation and knowledge and the way in which these factors bring stakeholders together (Fabry and Zeghni, 2012; Fabry, 2015). From this perspective destinations may 'host' clusters but not all destinations necessarily do so as they are not organized in this way or to this extent. A similar view is expressed by Clergeau and Violier (2013, p. 57) who see a cluster as a geographically anchored production process but stress that geographical proximity does not automatically give rise to organized proximity: 'Co-location is not a synonym for interaction.' That is, just because they are located close to each other firms do not always work together and interact cooperatively and competitively in the manner suggested by Porter.

In seeking to clarify whether it is appropriate to apply the concept of clusters to destinations, particular attention has been given to the nature of tourism demand, the tourism value chain and to the relationships between the actors involved in the production of tourist experiences. Tourism clusters differ from industrial ones by being areas where production and final consumption generally occur simultaneously and relatively few true forward linkages are generated (Gollub *et al.* 2003; Miller and Gibson, 2006; Fabry, 2009). Moreover, the pattern of demand for tourism may differ from that postulated by Porter (2000) who underlined the importance of the quality of local demand for stimulating differentiation and product improvement (Miller and Gibson, 2006). Variations in the quality and type of services such as restaurants, cultural attractions and events, outdoor attractions and local transport will occur from destination to destination as a function of both local and external use; connecting transport services will involve a mix of inward and outward traffic but the accommodation sector is essentially dependent on external demand.

Consideration also has to be given to the characteristics of the production process in tourism compared to manufacturing and whether this process is viewed from the perspective of the firm or of the end user, the tourist. In manufacturing, horizontal clusters occur where co-located firms produce substitute goods while in vertical clusters they produce intermediate goods which are subsequently assembled into finished products (Andria *et al.*, 2015). In tourism, the tourist experience effectively constitutes the end product. In general terms, a tourist's experience involves some combination of travel to, within and subsequently from a place or places, and an activity or set of activities, accommodation and hospitality while there. At issue here are the stages in the creation of this experience – what they are and whether or not there is a common sequence – and who is involved in the creation process.

From the firm perspective three principal types of tourism clustering have been identified: horizontal, vertical and diagonal (Michael, 2003, 2007). Horizontal clustering occurs 'where similar or like firms from the [same] stage in the value chain for the same industry co-locate in a geographic area' (Michael, 2007, p. 25). Held (1996, p. 25) cited by Miller and Gibson, 2006, p. 58) defines horizontal clusters as 'those whose component industries have a common resource base, but few logical buyer–supplier linkages.' Urban hotels, for instance, are commonly clustered together in various locations throughout the city, groupings of restaurants are often located in particular neighbourhoods, souvenir shops are concentrated close to major attractions, etc. (see Chapter 7). There may be little buyer–seller interaction within and between these sectors and firm behaviour remains competitive as each strives to attract its share of the destination's market. At the same time, horizontal clustering can bring economies of scale and induce cooperative behaviour, particularly in terms of joint marketing and creating a stronger image in the market, lobbying for infrastructure or land-use zoning, information sharing, access to labour or the development of training programmes.

Vertical clustering refers to the 'co-location of firms operating at different stages in an industry's supply chain' (Michael, 2007, p. 26). Michael does not provide a tourism example but creation of the tourist experience may engender the co-location of certain services at different stages of a sequence, particularly in terms of transport and accommodation (e.g. the grouping of hotels and rental car offices at or close to airports, terminus hotels where rail or ferry travel is involved) and between accommodation and attractions (e.g. hotels and restaurants at theme parks or close to natural attractions). In some cases ownership or strategic links might exist between these co-located businesses (e.g. railways and terminus hotels, airlines and rental car companies).

Michael (2003, 2007) uses the concept of diagonal clustering to take account of the heterogeneity of products and services which might make up the tourist experience and the complementary nature of these. Diagonal clustering:

> occurs where firms working together create a bundle of separate products and services that the consumer effectively purchases as a single item. The situation is common in many tourism destinations, where separate firms with separate production processes supply activities, transport, hospitality, accommodation, etc. The co-location of complementary providers adds value to the tourism experience; while, conversely, the absence of key services will probably limit growth.
>
> (Michael, 2007, p. 26)

Michael (2003, p. 138) contends that each form 'is only a category of the production linkages between firms' and that an optimal outcome would be one where firms operate across all three forms of clustering to maximize the gains from co-location.

For Clergeau and Violier (2013, pp. 60–61), however, the tourist, not the firm, is the 'essential actor' in the co-creation and co-production of the tourist experience:

> An experience which, undergone in situ, is not the result of a sequenced chain of events designed to produce a holiday offer for the market (Lanquar, 1994) but instead of a mobilization, instigated by the tourist, at a given time, in a given location; of a constellation – not a chain – of services destined for the co-production of an experience ... Which leads us to consider that a tourist destination is, in the words of Normann and Ramirez (1993), a *localized value constellation*, created with and for the tourist.
>
> (Emphasis in original)

Similarly, Fabry and Zeghni (2012) argue that tourists become co-producers of a service each time they buy one. Earlier, Fabry (2009) emphasized the need for tourism clusters to have external linkages through value chains which connect providers to customers in the market.

Different types of tourism clusters have been identified (Erkuş-Öztürk, 2009; Dragusin *et al.*, 2010; Ivanis, 2011):

- local/regional/cross-border;
- general (offer diverse tourism products with broad market appeal)/specialized (in a particular segment or product, e.g. shopping tourism, medical tourism, upmarket hotels);
- agglomerated (large number of firms)/non-agglomerated (small number of firms); and
- spontaneous (have developed organically)/planned (are the result of explicit public/private-sector policies).

It should also be noted that that the term 'cluster' may be used in a purely spatial sense to refer to close geographical groupings of phenomena (Jansen-Verbeke and Lievois, 1999; García-Palomares *et al.*, 2015; Sarrión-Gavilán *et al.*, 2015) or in a statistical sense in relation to objects which are more similar to each other than to other groups or clusters (Blasco *et al.*, 2014) without implying in either case any of the organizational features discussed in this section. However, statistical cluster analysis can be used to identify spatial clusters which might be interpreted in terms of clusters or industrial districts (Capone and Boix, 2008).

Networks

Network analysis is widely used in such fields as mathematics, social network analysis (SNA), policy network analysis and management and is increasingly being applied in tourism (Scott *et al.*, 2008a; Pearce, 2012; van der Zee and Vanneste, 2015). Usage and definitions of the term network vary from a loose reference to social connections to a more precise mathematical meaning (Scott *et al.*, 2008a). According to Rocks *et al.* (2005, p. 82): 'The basic components of a network are nodes and connections.' In a social science context, nodes are replaced with actors and connections with social ties and bonds (Davern, 1997). Three fundamental types of network are commonly recognized: exchange, communication and social (Szarka, 1990). The concern of network theory is with the collective, rather than the individualistic or dyadic nature of organizational action, behaviour and constraints (Pavlovich, 2003). Network characteristics commonly analysed include density, centrality, cohesiveness, size, prominence, level of formality, level of stability and level of flexibility (Rocks *et al.*, 2005; Scott *et al.*, 2008a).

In adopting a network perspective tourism researchers explicitly recognize the multiple stakeholders involved in destinations and the importance of understanding the relationships among them, especially in terms of managing destinations and delivering the tourist experience (Pavlovich, 2003; Nordin and Svensson, 2007; Camprubi *et al.*, 2008; Scott *et al.*, 2008b; Denicolai *et al.*, 2010). Camprubi *et al.* (2008, p. 48) consider destinations as 'a network of relations, links, ties, or contacts between multiple and diverse actors that, together, create the tourist product.' Scott *et al.* (2008b, p. 171) see destinations as 'collaborating networks of complementary organizations.' Baggio and Cooper (2010, p. 1768) define destinations as 'networks of organizations and communities where the destination functions by the movement of resources, such as information or investment, through the network.' Pavlovich (2003, p. 203) stresses the need to understand the architecture of linkages between 'a diverse and highly fragmented supply structure' and 'the strong market interdependence [that] forms as suppliers pass customers from one organization to another in order to provide a comprehensive tourist experience ... that acts to integrate the various components into a destination structure.' Figure 3.1 depicts the structure of relations in 2000 between organizations in Waitomo, a small rural destination in New Zealand renowned for its limestone caves, derived from network analysis.

The type of network examined and the actors included are influenced by the issues being addressed. The focus on firms and organizations reflects a view that a network perspective is particularly relevant for examining issues of governance and management (Bodega *et al.*, 2004; Merinero Rodríguez, 2008; Timur and Getz, 2008; Merinero Rodríguez and Pulido Fernández, 2009; Erkuş-Öztürk and Eraydin, 2010; Franch *et al.*, 2010; Meriläinen and Lemmetyinen, 2011). The collective nature of destination network management is emphasized by Meriläinen and Lemmetyinen (2011, p. 28), who argue that 'the coordination of cooperation seems to focus on the actor whereas strategic management approaches the action from the perspective of the whole network.' Others stress the role of networks in facilitating knowledge transfer among firms which in turn is said to foster innovation and competitiveness (Baggio and Cooper, 2010) or underline the ways in which network analysis

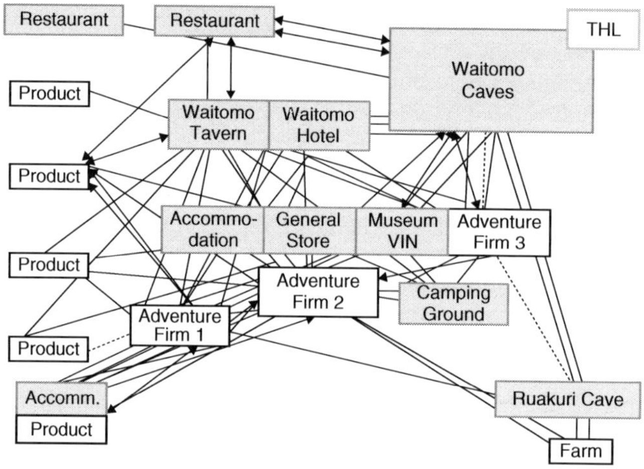

Fig. 3.1. The relational structure within Waitomo, 2000. From Pavlovich (2003), Figure 5.

provides insight into how the relationships among the actors affect a destination's development and performance (Gibson *et al.*, 2005; Presenza and Cipollina, 2010). The impact of networks on demand within destinations has attracted rather less attention, even though the notion of the co-creation of experiences by tourists is recognized.

Baerenholdt *et al.* (2004) contest the notion that destinations are networks. They identified six types of networks:

1. vertical networks based on the delivery of raw materials and services such as cleaning and washing;
2. horizontal networks and chains, for example of hotels;
3. additive networks in marketing managed by local and regional marketing organizations;
4. informal cross-cutting social networks among locals involved in tourism but whose social relations are in other arenas; and
5. networking of political connections;
6. networking with tourists.

Based on three case studies in Denmark, Baerenholdt *et al.* (2004, p. 25) then argued that tourism businesses might simultaneously belong to several types of network, not all of which are territorially bound and synomymous with 'destinations':

> Innovative horizontal networking practices among businesses and networking with tourists are mobile practices beyond the level of destinations, while practices in marketing, informal networks and political connections combine territorial and mobile networking to producing place images, cooperation in places and the materiality of tourist attractions and facilities in places. This combination of mobile and territorial networking is crucial. Tourist places are produced by the cooperative efforts of territorially defined relations and by the mobile interactions among tourist businesses, tourist organizations and tourists across place boundaries.

At issue here is the question of boundaries and which actors are included in the network. Destinations, they argue (p. 28), 'are first and foremost produced as images that are not territorially fixed.'

Scott *et al.* (2008b, p. 171) asserted that: 'Network analysis provides an alternative view to that of the "bounded" destination, as the degree of the links defines its spatial extent.' Pragmatically, however, Scott *et al.*, like many other researchers (Bodega *et al.*, 2004; Timur and Getz, 2008; Merinero Rodríguez and Pulido Fernández, 2009; Denicolai *et al.*, 2010) focused their empirical study on bounded destinations; in their case on the social networks of four Australian regions largely based on local government districts and marketing campaign regions.

More in line with the tradition of research on nodal regions and travel-oriented conceptualizations of destinations, other researchers have used SNA to analyse the spatial structure of both regional and local destinations. Now the nodes in the networks are the places visited by tourists whose movements constitute the linkages between them (Shih, 2006; Liu *et al.*, 2012; Aranburu *et al.*, 2016; Asero *et al.*, 2016; Piriou, 2016; Wang *et al.*, 2016). Such an approach appeals as a way of identifying, visualizing and measuring different structural characteristics, the strength of connections between places, the functions of different nodes and how these are influenced by their connectivity. The logic here is that: '[E]ach destination within a particular geographic region possesses development opportunities and constraints resulting from the influence of other destinations in the region' (Shih, 2006, p. 1038).

Studies which conceptualize destinations as networks tend to focus on analyaing and mapping the structural characteristics of the relationships between a set of actors or nodes, particularly those relating to density, centralization and tie strength (Pavlovich, 2003; Bodega *et al.*, 2004; Shih, 2006; Merinero Rodríguez, 2008; Scott *et al.*, 2008b; Timur and Getz, 2008; Merinero Rodríguez and Pulido Fernández, 2009; Liu *et al.*, 2012). Bodega *et al.* (2004) proposed a fourfold taxonomy of of interorganizational forms by crossing relational density with systemic centralization in a four-cell high/low matrix:

1. community model: low/low;
2. corporative model: high/high;
3. governed model: low degree of relational density/high degree of systemic centralization; and
4. constellational model: high degree of relational density/low degree of systemic centralization.

Favre-Bonté *et al.* (2016) characterize interorganizational networks along four dimensions:

1. The nature of the relationship between the members:
   i) horizontal, whereby relationships are built with competitors to share the same resources;
   ii) vertical, in which the aim is to achieve a transfer of additional resources, between client and supplier;
   iii) cross-industry between potentially complementary, non-competing organizations.
2. The mode of regulation through formal (e.g. written contracts) or informal (e.g. establishment of joint teams) mechanisms for coordination.
3. Structure or architecture which may take the form of centralized 'star' networks or decentralized community-based network.
4. The geographical scope of the network: local, regional or national. As in some of the cluster literature, geographical proximity and the enhanced frequency of interaction which it promotes is said to improve the functioning of the network.

Their application of this framework to French ski resorts is discussed in Chapter 9.

Most network studies depict relations within a destination at a particular point in time but a few acknowledge their dynamic nature and outline how destinations may evolve (Pavlovich, 2003; Gibson *et al.*, 2005; Merinero Rodríguez and Pulido Fernández, 2009). Figure 3.1 represents a much more complex destination in 2000 compared to the smaller and much simpler structure when tourism first developed at Waitomo a century or so earlier (Pavlovich, 2003).

Systems

The term system is used with varying degrees of rigour but might be thought of as 'a group of inter-related, interdependent and interacting elements that together form a single functional structure' (Weaver and Oppermann, 2000, p. 23). The basic concept derives from von Bertalanffy's general systems theory. As Leiper (2000, p. 570) notes:

> The distinctive purpose of general systems theory is to deal with anything or any idea that seems complex. The aim is to reduce complexity . . . The strategy for achieving this is, broadly speaking, breaking a whole (thing or idea) into its elements, and then identifying the crucial elements to see how they are connected.

Systems may be hierarchical in nature, being composed of a set of ordered sub-systems. They may also be open or closed depending on whether account is taken of the environments within which the system is embedded.

Following Cilliers (1998), Baggio (2008, p. 5) lists the characteristic properties of complex systems as:

> a large number of elements form the system; interactions among the elements are nonlinear and usually have a somewhat short range; there are loops in the interactions; complex systems have a history, the 'future' behavior depends on the past one; each element is unaware of the behavior of the system as a whole; it reacts only to information that is available to it locally.

Baggio (2014, pp. 446, 448) suggests that, by their nature, tourism destinations can be seen as complex adaptive systems:

> A destination has the properties of a system: an organized assembly of elements or parts (components) connected to each other with some defined relationship, and having the general objective of accomplishing a set of specific functions, or achieving particular goals ... the identification of destinations as systems is a useful analytical approach, but stimulates further questions on what type of system they are, what are the components and how their interactions affect their overall dynamic behaviour.

Destinations are commonly portrayed as some form of tourism system without reference back to a particular systems model or set of system characteristics. The use of systems allows for a more comprehensive conceptualization of destinations and a more explicit depiction of a wider range of factors and the relationships between them. The frequent lack of reference to a specific systems model means different elements and types of relationships are emphasized and they may be expressed more directly in tourism terms. As a result, destination systems are less uniform and more varied than the concepts of networks and clusters but might be categorized into two broad types: supply–demand systems and spatial systems.

Supply–demand systems

Studies portraying destinations as supply–demand systems focus on the relationships among the providers of goods and services in a particular location, the consumers of these (the visitors), the mechanisms by which the goods and services are promoted and delivered and the context in which this occurs (Laws, 1995; Machiavelli, 2001; Ejarque, 2005; Manente and Minghetti, 2005; Terhorst and Erkuş-Öztürk, 2011; Fayos-Solà *et al.*, 2014). Variations occur in terms of who or what is included, how the elements and actors are linked, and the importance of territory. Particular emphasis is given to the need to deliver an integrated set of goods and services. Machiavelli (2001, p. 8), for example, asserts: 'Tourists perceive their destinations as an integrated supply system, a "unicum" with which they relate and of which they evaluate the ability of providing satisfaction if all the main components find expression in a coordinated and synergetic approach.'

In one of the earlier models, Laws (1995) depicts a tourist destination as an input/output system in which inputs such as skills, resources and tourist expectations transform primary and secondary elements through various processes into a set of outcomes for each stakeholder group, for example, satisfaction for tourists, remuneration for staff and benefits or problems for residents. Understanding the way changes occur throughout the system are needed for effective destination management.

A similar model is proposed by Manente and Minghetti (2005, p. 230) who define a destination system as 'a group of actors linked by mutual relationships with specific rules, where the action of each actor influences those of the other so that common objectives must be defined and attained in a co-ordinated way.' Like Laws, they portray their destination system as an open one in which destination actors need to take account of their competitive environment (Fig. 3.2). Products are matched to visitor segments by the strategies and policies of local (including community) and non-local (e.g. tour operators) actors, and the behaviour of the visitors.

According to Bonetti *et al.* (2006, p. 112):

> assuming the territorial area as a starting unit of analysis, we can state that a tourism system does exist when all resources

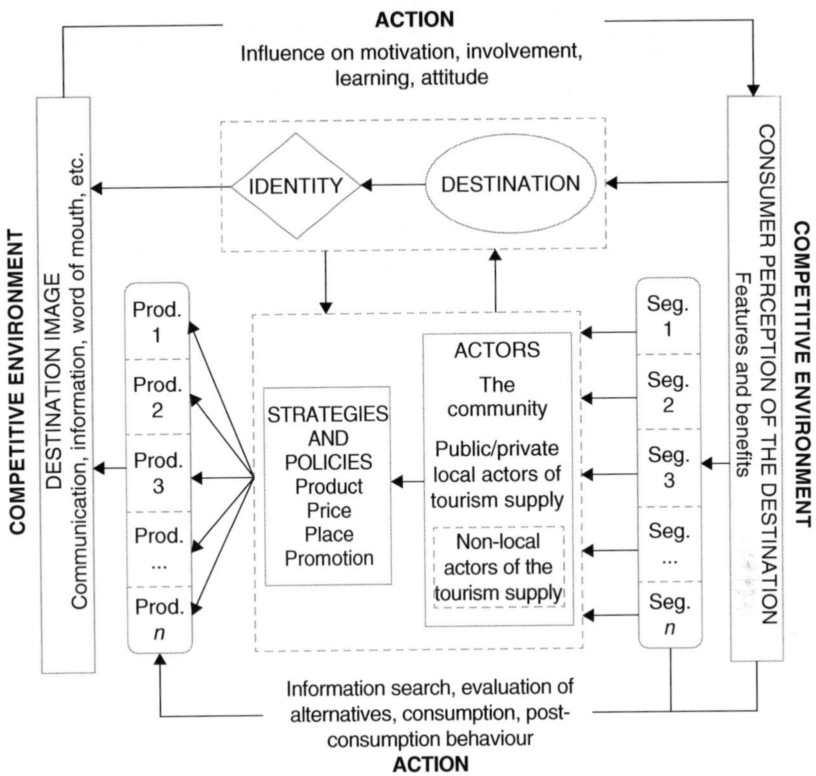

Fig. 3.2. The destination system. From Manente and Minghetti (2005), Figure 25.1.

> involved in the tourism activities are mutually linked by stronger relationships, compared to other territorial resources with different economic purposes.

They then focus on the relationships by which the tourism activities are linked in terms of two variables:

- the degree of interdependence among the system's stakeholders as determined by the density of relationships, the degree of willingness and the degree of mutual trust; and
- the degree of centralization of the system governance functions which is affected by recognition of the governance body, the degree of strategy formalization and the degree of centrality of the organizations.

Bonetti *et al.* cross these two variables in a quadrant matrix to produce a typology of 'four ideal types of tourism systems', namely:

1. market cluster (low interdependency/low centralization);
2. district (high interdependency/low centralization);
3. constellation (low interdependency/high centralization); and
4. tourism local system (high interdependency/high centralization).

Each of these types has a particular configuration (Fig. 3.3) but none has been shown empirically to be superior to the others with regard to sustainable development. Account needs to be taken not just of the system configuration but also factors related to the territorial area and the competitive context.

Prats *et al.* (2008, p. 182) 'understand *territory as a* collective *construction,* rather than simply as geographical proximity' and offer a systems model composed of four interconnecting blocks or major components: tourism agents, relational elements, the

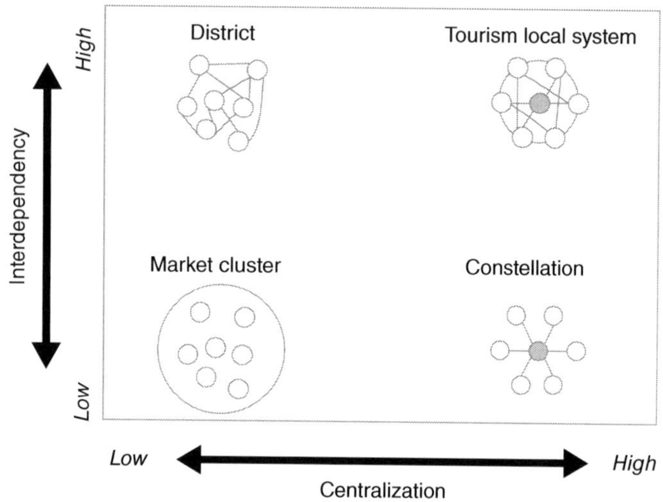

Fig. 3.3. Possible configurations of tourism systems. From Bonetti *et al.* (2006), Figure 8.4.

macro-environment and outcomes. In this way they bring out not only the interaction amongst the agents but also the impact of macro-environmental factors and the way in which the components influence the capacity for innovation.

Ejarque (2005) proposed three types of destination system based on the way in which and the extent to which the different products and services offered are integrated:

- the diffuse destination system in which there is little coordination and collaboration amongst providers;
- the dependent destination system typical of mature coastal resorts where integration occurs through the actions of external operators and there is little local autonomy in the way in which the products and services are offered in the market; and
- the integrated destination system where joint action has been required to combat a pronounced decline in demand.

In the FAS model of Fayos-Solà *et al.* (2014) a destination is presented as a system formed by the complex interaction of three sub-systems: productive Factors, Attractors and Support services. They provide a detailed list of the 27 dimensions by which these three sub-systems can be evaluated and emphasize that such evaluation should be carried out systemically and individually. Fayos-Solà *et al.* also stress (p. 71) that

> a destination is a system whose dynamics are formed by multiple agents interacting among themselves. These agents hold and control rights over different forms of capital, in the framework of existing institutional setups. Within a multi-agent system, the stakeholders develop their strategies in awareness of the physical and institutional environment.

They may be internal or external. Three groups of endogenous agents are recognized (local government, multiple private-sector agents and residents, professional and labour resources), together with three sets of exogenous agents (customers/tourists), intermediaries and competitors external to the destination and supra-local government (regional, national, international)).

Spatial systems

Rather than emphasizing relationships between actors and other factors within a destination, tourism spatial systems focus on linkages between places. Many spatial systems adopt an origin–linkage–destination structure whereby destinations are presented as a counterpoint to origins or markets (Pearce, 1995a, 2012).

The emphasis here is on destinations forming part of larger systems and on the way in which they are linked to origins by routes, transport services, distribution channels, flows of information and tourists. Leiper's (1979) model is one of the more widely known. He takes a whole systems approach and shows destination regions linked to generating regions by flows of departing and returning tourists. Overlaid on this is the tourist industry, depicted graphically in very general terms in the model but elaborated on in the accompanying text as comprising six sectors: marketing, carriers, accommodation, attractions, miscellaneous services and regulation. Leiper's model also shows the tourism system operating within a broader set of environments.

Again, different researchers emphasize particular elements and relationships in their systems (Pearce, 1995a, 2012). Some stress the structural mechanisms which influence the flows between origins and destinations and the form development takes in the latter (Hills and Lundgren, 1977; Britton, 1982). Others highlight the fact that many places are origins as well as destinations, that reciprocal flows occur between them and that a hierarchy of destinations may exist (Lundgren, 1982). Campbell's (1967) seminal model of travel patterns distinguishes between various groups on the basis of the relative importance of the travel and stay components of their trip. According to his model, 'recreational' travel is scattered radially from the city origin; 'vacational' travel is essentially linear and highway oriented; and that of an intermediate group of 'recreational vacationists' involves a hub-and-spoke pattern of side-trips from a regional base. The model also shows not all places that lie on a vacational path are visited; some are just passed through. Later models of travel patterns offer variations on this theme with regards to the ways in which multi-destination travel links multiple destinations in various combinations. Lue *et al.* (1993) proposed a fivefold typology of the spatial patterns of pleasure vacation travel: (i) travel to a single destination; (ii) en route; (iii) base camp; (iv) regional tour; and (v) trip chaining. Oppermann (1995) suggested the base camp pattern might be more appropriately seen as a form of travel associated with a single destination and then differentiated multi-destination travel into the following types: (i) stopover; (ii) full loop; (iii) destination area loop; (iv) jaw loop; and (v) multiple destination areas loop.

Dredge (1999) built on the work of Gunn (1972), Leiper (1979), Lue *et al.* (1993) and others to develop a normative planning and design model combining a destination region or regions; tourist generating markets; primary or secondary nodes comprised mainly of attraction complexes and service components; districts characterized by a different tourism emphasis; circulation routes and gateways through which visitors enter the region. According to this model, destination regions may have one or multiple nodes and be chained together by patterns of tourist movement. While it sets out some useful relationships and functions this model overlooks others. In particular, it does not take account of the issue of multi-functionality discussed in Chapter 2. The assumption that 'tourist generating markets and destination regions are separate geographical entities' (Dredge, 1999, p. 781) might be challenged (Lundgren, 1982). Primary nodes such as major cities are also often key gateways due to their transport infrastructure (Fig. 2.1). Moreover, the notion that 'a destination region is a location that a person chooses to visit for at least one night' (p. 781) does not deal adequately with aggregate travel patterns and overall demand which determine the relative importance of nodes and the links between them.

Miyakuni and Vander Stoep (2006) also take a planning perspective and suggest that rather than dealing with isolated components a tourism system can be analysed using a linkage lens to connect resources in an integrated system and thereby enhance the tourist experience. They broaden and specify the ways in which resources might be connected by identifying six overlapping types of linkages – physical, visual, transportation, promotional/economic, interpretive and informational – which they illustrate with a case study of a tourism corridor in Shuri, Okinawa.

Barrado Timón (2004) takes a different geographical perspective and emphasizes the way in which tourism is inserted into or developed within a certain region or territory (Fig. 3.4). Here a destination is portrayed as a sub-system which constitutes the area of overlap between two larger systems, one sectoral (tourism), one territorial. Parts of the tourism sector (e.g. most of the marketing and distribution) occur in the market. Likewise, not all of the territory in which a destination develops is taken up by tourism; space is also occupied in varying degrees by other functions and activities (e.g. general infrastructure, residential and industrial areas, etc.). The destination is that sub-system where elements of both systems coincide, that is, that part of a territory given over to the accommodation, infrastructure, facilities and resources used by tourists. Not only is it necessary to take account of the diverse relationships between these elements within the destination but also the relationships with the larger territory and the rest of the sector. While acknowledging the external links, Barrado Timón places particular emphasis on the spatial structure of destinations and the regions within which they develop, factors essentially neglected in the other concepts discussed. Variations in the nature, morphology and organization of destinations are explained by such factors as the perception of visitors and the scale and density of development.

A further perspective is provided by Violier (2009) who proposed an analytical model of the evolution and management of tourist places in which the spatial relationships between the 'world' and the destination are expressed in terms of different types of proximity. Violier distinguishes between time distance and cultural difference and argues that tourists will make greater use of intermediaries to arrange their travel to a destination when one or both of these forms of proximity increases. Intermediation will be greatest for destinations characterized by both cultural and time distance, for example, travel from France to China. He also argued that the co-presence of tourists, residents and public and private actors results in different forms of interaction according to the type of place and the relative role there of tourism. In particular, in tourist cities where tourism is subordinate to other functions interaction between these groups will be more complex than in those places variously labelled resorts, tourist sites, resorts, '*comptoirs touristiques*' (distinguished from resorts by a closed form of

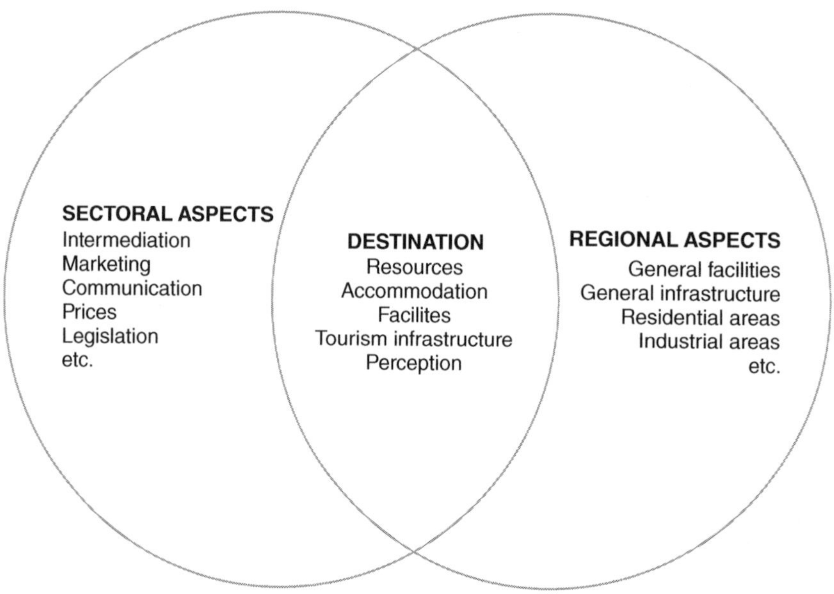

Fig. 3.4. Concept of a tourist destination. Redrawn from Barrado Timón (2004), Grafico 1.

management) and 'touristified' cities. Given these differences, different types of destinations might be expected to follow different development trajectories.

Fistola and La Rocca (2017) refer to the 'tourist essence' of cities; they portray urban tourism as a sub-system in cities viewed as urban systems and show how the nature and spatial extent of tourism vary according to the type of tourist cities they identify: emblem cities, scattered supply cities, mono-polar cities and niche cities. Moreover, from a systems perspective an entropic state in the tourism sub-system may produce disruptions in the larger urban system, for instance when pressures arise due to excessive numbers of tourists invading the urban space.

Social Constructs

Other researchers, especially social scientists, frame destinations as social constructs. Iwashita (2003, p. 331) summarizes social constructionism as 'the principle that social and cultural phenomena and individuals are social constructions produced and reproduced through historically and culturally situated interactions of people and human activities in society. Objects are not given in the world but constructed, negotiated, formed and fashioned by human beings through their social interaction and practices.' Social constructionism emphasizes transformation, involvement and the multiple meanings people give to their lives and the places they inhabit.

Ringer (1998, p. 2) and contributors to his volume *Destinations: Cultural Landscapes of Tourism* conceptualize destinations, or more specifically destination communities, as social constructs, as 'groups of people and their places of lived experiences whose cultural landscapes and local economies increasingly exhibit the influx of new ideas and changing patterns of social interpretation and communication associated with tourism's progress.' This stance reflects the view (p. 1) that 'tourism is a cultural process as much as it is a form of economic development' and stresses the multiple forces and factors that continually transform the socio-spatial contexts that constitute destinations. Saarinen (2004, p. 169) argued:

> As a social construct, the identity of a destination is a constantly changing 'outcome' of the process of transformation and the discourses constructing it, identifying the destination as a tourist space with both similar and distinctive elements relative to other tourist destinations.

Moreover, 'a destination as a social construction can be perceived and interpreted differently by different groups and individuals because social and cultural meanings are subjective and differ between individuals' (Iwashita, 2003, p. 336). One consequence of these multiple interpretations is that the boundaries of the destination are hard to define: 'The destination many expand or shrink and it may receive deeper or superficial meanings' (Saraniemi and Kylänen, 2011, p. 139). Similar ideas are expressed in Lemasson and Violier's (2009) volume where the notion of co-presence in tourism spaces recognizes that these are spaces shared, and changed, by tourists and residents.

Viewed from a social construct perspective stressing cultural as well as economic processes, change is a core attribute of destinations as they are made and remade by multiple stakeholders (Ringer, 1998; Saarinen, 1998; Lemasson and Violier, 2009; Saraniemi and Kylänen, 2011). According to Saarinen (1998, p. 160):

> the destination is not a stable, permanent socio-spatial structure, but a cultural landscape subject to continual transformation and reformation, in which it emerges, changes, disappears and re-emerges in varied forms. In this manner, destinations are institutionalized and de-institutionalized, reproducing social structures and meanings through the adopting and adaptation of previous ones.

In reviewing destinations as social constructs from a marketing perspective, Saraniemi and Kylänen (2011) give particular weight to the role of tourists as producers, as co-creators of their own experiences so that destinations might also be viewed as markets. They contend (p. 139) that:

> The success of destinations no longer lies in the bundle of products offered by different service providers to customers but in the competitive processes that enable consumers to immerse themselves into and find the elements that they seek in (re)presenting and (re)producing their fragmented identities.

Assemblages

Briassoulis (2017) challenges the notion of what she refers to as 'unitary whole (systems)' characteristic of conceptualizations of destinations as an amalgam of a set of supply-side components (Chapter 2), actor networks, industrial districts, clusters and these integrated supply systems. Instead, Briassoulis (2017, p. 310) argues that destinations are more appropriately considered as assemblages:

> Tourism demand activates spontaneous and planned processes of provisionally and contingently assembling biophysical and human, material and immaterial, mobile and immobile, existing and new components of a SES [socioeconomic system], other SESs, and spatial levels (the tourism supply) to meet this demand in a period. The resulting place- and time-specific, dynamic, and multi-level arrangement is a tourism assemblage (TA), an individual singularity, which weds tourism supply with tourism demand, turning, in the process, a place into a destination. The TA is place bound.

Rather than consisting as a unitary whole, destinations from this perspective are considered to be formed by multiple tourism assemblages. The components making up assemblages are rather more wide-ranging and more explicitly classified than in many of the other frameworks. In addition to the biophysical (e.g. beaches and forests) and human-made/material (e.g. sites of interest, tourist accommodation), Briassoulis also includes the immaterial (e.g. entrepreneurship, legislation and customary rules) and distinguishes between the collective (e.g. firms, public bodies and NGOs) and individuals (e.g. locals, entrepreneurs and tourists). Taking a broader, integrative perspective (p. 310), multiple 'TAs coexist and cofunction with numerous assemblages serving other human purposes but do not necessarily coincide with them spatially and/or temporally.' Furthermore (p. 311):

> The properties of the emergent TAs (extent, duration, distributions of characteristics, diversity, resilience, carrying capacity, etc.) are not the sum of the properties of their components but emerge from complex, contextual, and contingent interactions among them that depend on their capacities, especially of the critical or limiting components that deeply influence tourism development such as attractions, water, climate, culture, tourism demand, governance, and capital, several of which are shared with other assemblages.

Bruttomesso (2018) interprets urban tourism conflict in Barcelona through the lens of assemblage thinking. In one of the few empirical tourism studies that have taken this approach she examines the time- and space-specific events and processes associated with resident and urban activist attempts to reclaim a square in the old part of the city which has come under pressure from the expansion of tourism. Cohen (2018) suggests the concept of assemblage could provide an alternative way of investigating urban backpacker enclaves.

Discussion and Conclusions

The systematic juxtaposition and review of these six sets of frameworks further underlines the multi-faceted nature of destinations. It also shows that none of the frameworks addresses singlehandedly all of the features of a destination that emerge from this overview and the discussion in Chapter 2. Rather, each framework focuses on and attempts to account for particular features. Certain features recur; others are emphasized in particular frameworks and ignored or given lesser weight in others. Differences in interpretation do occur and some theoretical disagreement does arise not only because of the various lenses through which they have been viewed but also because the particular problems being addressed differ as do destinations themselves. Taken together, however, the different features of these frameworks are on the whole complementary and compatible rather than in conflict.

The recurring and interrelated features that emerge from this review can be grouped on three major dimensions: geographic, mode of production or organization and dynamic (Table 3.1). Each dimension is comprised of two or three constructs, each of which in turn is made up of a number of variables. In elaborating on this table below, points made by specific studies are referenced; more general ones drawn from the review above are not.

Table 3.1. Destination dimensions, constructs and variables.

Geographical dimension		
	Place	
		Contextual factors: political, economic and competitive environments
		Cultural characteristics and social embeddedness
		Geographical embeddedness
		Tourism resources
	Space	
		Spatial concentration and co-location of actors
		Boundaries and geographical extent
		Spatial scale
		External links
Mode of production or organization		
	Structure	
		Organizational configuration
		Interdependence: technical complementarity
		External linkages
		Coordination: institutionalization intermediation
		Co-creation
	Behaviour	
		Cooperative/competitive: complementarity, trust, economies of scale/efficiencies
	Actors	
		Sets of actors
		Meta-agent
Dynamic dimension		
	Structural evolution	
		New nodes, network density
	Changing places	
		Place characteristics
	Driving factors	
		Innovation and competitiveness
		Continuous adaptation

Geographic dimension

The geographic dimension of destinations discussed in Chapters 1 and 2 is expressed in terms of a range of variables accorded differing levels of significance in the different concepts and frameworks reviewed in this chapter. These variables can be grouped into the constructs of place and space but there is little consideration of the links between the two.

Place

Local and broader contextual factors are incorporated in some of the destination systems, particularly those taking an open systems approach. These factors include the broader political, economic and competitive environments. Cultural characteristics and the degree of social embeddedness – the strong links between tourism and the community that give rise to a particular milieu – are stressed by social constructionists and those adopting an industrial districts approach. Such attributes might also be acknowledged occasionally in some network studies but generally are not. Some cluster and systems studies refer to the importance of localized tourism resources. Similarly, the make-up of assemblages includes biophysical and human-made/material components. Other cluster studies and those based around networks, however, often omit resources. This may in part be due to the proposition advanced by Porter (1998) and other cluster and network proponents that success is no longer based essentially on resources and other comparative advantages but on the way in which these are exploited in a competitive manner which puts the focus on the actors involved in this exploitation. However, to be exploited competitively potential resources must first exist as these cannot as a general rule be re-located (Barrado Timón, 2004). This factor, coupled with most production and consumption of tourism occurring *in situ*, underlies the place-based, geographically anchored, territorially bound nature of destinations.

Space

Spatial concentration and the co-location of specialized tourism firms and activities are key characteristics of tourism clusters and industrial districts. Following Porter (2000), the benefits of co-location seen in terms of spatial proximity are used to explain the functioning of tourism clusters. Certain French studies take a more nuanced approach to proximity, distinguishing between geographic proximity, organized proximity and cognitive proximity (Bocquet, 2008; Marcelpoil and François, 2008; Clergeau and Violier, 2013). The focus here is on the relationships between firms and other actors within a particular locality or region, generally without any consideration of their specific location within the destination. The interdependence between the various physical elements providing for the tourists' 'total experience' – the attractions, accommodation, restaurants and so forth – is also recognized but again little or no attention is paid to where these are actually located. The 'mapping' of clusters and networks has generally involved linking the actors rather than depicting their spatial location or that of the physical supply-side elements such as accommodation and attractions. Except for some of the spatial systems models and the place networks these frameworks do not deal with relationships between destinations.

Differences occur with respect to the geographical extent of such concentrations of actors and physical elements, that is, whether they are 'bounded' in particular locations as portrayed in the conceptualizations of industrial districts (Becattini, 1990; Gaido, 2002); defined in cluster and network analysis (Porter 1998; Scott *et al*., 2008b) by a set of linkages; correspond to a production-consumption unit (Barrado Timón, 2004); or are seen to be inherently variable due to the multiple interpretations of places emphasized by the social constructionists (Saraniemi and Kylänen, 2011) or the place- and time-specific arrangements of multiple assemblages (Briassoulis, 2017).

Despite this concern with boundaries most of these frameworks do not deal directly with the issues of spatial scale which were shown to be important in Chapter 2. The hierarchical nature of spatial sub-systems draws attention to the destinations being conceptualized at different spatial scales (Pearce, 1995a, 2012). The concept of assemblages also incorporates the notion of panarchy which recognizes the 'influence of systems at scales above and below the focal scale' (Briassoulis, 2017, p. 29). By their very nature the focus of industrial districts and clusters is on a sub-national or sub-regional scale.

External linkages are a vital feature of system frameworks in which destinations are depicted as sub-systems and need to be understood as counterparts to origins or markets (Leiper, 1979; Pearce, 1995a, 2012; Manente and Cerato, 2000a). Reciprocal flows of tourists occur in such systems with the result that most places are origins as well as destinations but in general the notion of places having multiple functions is not explicit or developed along the lines discussed in Chapter 2. External linkages are less evident in the other frameworks (Fabry, 2009). For tourism it is not so much a question that the surplus of production cannot be sold locally (Becattini, 1990) but rather that external linkages by way of image creation, distribution channels and transport connections are essential to generate demand and encourage flows of tourists to the destination as most production and consumption incurs *in situ*. Moreover, as some of the network studies indicate, external network ties are also important for accessing information and as a source of innovation (Pavlovich, 2003; Sundbo *et al*., 2007; Camprubi *et al*., 2010).

Mode of production or organization

Structure

The use of frameworks of industrial organization has resulted in destinations being portrayed much more explicitly as modes of production or organization. A first concern has been with characterizing and analysing the structural elements of production, that is, the organizational configuration of the destination and the nature of the relationships involved. The underlying factor here shared by the majority of the studies cited is the mix of products and services needed by tourists: they require accommodation, attractions, transport, hospitality and so forth. Moreover, these products and services are largely consumed *in situ* (Gollub *et al.*, 2003; Fabry, 2009; Chhetri *et al.*, 2017). This technical complementarity (Terhorst and Erkuş-Öztürk, 2011) in turn results in the interdependence of multiple suppliers, those firms and organizations providing the products and services sought by the tourist. The relationships arising from this interdependence are variously described as horizontal, vertical and diagonal (Michael, 2003). As shown in Figs 3.1–3.4, the way in which these relationships are configured varies depending on the framework adopted. The relationships may be both formal and informal.

Interdependence implies a need for coordination. This may occur between two or more suppliers contractually or informally but some form of institutionalization through the intervention of destination management or marketing organizations (DMOs) and/or other organizations is often needed to coordinate and undertake destination-level activities such as joint marketing or advocacy for the tourism sector. As a result, the nature and degree of institutionalization may play an important role in the functioning of a destination. Coordination to deliver an integrated set of products and services to tourists can also occur in other ways, particularly through intermediation whereby packages including two or more products are put together and sold by tour operators and other intermediaries. The extent to which this occurs varies from destination to destination (Ejarque, 2005; Violier, 2009; Clergeau and Violier, 2013). Moreover, various combinations of products and services may be packaged, or, seen through the assemblage lens, various assemblages exist rather than a 'unitary whole' (Briassoulis, 2017). These structural elements are essentially seen from a supply-side perspective, especially that of the firm. Supply–demand systems emphasize the need to develop products and services sought by tourists (Fig. 3.2) and some may give greater weight to the active role tourists have in shaping what is offered (Fayos-Solà *et al.*, 2014). Others go further and argue that the tourist experience is co-created by the tourists themselves (Saraniemi and Kylänen, 2011; Fabry and Zeghni, 2012; Clergeau and Violier, 2013). Clergeau and Violier (2013) asserted that the extent to which tourists co-create their experiences is conditioned by the otherness of the places visited: they will play a greater role in shaping their own visits and experiences in familiar places and be more dependent on intermediaries in more exotic locations.

Behaviour

Attention is also drawn to the behaviour of firms within the destination. Although most of the authors do not draw directly on the wider concept of 'co-opetion' (Brandenburger and Nalebuff, 1996; Bengtsson and Kock, 2000), such behaviour is generally portrayed as cooperative but competitive. According to the frameworks and theories applied, cooperation is a function of complementarity, trust and economies of scale. The complementary nature of products encourages firms to cooperate as they recognize the benefits of working together to deliver a complete and satisfying experience to the visitor. In this context Maulet (2006) refers to the degree of systemic consciousness amongst firms while Bodega *et al.* (2004) emphasize the need for systemic rather than just individual success. Trust results from a shared culture and sense of belonging to a district. Economies of scale also play a significant role in fostering cooperative behaviour, particularly through the greater efficiencies that may be achieved. Infrastructural developments are also commonly a question of scale (Brown and Geddes, 2007). At the same time, following Porter's (1998) cluster arguments, the behaviour of firms remains competitive as each strives to attract its share of the destination's market.

Actors

As this discussion shows, different frameworks tend to stress the role of particular sets of actors involved in these modes of production. Most focus explicitly on tourism firms and perhaps also local authorities and other public institutions. Tourists and residents are included less frequently. They feature in some of the cluster and systems conceptualizations but it is with the portrayal of destinations as social constructs that tourists and residents are most directly and strongly acknowledged as active participants, not just in the co-creation of the tourist experience but also in that of the destination as a whole (Ringer, 1998; Lemasson and Violier, 2009; Saraniemi and Kylänen, 2011). But it is the collective nature of the actors and the way in which they interact to form the destination which is a common point in most of these studies.

Dynamic dimension

Application of these different frameworks to the study of destinations highlights their inherently dynamic nature. Some attention is given to the structural development of destinations in network analysis and through the growth of clusters or industrial districts in which changes in organizational structure are analysed with reference to factors such as the addition of new nodes and increasing density of relationships. The social constructivists in particular highlight changes in place characteristics as destinations are made and remade.

Two broad sets of factors are seen to drive the dynamism of destinations. Studies using network, cluster or industrial district frameworks stress the importance of innovation and competitiveness which are said to result from the flow of information, knowledge transfer and labour mobility which derives from co-location and external ties. Others, particularly those adopting a social constructivist approach, see destinations as places that are continually adapting and changing as a result of social as well as economic processes. In portraying destinations as complex adaptive systems, Baggio (2008) stresses their dynamic nature in terms of self-organization and self-similarity, robustness and resilience, and 'edge of chaos' behaviour. Similarly, tourism assemblages are also portrayed as being inherently unstable, 'adapting to sudden or gradual changes in components and/or their relationships' (Briassoulis, 2017, p. 312). Factors that may activate such changes include competition for resources from tourism and other assemblages, that is, this framework incorporates the impact of different forms of tourism and changes in the broader environment, one of the few to do so. Tourists may also activate tourism assemblages (Briassoulis, 2017) and clusters may be portrayed as demand driven (Chhetri *et al.*, 2017) but overall there is relatively little weight given explicitly to growth or change in demand as a driving factor. Violier (2009) also draws attention to the different trajectories that destinations may follow depending on the interaction between different groups of actors and the relative role of tourism in different places.

Table 3.1 does not imply that all destinations share the same characteristics. Rather, they vary. In terms of structure, for example, the level of organizational complexity may vary depending on the degree of technical complementarity and the number of owners; destinations may be more or less coordinated and have differing levels and forms of institutionalization and intermediation; co-creation may occur to a greater or lesser extent. The nature and degree to which destinations differ on these and other variables will also determine whether or not they conform to the conditions of particular frameworks, that is, whether or not they might be thought of as clusters, industrial districts, networks, some form of system, a social construct, an assemblage, or seen in some other way. While some of these authors have applied their framework empirically, others have just discussed them conceptually. What is needed now is a systematic review and evaluation of the extent to which the concepts discussed here and in the previous chapters are reflected in the empirical research on destinations. Of particular interest is the extent to which these frameworks elucidate the relationships between spatial and organizational structure given the dimensions and constructs summarized in Table 3.1. After the fuller discussion of the frameworks in this chapter further consideration can be given to any links between them and other ideas introduced in Chapters 1 and 2. With their emphasis on similar, small and medium-sized businesses, for example, the Chapter 3 frameworks generally say little about the role of large enterprises, issues of supplier concentration or product differentiation, key aspects of market structure. First, however, methodological issues relating to operationalizing these and other concepts and frameworks need to be addressed.

4 Analysing the Structure of Destinations

The diverse definitions and conceptualizations of destinations, together with the range of issues being examined, mean there are many different approaches to analysing the structure of destinations. Much of the discussion of organizational matters has been conceptual in nature and many challenges arise in operationalizing the various frameworks discussed in Chapter 3. The analysis of both spatial structure and organizational structure involves common issues of the interdependence and interconnectedness or interaction of multiple phenomena or actors but the two are generally treated in separate streams of research. SNA is one technique which has been used to examine aspects of both forms of destination structure but beyond this and issues of case studies or comparative research there is relatively little overlap in methodological matters.

Destinations do not develop in a vacuum. Studies of both spatial and organizational structure need to take account of the context and environment in which destinations have developed and exist, that is, the characteristics of the place will influence its structure and evolution. This point is emphasized in particular in the industrial district research and studies which stress the importance of milieu and social capital (Gaido, 2002; Mottiar and Ryan, 2006, von Friedrichs Grängsjö and Gummeson; 2006), studies which frame destinations as systems, particularly open systems (Barrado Timón, 2004; Manente and Minghetti, 2006; Prats *et al.*, 2008; Fayos-Solà *et al.*, 2014; Flores Ruiz, 2014) and research by social constructivists (Ringer, 1998). Barrado Timón (2004) and Flores Ruiz (2014) situate the destination in a territory and distinguish between broader environmental characteristics and those relating directly to the tourism sector. Gordon and Goodall (2000, p. 291) contend that:

> in 'real' places the [tourist] experience is conditioned by a range of services independently provided within a physical, social, and cultural environment (in both intended and unintended ways) by past patterns of development, as well as by the presence of visitors arriving independently.

What then is the economic, social, political and historical context in which the destination is set, what are its physical site characteristics and how do these factors affect form and function? In addition to competition within sectors, account has also to be taken of competing or complementary forms of land use (residential, commercial, industrial, conservation, etc.); of physical factors which encourage or constrain tourist-oriented services, facilities and activities; and of administrative structures and processes which shape the form and function of destinations (e.g. unitary or fragmented administrative units; areas with strictly planned or *laissez-faire* approaches to land use). The relative size and importance of other sectors will influence the competition which tourism faces in acquiring and developing sites and the demand generated from them, especially for business travel and visits to friends and relatives. Morphological differences might be expected between those destinations which form part of a larger settlement and those developed as enclaves or on green-field sites which are less constrained by land-use competition.

Destination landscapes will also reflect the time at which tourism initially developed and subsequent stages of growth occurred. In particular, they may be shaped by the dominant mode of transport which prevailed when the destination began to develop or experienced major periods of growth. Railways, trams, motor cars, planes and cruise ships all offer different levels of accessibility and mobility and have been important in varying degrees at different times in different places. Thus the location of some present-day accommodation may be the legacy of a bygone era when it was closely linked to a mode of transport which has diminished in importance, for instance, terminus hotels adjacent to major railway stations. Attitudes toward tourism by national governments, local authorities, planners, investors and residents also change over time as do the tastes of tourists. Periods of boosterism may be succeeded by phases of more controlled growth; rampant consumerism may be followed by a growing concern to act responsibly; fixed stays at beach resorts may give way to or be complemented by multiple, short city breaks as fashion or society changes. Other land uses in or around the destination will also evolve and their decline or increase will open up opportunities for or put pressure on the tourism sector. Each of these changes will impact on the supply and demand for products and services in destinations and shape the way the different physical elements are arranged in space and influence the relationships between actors.

Subsequent sections in this chapter deal in turn with methodological matters that are more specific to the analysis of the spatial and then the organizational structure of destinations. Conclusions are then drawn. More specific technical matters are dealt with as they arise in Chapters 5 to 9.

Analysing Spatial Structure

Analysis of the spatial structure of destinations focuses on the location, distribution and interaction of phenomena. The scale of analysis has a large influence on the data and methods used.

What is being examined?

Given the multi-faceted nature of tourism a wide range of potential variables in terms of both demand and supply may by analysed to determine the spatial structure of a destination. Some studies incorporate both demand and supply variables, many are limited to one perspective only and the set used is often far from comprehensive. What data are used will depend on the particular purpose of the study, the techniques employed and availability. Decisions over what is included will impact on the interpretation given and the insights gained.

Demand

Demand can be expressed with regard to such variables as visitor numbers, bednights and expenditure, each of which may be further categorized into segments according to such variables as origin, demographics, purpose of visit, motivation or category of expenditure. Often research is restricted to one segment, for example either international or domestic tourists but not both (Pearce, 1996a; Bowden, 2003; Brida *et al.*, 2017; Koo *et al.*, 2017). Just who is included is particularly important in terms of interpreting patterns of demand as this may vary significantly from study to study, from one destination to the next and between products and services. Many of the services and spaces used by visitors may be shared by local residents whose demand may be a more significant location factor for restaurants, entertainment, shopping and public transport, especially in cities where the local population outweighs the influx of visitors. Visitor demand associated with attractions will come primarily from leisure travellers while that for commercial accommodation and transport commonly also includes other visitors, such as those travelling for business. As the motivations and activity patterns of these segments may vary, careful attention needs to be given to who is included.

Supply

Differences in the mix of supply-side elements will occur from destination to destination due to their mutual technical complementarity (Terhorst and Erkuş-Öztürk, 2011). As these authors note, sun-sea-sand destinations show a high degree of technical complementarity of a limited range of goods and services while much less technical complementarity occurs in cities where many goods and services, particularly attractions, are substitutes of each other. Destinations having a large share of non-hotel accommodation will generate demand for more supporting services than resorts characterized by all-inclusive hotels (Pie i Ninot, 2013).

Moreover, supply-side sectors might be further sub-divided along various dimensions. Accommodation might be analysed by type of accommodation (hotels, motels, bed and breakfasts, backpackers, camping grounds); ownership and management (independent, chain or state-run); quality (e.g. according to some star rating system); size (number of rooms); whether it is formal (registered) or informal (permanent residences, second homes, Airbnb). Transport includes services and infrastructure to, from or within a destination; different modes (road, rail, air, water); public and private transport; public- or tourist-oriented. Other infrastructural elements such as water supply and sewerage systems are fundamental to the functioning of destinations but are generally overlooked elements in studies of spatial structure. Decisions therefore have to be made about what sectors are to be included and how they are to be measured (Majewska, 2015; Parrilla-González and Pulido-Fernández; 2017). Sarrión-Gavilán *et al.* (2015, p. 42) contend that statistics on bed places in regulated accommodation can be considered as 'a proxy indicator of tourism importance at municipal level'. However, as Sánchez Martín *et al.* (2017) point out, the distribution of accommodation is not always a reliable indication of the importance of tourism as existing bed-space capacity might not match an area's potential as determined by resources and location. In other words, accommodation might not be in the right place, some areas may be saturated while others are relatively under-served with respect to potential resources. This was found to be the case in their study of Caceres in Spain. Boundaries between classes of elements may also be blurred; some might have multiple functions or be seen and used by different visitors in different ways. Promenades in traditional coastal resorts, for instance, facilitate beach access but may also have an important social function, the *m'as-tu-vu* component (Clary, 1977), that contributes to the overall attractiveness of the destination.

Attractions are a key set of elements inasmuch as they underpin demand and provide the 'energizing power' of tourism systems (Gunn, 1979). As such, the location and distribution of attractions are likely to be key determinants of the spatial structure of destinations at a range of scales from the local to the national. Attractions are numerous and varied and are especially challenging elements to deal with comprehensively. Attractions might be classified as cultural or natural; for entertainment, education or both; organic or purpose built (cathedrals/theme parks); primary or secondary; open access or paid admission. Considerable debate has occurred over what constitutes an attraction, with little consensus on what should or should not be included (Lew, 1987; Leask, 2010). Gunn (1979) differentiated between attraction defined as experience and attraction defined as physical entity. P.L. Pearce (1991, p. 46) defined an attraction as 'a named site with a specific human or natural feature which is the focus of visitor and management attention'. MacCannell's (1973, p. 597) concept of an attraction was that of 'an empirical relationship between a *tourist*, a *sight* and a *marker* (a piece of information about a sight (e.g. guidebooks, slideshows)'. MacCannell's ideas on the structure of an attraction were developed by Leiper (1990, p. 187) who adopted a systems approach in which, following Gunn (1979), he substituted nucleus for marker and stressed the functions of different types of markers – generating, transit and contiguous – which 'act as the catalytic element, linking the human and nuclear elements of an attraction system'. Richards (2002, p. 1048) reported survey results which provided empirical support for Leiper's model, showed the use of different markers varied by the type of holiday taken and suggested that the notion that 'tourists are "pushed" towards attractions by their motivations'. In her review, Leask (2010) essentially took a site-based approach, with an emphasis on permanent features subject to visitor-oriented management. Such an approach excludes many of the less tangible characteristics which Matley

(1976) and Karski (1990) attribute to the attractiveness of many cities as well as interest in the culture of particular communities or ethnicities associated with people rather than physical phenomena. The nature of attractions will also change with scale: at the resort level the concern may be with beaches and ski slopes; at the regional or national level this will shift to coasts and mountain ranges.

A lack of clarity or completeness also occurs with respect to the use of the terms attractions and activities. In some cases these two terms are used synonymously or intermingled; in others an explicit or not-so-clear distinction is made between them. Most commonly, attractions are portrayed in terms of designated sites/sights whereas tourist activities are what tourists do, what they engage in, how they 'consume' the city. Thus the Eiffel Tower is an attraction, a site/sight, while someone going up it or admiring it from below is engaging in an activity, generally sightseeing. Likewise, shops might be attractions and shopping a tourist activity. Attractions as sites/sights or physical entities are fairly readily inventoried and mapped, particularly if they are seen as permanent features subject to visitor management. Activities present greater challenges in terms of coverage and their spatiality. Unlike attractions fixed in space, mobility is a key characteristic of many tourist activities such as sightseeing. Analysis of both attractions and activities is essential in building up a more comprehensive picture of the structure of destinations. These two strands of research are closely intertwined; one may inform the other but often they are carried out separately and the resultant studies offer only a partial view of the spatial structure of destinations.

Data availability

In practice, comprehensive sets of relevant, consistent data are rarely available and are also difficult to generate independently. Much destination research is based on official statistics. These are commonly derived from two sources: supply-side censuses of businesses, especially of accommodation, and visitor statistics derived from surveys, arrivals at borders, hotel registrations and the like. Official statistics are generally limited to the formal rather than informal dimensions of destinations. As a result, analyses of accommodation tend to focus on commercial establishments not stays in second homes or with friends and relatives; attractions which record entries and/or paid admissions are included more frequently than those that do not; public transport usage is emphasized over private. Statistics on supply are commonly more readily available and analysed more frequently than those on demand (Sarrión-Gavilán *et al.*, 2015; Hernández-Martín *et al.*, 2016). Distinguishing visitor demand from that by local residents is also a challenge. Commercial accommodation is essentially the domain of visitors but other facilities, services and spaces are frequented in varying degrees by visitors and residents alike: public open spaces, cultural and natural amenities, restaurants, shops, transport infrastructure and services. Stays with friends and relatives by definition involve private residences shared with guests. The advent of platforms such as Airbnb adds to this mix.

Other complementary sources of information are also employed. Field observations, aerial photographs and land-use zoning maps are commonly used in morphological studies (Pearce, 1998a; Longjit and Pearce, 2013; Pie i Ninot, 2013). These too have their limitations. Field observations and aerial photographs may enable land use to be mapped but, as is also the case with most statistical sources, by themselves they do not reveal patterns of ownership or jurisdictional responsibilities. Knowing how factors such as ownership and jurisdictional responsibility are configured on the ground becomes crucial in terms of explaining spatial structure and developing appropriate destination management structures and strategies. Visitor surveys and qualitative studies of demand and visitor behaviour are increasingly being complemented by technological advances enabling the tracking of visitors using global positioning system (GPS) devices or determining their spatial behaviour or through the analysis of 'big data' sets such as geotagged photos or cell phone calls (McKercher *et al.*, 2012; García-Palomares *et al.*, 2015; Hu *et al.*, 2015; Aranburu *et al.*, 2016; Raun *et al.*, 2016). These technological advances have enabled more spatial data to be captured and analysed but are generally practical only at a local scale or in small countries (Chapter 7).

The spatiality of the data needed will depend on the scale at which the destination is being examined (Pearce, 1997a, 1999). When the emphasis is at a regional or national scale it may suffice to know the total number of hotel rooms in a city, but where the focus is at the local level more detail will be needed on where those rooms are located, something which is not always readily available. If the data are available in areal units the question of spatial scale again arises: for example, are Parisian hotel data recorded at the level of the *arrondissement* sufficient or is a more detailed breakdown needed to fully understand the impact of locational factors and any relationships with other tourism elements or forms of land use? If data are available or can be collected on individual items (e.g. hotels or attractions) are they to be depicted discretely (e.g. dot distribution maps of hotels) or aggregated in some way? Moreover, as Wall (1997, p. 242) observed, attractions occur and can be considered as points, lines and areas and these can be viewed at different scales:

> at the scale of a country, destination areas, such as coastal resorts or national parks, may be viewed as a series of points. On the other hand, a single destination area may be viewed as a combination of points, lines, and areas, or as a series of nodes and links.

Truchet *et al.* (2016) refer to local, spread and diffuse attractions. With big data sets point data are statistically clustered into larger spatial units represented as various sized hexagons and polygons which may be further refined into areas of interest with irregular boundaries (García-Palomares *et al.*, 2015; Hu *et al.*, 2015; Vu *et al.*, 2015). GPS data may also be aggregated into cells or grids to define nodes or hotspots of tourist activity (Kadar, 2014).

Consistent time-series data are also required if the way in which a destination has evolved is to be examined (Baum and Haveman, 1997; Hernández-Martín *et al.*, 2016; Parrilla-González and Pulido-Fernández, 2017). These are often not readily available due to changes in definition and boundaries, differences in data collection methods and institutional changes.

Location and distribution

Common steps involved in analysing location and distribution involve identifying, compiling, plotting and grouping phenomena through cartographic representation and/or statistical analysis. Maps are fundamental tools used to establish the morphology of destinations and depict the distribution of phenomena at local, regional and national scales. Plotting the distribution of particular tourism elements (hotels, attractions, transport infrastructure) or land-use zones, a process assisted in recent years by the use of geographic information systems (GIS), enables spatial patterns in the target variables to be identified. Where multiple variables are included this in turn allows the spatial differentiation or co-location of the various elements to be established with varying degrees of detail and enables aspects of the relationships between them to be examined. The visual interpretation of patterns and qualitative accounts of observations in the earlier morphological studies have been augmented more recently by the use of cluster analysis and other statistical techniques to explore the strength of the relationship between the variables.

Lorenz curves are an effective means of depicting the extent of concentration or dispersion of tourism attributes across countries or regions (Pearce, 1995a; Dean Runyan Associates, 2013). A first step in the construction of Lorenz curves is the derivation of location quotients (LQs) for each areal unit (e.g. regions or counties). These are calculated by relating the percentage of the target variable (e.g. tourism-related employment or bednights) in an areal unit to the percentage of the countrywide or regional base variable (e.g. total employment or land area). Values greater than unity indicate a region is receiving relatively more tourism-related employment or bednights or than the national or regional average; values less than 1 show that it is below the average. Lorenz curves are constructed by ranking the areal units in descending order of their LQs and then calculating and plotting the target variable against the base variable. If there were a perfect correspondence between the distribution of the target variable and the base variable, that is there is a uniform distribution, then a straight diagonal line would result. The greater the divergence between the Lorenz curve and this diagonal, the greater is the degree of concentration of the target variable. Only general comparisons can be made across countries where the number and size of the areal units vary from country to country as the form the curves take is in part dependent on the number of units used. While Lorenz curves show the extent of concentration they do not depict where this concentration occurs. However, LQs can be mapped to depict those areas where tourism is more concentrated than others but this also requires decisions about category boundaries (Capone and Boix, 2008; Segarra-Oña *et al.*, 2012; Majewska, 2015).

Conventional mapping of the distribution of supply and demand, depicting, for example, the distribution of hotel beds or second homes by standard administrative units (Khiati, 2018), has been complemented by the use of Exploratory Spatial Data Analysis (ESDA) techniques to identify patterns of spatial clustering and association of phenomena at scales from the local through to the national (García-Palomares *et al.*, 2015; Hu *et al.*, 2015; Sarrión-Gavilán *et al.*, 2015; Vu *et al.*, 2015). Here the basic aim is to establish the nature and extent of spatial association, that is, the degree to which the characteristics of a spatial unit are shared by neighbouring units. Of particular interest in the present context are Local Indicators of Spatial Association (LISA), the most frequently used being the local Moran's I statistic (Anselin, 1995). LISA enables concentrations of high and low values to be assessed and spatial outliers to be identified by distinguishing between four types of local spatial association:

- high–high (HH): units with a high value surrounded by others with high values, often referred to as 'hotspots';
- low–low (LL): units with a low value surrounded by others with low values;
- high–low (HL): units with a high value surrounded by others with low values (outliers); and
- low–high (LH): units with a low value surrounded by others with high values.

Procedures for doing this are discussed in detail by García-Palomares *et al.* (2015), Majewska (2015) and Sarrión-Gavilán *et al.* (2015).

These techniques have been used for a variety of purposes and different types of data such as geotagged photos as an indicator of the clustering of attractions at the local level (García-Palomares *et al.*, 2015), a range of relative indicators to analyse the spatial distribution of supply at a regional level (Sarrión-Gavilán *et al.*, 2015) and interregional effects of agglomeration on a national scale (Majewska, 2015).

Three supply-side studies of the distribution of different facets of tourism in Andalusia demonstrate different sources of data and quantitative techniques which might be used to analyse the structure of a regional destination (Table 4.1). As the authors of the three studies note, Andalusia provides a very appropriate region in which to test a range of techniques: it is one of the most important regional destinations in Spain with a varied geography and a mix of coastal, urban and rural municipalities. Moreover, with the devolution of greater responsibility for tourism following the creation of the Autonomous Communities in Spain (Pearce, 1997b) greater attention has been given to issues of sustainability and to the spatial aspects of planning in Andalusia. This has brought a sharper focus on tourism at the municipal level and on the need to consider local characteristics and differences throughout the region. As Table 4.1 shows, the three studies aim to demonstrate the validity of particular methodologies applied to different intraregional issues (identifying potential tourist resources; analysing economic dynamism and monitoring spatial changes); they are set

Table 4.1. Characteristics of studies analysing aspects of the regional structure of tourism in Andalusia (author's summary).

Article	Merinero-Rodríguez *et al.* (2014)	Parrilla-González and Pulido-Fernández (2017)	Sarrión-Gavilán *et al.* (2015)
Aims	To generate and apply a model to identify the characteristics of a regional tourism system relating to the tourist potential of localities in Andalusia	To demonstrate how supply-side variables can be analysed to identify and segment local level destinations in terms of their economic dynamism	To illustrate the combined use of GIS and ESDA techniques as valuable tools for monitoring changes in the spatial distribution of tourism, and discovering patterns of spatial association
Framework	Regional tourism system	Clusters	Not stated
Variables at the municipal level	17 variables incorporating natural and heritage resources relating to theoretical segments and 5 variables relating to 4 common resources (accommodation, hospitality, tourist information and tourism associations)	30 variables for 1996 and 2011 covering accommodation, services, transport, some attractions (museums and cinemas), tourist offices, a tourism index and unemployment	Bed places in regulated accommodation (hotels, apart-hotels, hostels and pensions, apartments, camping grounds and rural tourism establishments) for 2000 and 2011 plus population
Techniques	Factor analysis and cluster analysis	Latent class analysis	GIS and ESDA
Findings	Identification of 4 types of municipalities on the basis of their tourist potential: pure coastal; pure urban; mixed urban; and rural (effectively not tourism destinations)	Identification of 3 groups of municipalities in terms of their economic tourism dynamism: very dynamic destinations (44); dynamic destinations (56); and non-dynamic municipalities (660)	29 maps showing much detail of differing patterns of distribution and spatial association, including: a trend towards an increase in the number and size of clusters; some decrease in pressure on local population but growth in environmental pressure in some mountain areas and on the Costa del Sol

against different conceptualizations of destinations, use different techniques and sets of variables and come up with varied but related patterns and findings.

Merinero-Rodríguez *et al.* (2014) employ the term regional tourism system, a term also used, if superficially, in the new spatial planning policies and plans of the autonomous communities. They understand a regional tourism system to be the configuration of all the types of geographical spaces found in a given autonomous community based on tourist resource characteristics at a local/municipal scale. Merinero-Rodríguez *et al.* derive a classification of municipalities but there is no subsequent attempt to show any links between these different spaces, something which might be expected in a systems approach. Parrilla-González and Pulido-Fernández (2017) situate their study in a broader discussion of destination concepts, including notions of complementarity associated with clusters, the idea that destinations are dynamic by nature, and the love of variety and coordination theorems of Candela and Figini (2012). They then draw on Candela and Figini's ideas and Cocossis *et al.*'s (1991) pentagonal model to develop a framework of tourism dynamics consisting of five components: infrastructure, services and tourist attractions; accessibility; coordination and auxiliary services; governance and tourism management; and knowhow. In contrast, Sarrión-Gavilán *et al.* (2015) do not set their spatial statistical analysis against any destination concept; they identify spatial clusters but deal with these only in methodological not conceptual terms.

The selection and availability of relevant region-wide data at the local level is raised in each study. Sarrión-Gavilán *et al.* (2015) focus on statistics on bed places in regulated accommodation. They also include permanent population figures which they relate to tourist bed spaces available to derive Defert's Tourist Function Index (DFTI) as a measure of the pressure tourism exerts on municipalities and calculate the number of bed spaces per square kilometre (TSI) as a proxy indicator of pressure on the environment. With their focus on tourist resources, Merinero-Rodríguez *et al.* (2014) began with a much wider set of variables which they subsequently reduced to two smaller sets: 17 natural and heritage resource variables relating to the ten theoretical tourist segments outlined in the regional tourism plan (e.g. sun, sand and sea; cultural-historical, beauty and well-being) and five variables relating to the four common resources: (i) accommodation (total beds); (ii) hospitality (total places); (iii) tourist information (total number of tourist offices and total number of webpages); and (iv) tourism associations (total number of associations related to tourism). Reflecting the idea of complementarity inherent in clusters, Parrilla-González and Pulido-Fernández (2017) include the widest range of data, 30 variables for the years 1996 and 2011 covering accommodation, services, transport, some attractions (museums and cinemas), tourist offices, a tourism index and unemployment. No measures of two of the five pentagonal components (governance and tourism management and knowhow) were used because of the non-availability of data. They thus limit their focus to the economic dynamics of tourism rather than a more complete picture of tourism dynamics. To reduce the number of variables in order to group the municipalities by cluster analysis, Merinero-Rodríguez *et al.* (2014) undertook a two-step factor analysis; first of the natural and heritage resource variables which resulted in ten indicators of tourist potential, and second of those indicators which in turn yielded three tourist resource factors: urban, coastal and nature. The five common variables were used as controls. Cluster analysis subsequently resulted in the identification of four types of municipalities on the basis of their tourist potential.

At the local level much effort has been given to identifying and quantifying the importance of attractions and mapping their distribution. Ashworth and de Haan (1985) and Jansen-Verbeke and Lievois (1999) identified then plotted listed buildings and heritage attractions in Norwich and Leuven respectively through the use of markers, namely mentions in guide books or promotional material. Ashworth and de Haan (1985) attributed a value to each attraction based on the length of entry in the publications used. Jansen-Verbeke and Lievois (1999) devised a typology of heritage clusters formed by two dimensions which influenced their role as tourist attractions: a morphological positioning index and a functional index. Pearce (1998a) compiled published and unpublished entry data on a range of sites in Paris from agencies responsible for managing or monitoring monuments, museums and other sites known to attract visitors, such as the sewers. When mapped, these data show not only the location of the attractions but also the number of visitors, segmented in many cases into domestic and international visitors (Fig. 7.1). However, a varying proportion of the entries, especially to the religious sites, are Parisians and not visitors to the city. Visitors might appreciate some attractions such as the Eiffel Tower or the Arc de Triomphe without actually entering them and thus being recorded in the official statistics. Moreover, these attractions are primarily cultural venues and while Paris is undoubtedly known for its monumentality and museums, visitors to Paris do visit other attractions and engage in activities not recorded in entry data such as going for a walk, visiting gardens or taking a river cruise on a *bateau-mouche*

(Pearce, 1997a). In a survey of international visitors to Paris, Freytag (2010) included both 'sights to be visited' and 'very important activities' but no data are provided on where the latter (e.g. taking pictures, spending time with travel partners and escaping from daily life) occurred. Surveys are commonly carried out to identify the features of a destination that tourists are aware of or visit and the activities they engage in, either as the main focus of a study or a complementary aspect of it (Ashworth and de Haan, 1985; Hartmann, 1988; Keul and Kühberger, 1997; Jansen-Verbeke and Lievois, 1999; Shoval and Raveh, 2004; Freytag, 2010; Dejbakhsh *et al.*, 2011; Budović and Ratkaj, 2018). In particular, surveys enable profile and other data on the visitors to be combined with characteristics of the attractions visited so that patterns of spatial behaviour of different segments can be established.

The advent of photo-sharing on websites such as Flickr and Panoramio has led to the analysis of geotagged photos as a means of identifying tourist hotspots or areas of interest in cities (Kadar, 2014, 2015; García-Palomares *et al.*, 2015; Hu *et al.*, 2015; Vu *et al.*, 2015; Zhou *et al.*, 2015; Gutiérrez *et al.*, 2017; Encalada *et al.*, 2018). The underlying assumption here is that the photos taken by tourists represent the urban features they are interested in, that is, they constitute the attractions of the city. Distinctions between photos uploaded by tourists and residents have been made through the use of owner IDs (Vu *et al.*, 2015) or the period during which the photos have been taken (García-Palomares *et al.*, 2015). Not all tourists, of course, take photos let alone share them publicly on Flickr and Panoramio and not all attractions and activities are photographed, especially nocturnal and indoor ones. However, vast numbers of geotagged photos can be located accurately as point data before being clustered objectively using a variety of spatial correlation techniques to produce maps of tourist space usage in urban areas. Comparisons of the rankings of images with the rankings of attractions on Planetware, a professional travel website (Zhou *et al.*, 2015), and assessment of the degree to which 'the spatial clusters [of photos] coincided with the cities' main attractions' (García-Palomares *et al.*, 2015, p. 414) have shown good levels of consistency. Geotagged photos can also reveal areas of interest to tourists outside known landmarks or aesthetic areas (Hu *et al.*, 2015). In general, however, the emphasis has been on where the photos have been taken rather than what they depict. In the present context, a major advantage of the application of the same techniques of data collection and analysis across multiple cities is that it greatly facilitates inter-city comparison of the extent to which common spatial patterns can be found in the way in which tourists photograph, and, by extension, use cities (Kadar, 2014, 2015; García-Palomares *et al.*, 2015, Hu *et al.*, 2015). Analysis of photos taken in different years also provides a means of analysing changes in spatial patterns over time and enabled Encalada *et al.* (2018) to identify major monuments and other attractions in Lisbon that remained stable over the period 2007–2014, emerging tourist hotspots and areas where visitor interest had waned.

Spatial interaction

Other techniques are employed to examine patterns of spatial interaction which provide important insights into the spatial structure of destinations and contribute to differentiating between destinations and the various functions they perform. In the vast majority of these studies tourists and their travel patterns are at the heart of the analysis. Economic flow data may also be used.

The complexity, challenges and opportunities of analysing international tourist flows by aggregating and abstracting individual itineraries into larger patterns of multi-destination or circuit travel are extensively reviewed by Forer (2005). Drawing on the space–time aquarium of Hägerstrand, Forer suggests itineraries can be conceptualized in terms of space, time and activities (*x,y t,a*). 'Working with tourism', he states (p. 30), 'requires a sound data model, robust analytical tools and clear visualisation'. He then goes on to demonstrate the various ways by which data from New Zealand's international visitor survey and from customized regional surveys can be analysed using GIS to produce maps showing not only the networks of different market segments (e.g. by origin or mode of travel), both as desire line flows or as 'grounded' flows, but also maps portraying which days on a trip particular destinations are more likely to be visited. Other maps show the average duration of stops and arrival times at destinations. Desire line maps depict networks of straight line flows between origins and destination while grounded maps portray flows interpolated with reference to actual road networks (Figs 5.13 and 5.14).

The basic structure of domestic tourism can be derived by determining the direction and intensity of movements between sets of regional origins and destinations based on the analysis of a regional origin–destination matrix which shows the destinations visited by tourists from each originating region and the level of demand recorded in the destination regions from each regional origin (Pearce, 1993b, 1996a; Guardia Galvez *et al.*, 2014). Such a matrix indicates that at a national level domestic tourism is made up of interregional or intraregional flows (those occurring within the home region, i.e. regions falling along the diagonal of the matrix) and that the regions are origins as well as destinations. This latter point is often neglected in destination studies although it is explicitly recognized in several early models such as Fig. 2.1 and that of Lundgren (1982). The net inflow or outflow of tourists reveals whether a region might primarily be considered an origin or a destination but this is a question of relativity; to a greater or lesser degree regions will have both functions.

In principle, all possible flows in an origin–destination matrix can be taken into account. In practice, the detail involved in mapping all flows is complicated and creates issues of legibility and comprehension if some form of synthesis is desired. Dominant flow analysis, which focuses on the two largest flows into and out of each region, is a useful approach to identify basic national patterns, particularly when those flows account for a large proportion of demand (Pearce, 1993b, 1996a). Analysis of these patterns of actual flows can be usefully complemented by use of the Relative Acceptance (RA) Index which measures the magnitude of such flows against some predicted value. The RA Index is obtained by the formula:

$$RA_{ij} = \frac{A_{ij} - E_{ij}}{E_{ij}}$$

where RA_{ij} is the relative acceptance from origin *i* to destination *j*; A_{ij} is the actual flow from origin *i* to destination *j*; and E_{ij} is the expected flow from origin *i* to destination *j*.

More recently, monetary flows resulting from domestic tourism have been analysed using various econometric approaches (de la Mata and Llano-Verduras, 2012; Haddad *et al.*, 2013; Álvarez-Díaz *et al.*, 2017).

A regional origin–destination matrix for domestic tourism is generally constructed from data drawn from national household surveys (Pearce, 1993b, 1996a; Instituto de Turismo de España, 2013; Dominguez Perez, 2016; Álvarez-Díaz *et al.*, 2017; Khiati, 2018). Such surveys are resource-intensive as a large sample is needed to ensure reliable information is provided on detailed patterns of demand. As a result, few countries carry out this research on a regular basis and the amount of detail published also varies. The scope for generalization is therefore limited.

Key concepts drawn from SNA and applied in flow-based network studies to identify nodal functions and the spatial structure of destinations include different measures of centrality: degree centrality, closeness centrality and betweenness centrality (Freeman, 1978/79; Scott, 2000). Degree centrality measures the number of direct ties which a given node has to other nodes in the network. In-degree centrality relates to the sum of inward ties and concentration while out-degree centrality refers to the number of outward ties and dispersal. Together the two measures provide a means of assessing a node's function in the network, for example as an entry or exit gateway, hub or end point. Closeness centrality is based on geodesic distance and measures how close a node is to all other nodes. Higher values for closeness centrality indicate greater transferability and the likelihood the node is at the core of a network (Liu *et al.*, 2012). Betweenness centrality reflects a node's indirect links to other nodes in the network. Applied to destinations, a node with high betweenness centrality means 'that it is a highly critical intermediary between pairs of other destinations, since most tourists will stop at this destination while traveling between other various destinations' (Shih, 2006, p. 1032).

Other techniques have also been used to operationalize and assess the different and multiple roles which nodes may have. Pearce and Elliott (1983) examined the role of a given node in a total trip by use of the Trip Index (*Ti*). This relates the number of nights spent at the destination (*Dn*) to the total number of nights on the trip (*Tn*) according to the formula:

$$Ti = \frac{Dn}{Tn} \times 100$$

A *Ti* value of 100 shows the entire trip is spent at the destination, that is, the place is the focus of a single destination trip. An index of 10, on the other hand, would indicate it accounted for 10% of a multi-destination trip. A value of 0 shows no overnight stay was made at the node and that it was being visited either in transit or on a day trip. Pearce and Elliott (1983) demonstrated how the Trip Index distinguished between visitation patterns to a major urban centre, Christchurch, and a national park in the South Island of New Zealand. The latter was almost exclusively visited as part of a larger circuit whereas the proportion of time spent in Christchurch was greater, including a significant segment of visitors who made it their sole destination. Differences also occurred by origin and purpose of visit, demonstrating nodes may simultaneously have several roles of greater or lesser importance. Murphy and Keller (1990) incorporated seasonal variations in their application of the Trip Index to Vancouver Island, British Columbia, and were able to develop a functional classification of areas (e.g. year-round destination, mixed market, transit area) and identify a hierarchical spatial pattern influenced by the location of the island's two principal gateways. In South Carolina Uysal and McDonald (1989) found higher Trip Index values in coastal cities while other areas were more important as stopover and/or short trip destinations.

Lohmann and Pearce (2010) developed an analytical framework of nodal functions and applied it to ferry passengers going to two contrasting nodes in New Zealand linked by inter-island ferry services: Wellington, the capital and major urban centre; and Picton, a small town, railhead and port popular with summer visitors. Both these nodes typify the eccentric location at the end of a catchment area characteristic of many gateways (Burghardt, 1971). The framework incorporates the Trip Index but also retains the absolute number of nights and explicitly includes the main reason for going to the node rather than making assumptions about this (Fig. 4.1). Absolute length of stay was deemed to be important as travellers staying some time in a node are more likely to engage with the tourism sector and treat it as a destination, either primary or secondary. Three possibilities for the number of nights spent in the node were considered: travellers who do not stay overnight, those who spend one night and those who stay two or more nights. In terms of purpose of visit the focus was on determining whether or not the main reason for going to the node was to make a ferry connection, that is, the node was being used principally as a gateway to enter or leave the North or South Island. On this basis four classes of visitor were identified: gateway travellers; overnight gateway visitors; stopover visitors; and destination tourists. Wellington was shown to have a more evenly distributed range of functions than Picton dominated by its gateway role.

At the local level the focus is on the analysis of intra-destination movements, establishing the activity spaces of tourists and examining how the places they visit and the activities they engage in are linked together. In this way points of interest are identified from their movements rather than listed *a priori* as is often the case with visitor surveys. Tracking tourist movements in an urban environment is challenging due to the large and mobile populations found in cities and the resources and techniques required to collect and analyse detailed data on spatial

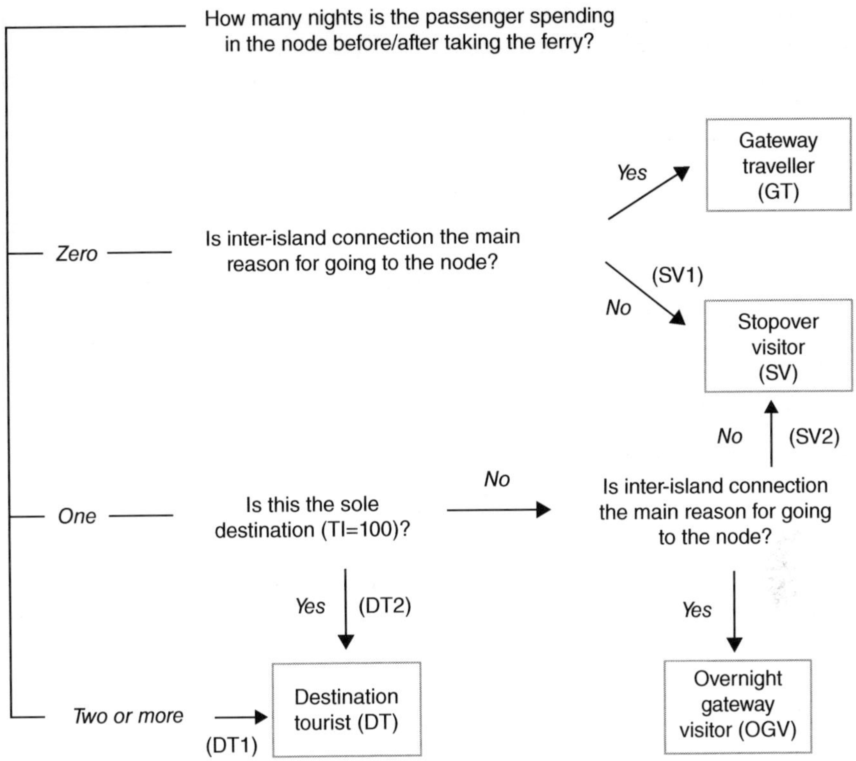

Fig. 4.1. Operationalizing nodal functions. From Lohmann and Pearce (2010), Figure 2.

and temporal patterns. Various approaches have been used to address this issue.

Early attempts involved observation, that is, following visitors around the city, supported in some cases with interviews, surveys or pedestrian counts (Murphy, 1980; Hartmann, 1988; Keul and Kühberger, 1997; Jaakson; 2004). McKercher and Lau (2008) examined the intra-destination movements of independent travellers within Hong Kong by combining the use of trip diaries with arrival interviews and a post-departure survey. Their study did not depict where these travellers went but they identified 11 movement styles and highlighted the role of territoriality (distance travelled) in influencing these:

> Daily territorial engagement can range from the highly constrained, at one end, where no movement occurs, to limited or convenience-based engagement in the immediate surrounds of the hotel or to deep engagement where the visitor moves freely throughout the destination.
>
> (McKercher and Lau, 2008, p. 372)

Dejbakhsh *et al.* (2011) asked international visitors to Melbourne to trace their movements around the city centre on a map and to a complete a questionnaire designed to test cultural differences in their movement patterns based on Hofstede's four dimensions: power distance; individualism/collectivism; masculinity/femininity; and uncertainty avoidance. Han *et al.* (2018) used SNA to explore the intra-destination movements of Japanese and Chinese visitors to Seoul based on an exit survey in which the respondents listed ten places they had visited in the Korean capital and the order in which they had visited them.

More recently, GPS techniques have been used to track tourist movements over time and space in cities such as Bilbao (Aranburu *et al.*, 2016); Freiburg (Bauder and Freytag,

2015); Lisbon (Caldeira and Kastenholz, 2018); Palermo (de Cantis *et al.*, 2016); Poznań (Kotus *et al.*, 2015); Melbourne and Sydney (Edwards and Griffin, 2013); and Hong Kong (Shoval *et al.*, 2011; McKercher *et al.*, 2012; Grinberger *et al.*, 2014). Due to technical and logistical issues, GPS data are usually only collected at a local level and for a single day's activity from each participant, group travellers are normally excluded and sample sizes may be small. GPS applications enable the individual paths of tourists to be tracked and aggregated as lines linking nodes, the nodes representing points of interest or hotspots of tourist activity. This reveals the intensity of movement along particular itineraries and visitation rates at particular places. Both types of patterns can be analysed to provide a picture of urban space usage by tourists (Bauder and Freytag, 2015) but the focus tends to be either on itineraries and the way places are linked together or on the intensity of visitation to particular parts of a destination. Vu *et al.* (2015) also illustrate the possibilities of mapping the trajectories of tourists according to the spatial and temporal sequence of geotagged photos and use Markov chain analysis to analyse the flows of Asian and Western tourists in Hong Kong.

Additional information relating to tourist profiles can be a collected by complementary surveys to segment visitors *a priori*. In contrast, de Cantis *et al.* (2016) segmented independent cruise passengers visiting Palermo, Sicily, based on movement attributes derived from GPS tracking data (speed, total length of tour, total duration of tour and number of attractions visited) and plotted the mobility patterns of the seven groups identified. In a multi-method study of Poznań, Poland, Kotus *et al.* (2015) combined GPS tracking of four groups of visitors with a short survey and a focus group study, an inventory and mapping of the spatial structure of attractions and other urban/tourism elements and analysis of the city's image based on photos uploaded on Flickr. Analysis of the resultant spatial patterns showed a large degree of consistency between the different techniques used.

Qualitative research also has a role in understanding the behaviour of urban tourists. Simon (2010) and Lepan (2013) carried out of in-depth semi-structured interviews with visitors to Paris and while their sample sizes were small and the generalizability of their results is limited, the richness of detail on what the visitors did, where and why provides a basis for new interpretations and reveals useful insight into attractions, activities and tourists' spatial behaviour in *la ville lumière*.

Analysing Organizational Structure

In broad terms organizational structure is about the relationships between sets of actors and their functions: '[T]o understand organizing we must start by finding out who the actors are, what they do and how they do it' (von Friedrichs Grängsjö and Gummeson, 2006, p. 63, citing Weick, 2001). Some characteristics are common to many of the frameworks discussed previously but methodological variations also occur because each framework shapes the way in which the research is undertaken and emphasizes particular points. Actors are given different weight in different frameworks and studies, relationships are analysed in different ways. Unlike the spatial structure studies, much of the organizational research does not deal directly with the issues of scale which were shown to be important in Chapter 2 and are more common in research on spatial structure. By their very nature the focus of industrial districts and clusters is on a sub-national or sub-regional scale: districts occur within a larger unit; clusters imply spatial concentration. Some network studies include national or even international ties but most deal with local or regional destinations as do those adopting a systems or social construct approach. As scale increases the number of actors and relationships expands rapidly and their analysis becomes more complicated whereas many measures of location and distribution can be readily applied across large territories and country-wide.

Networks

In empirical studies SNA, as outlined in the spatial structure section, is one of the most common means of analysing the organizational structure of destinations. In most network studies the emphasis is on analysing the relations between tourism firms, destination management or marketing organizations and to a lesser extent central or local government agencies (Timur and Getz, 2008; Tinsley and Lynch, 2008; Merinero Rodríguez and Pulido Fernández, 2009; Erkuş-Öztürk and Eraydin, 2010; Franch *et al.*, 2010). As Baerenholdt *et al.* (2004) have noted, who is included varies according to the type of network. Following Szarka (1990), Lynch and Morrison (2007) suggest that in exchange networks the actors are those companies and organizations with which the firm has commercial transactions; and in communication networks non-trading links providing offical or semi-offical information may exist with such actors as consultants, advisors and local and central government; social networks are made up of family, friends and aquaintances. Viken and Aarsaetter (2013, pp. 43–44) distinguish between 'informal, stable, and trust-based connections between tourism operators that exchange information, services in kind, and who practice reciprocity in commercial transactions' and governance networks in which local authorities and development agencies take the lead in encouraging tourism operators to form a network for marketing purposes.

Tourists are included less frequently as actors in a network. Denicolai *et al.* (2010, p. 261) define the tourist destination as 'a local network that creates value with and for the tourist; the latter is at the centre of the value creation process and not at the end, as in the traditional value chain process'. However, their subsequent empirical analysis of networked core competencies focused on tourism firms and public institutions. A similar value creation perspective is adopted and developed by Sfandla and Björk (2013, p. 498) who contend that 'a network without the existence of tourists is deeply meaningless in a tourism experience context'. They stress the role of tourists in the co-creation of their own experiences in a Tourism Experience Network (TEN), a network which involves value-in-use relationships between tourists as consumers and other actors and value-in-exhange relationships between between tourists and facilitators. The non-linear TEN connects to and complements a linear experience supply chain based on added-value exchanges between facilitators and other actors. The TEN proposed is wholly conceptual in nature and was not set explicitly in a destination context. Baerenholdt *et al.* (2004) provide some empirical evidence of networks with tourists, noting the close social interactions of repeat tourists with hosts of bed and breakfasts and guest houses on the Danish island of Bornholm and the role such hosts play in recommending activities and attractions to their guests. Such observations are not underpinned by detailed structural analysis of networks in the same way as networks of firms and organizations have been mapped (Scott *et al.*, 2008b; Merinero Rodríguez and Pulido Fernández, 2009). A major challenge here of course is the far larger number of tourists that would be involved compared to other actors.

Selection of the actors is an important initial step in empirical studies. In network studies this typically involves building up an initial list based upon information obtained from key informants and subsequent snowballing, from membership lists of DMOs or other relevant organizations, trade directories and websites (Scott *et al.*, 2008a; Timur and Getz, 2008; Merinero Rodríguez and Pulido Fernández, 2009; Pulido-Fernández and Merinero-Rodríguez, 2018). In some cases individual members or a sample of them will be surveyed; in others, the research will be limited to representatives of sector organizations. Community representation is generally through some community organization rather than from members of the community at large. Subsequent networks may display differentiated groups of actors: Prats *et al.* (2008) distinguish between research and development (R&D) centres, private firms, public administration and local community; Timur and Getz (2008) depict three groups: tourism industry, government and host environment.

Variations occur not only in the actors included in analyses of the architecture of destination networks but also in the nature of the relationships between actors and the various measures used. In Timur and Getz's

(2008) study the links between actors were operationalized by the existence of business contacts and the content of the links were joint tourism programmes or projects. Pulido-Fernández and Merinero-Rodríguez (2018, p. 142) conceptualized a tourist destination as 'a network of relationships established between the different tourism actors involved in the productive process of tourism in that destination' and had their respondents identify the actors they were involved with and the nature of their relations with them. They complemented this with documentation to check which stakeholders were involved in the different activities such as marketing or planning and development. Scott *et al.* (2008b) focused on the frequency of contact between the actors. All three studies measured centrality; Scott *et al.* (2008b) also took account of density and modularity; and Pulido-Fernández and Merinero-Rodríguez (2018) used indicators of density and intermediation as well. In other studies the focus has been on establishing the power and influence of different stakeholders in which measures of their centrality have been supplemented by information obtained through in-depth interviews and surveys (Beritelli and Laesser, 2011; Kennedy and Augustyn, 2014).

Pavlovich (2003) employed similar constructs (centrality, density, strong and weak ties) but took a very qualitative approach in her case study of the evolution of a small New Zealand destination (Fig. 3.1). Drawing on a mix of interviews, archival data and personal observations she mapped the changing pattern of network relations over a 100-year period and accompanied these with a richly descriptive narrative interpreted in terms of network theory. Other network studies have used different methods to examine factors which influence the nature of relationships between network members. Gibson *et al.* (2005) investigated the values associated with development–adaptation, cooperation–competition and commercial–non-commercial issues of the members of a local destination network. They combined the analysis of documents (e.g. meeting minutes) with observations at meetings and a questionnaire survey. In an inductive study of the mechanics of network collaboration in a local hotel network in Sweden, von Friedrichs Grängsjö and Gummeson (2006) carried out in-depth interviews of committee members, supplemented with information from member websites and data on the overall level of bookings and the level of external competition.

Kelman *et al.* (2016) argue that it is not an either/or issue in terms of the use of quantitative and qualitative techniques in network research; rather, the two approaches are complementary and both are needed to fully understand network structure and behaviour. In one set of destinations information obtained from interviews with network members provided confirmation of network patterns resulting from quantitative SNA and insight into their causality. In a second destination the interviews revealed weak and diffuse ties which had not been uncovered by the quantitative analysis.

Clusters and industrial districts

Although other stakeholders have been recognized in some cluster studies the focus in most cases has been on the range of firms involved in providing tourist services and supporting organizations and institutions (e.g. local government, universities, banks, consulting firms) and sometimes the local community (Gollub *et al.*, 2003; Miller and Gibson, 2006; Acuña Ortega and Brugnoli, 2007; Mazaro and Varzin, 2008; Kachniewska, 2013; Chin *et al.*, 2015; Perles-Ribes *et al.*, 2017). A distinction may also be made between firms or institutions and the leadership of key individuals, particularly in terms of the success of clusters (Hawkins and Calnan, 2009; Gardiner and Scott, 2014).

Empirical research on tourism clusters generally encompasses a wider range of factors than network analysis. These factors are generally framed in some explicit conceptualization of a cluster. In particular, many of these studies draw on Porter's (2000) diamond of sources of competitive advantage or develop some specific tourism extension or modification of it (Silva Lira, 2002; Jackson and Murphy, 2006; Miller and Gibson, 2006; Thomazi, 2006; Miller *et al.*, 2008;

Chin *et al.*, 2015). These frameworks suggest a list of related factors to be investigated. Varisco (2004), for instance, considered variables related to the business dimension, integration (vertical, horizontal and transversal), innovation, endogeneity and articulation (coordination and cooperation). Jackson and Murphy (2006) argued some characteristics of industrial districts are shared by clusters but the latter also have an additional set of defining features (Table 4.2).

Information on these factors is then compiled and analysed in greater or lesser depth through varying degrees of fieldwork. Often this includes a qualitative assessment of these factors based on compilation of basic data and documentation, observation and interviews with a range of actors. Chin *et al.*'s (2015, p. 8) study in Bali, for example, involved 'an intense three-week visit using a rapid rural appraisal type approach'. In other instances, relationships are examined in more detail though survey research with questions being asked directly about issues such as the degree of integration, challenges faced and attitudes towards cooperation and competition (Varisco, 2004; Jackson and Murphy, 2006; Thomazi, 2006; Perles-Ribes *et al.*, 2017).

Table 4.2. Characteristics of industrial districts and clusters. From Jackson and Murphy (2006), Table 1.

Characteristic	Industrial district	Cluster
Interdependence of firms	*	*
Flexible firm boundaries	*	*
Cooperative competition	*	*
Trust in sustained collaboration	*	*
Community culture and supportive public policy	*	*
Shared understanding of competitive business ethic		*
Private-sector leadership		*
Wide involvement of cluster participants		*
Appropriate cluster boundaries		*
Institutionalization of relationships		*
Social structure and attention to personal relationships		*
Life cycles		*

Perles-Ribes *et al.* (2017) took a very targeted approached in addressing the question of whether a cluster was a necessary condition for success in Benidorm. They focused primarily on private firms directly related to the tourism value chains on the grounds that prior research had established the importance of entrepreneurial activities and private-sector leadership. In terms of the composition of the target population of their survey they selected 'those activities which would not exist or which would not be so large without the tourism development of Benidorm' (Perles-Ribes *et al.*, 2017, p. 1585). Their semi-structured questionnaire survey focused on four themes: (i) competition in the destination and leading activities; (ii) existing interrelations (vertical and horizontal integration, formal and informal relationships); (iii) innovation behaviour related to types and sources of innovation; and (iv) actions that could contribute to improving Benidorm's competitive position as a tourist destination.

Marco-Lajara *et al.* (2014, p. 782) are quite explicit about what is needed:

> any author trying to identify the existence of an industrial district in a given territory must set out to prove that there is a high geographical concentration of firms associated with the same industry, that those firms are SMEs and, finally, that a community of firms and people exists as well.

With regard to the latter point they argue that the question of scale is critical and opt for the use of local labour systems.

Mottiar and Ryan (2006) incorporated triangulation in their case study of Killarney which provides a good example of the ways in which aspects of the industrial district concept applied to destinations can be operationalized. Data collection involved the use of key informant interviews, a survey of a spectrum of 81 local tourism suppliers, archival sources, field notes and observations. The information gathered was then analysed and interpreted in terms of five themes said to characterize industrial districts. These were similar but not identical to those shown in Table 4.2, namely: (i) geographical and sectoral concentration: firm size; (ii) strong inter-firm relations; (iii) competition and cooperation; (iv) social and professional milieu; and (v) innovation. In terms of interrelationships, most of the firms saw their main competition coming from local businesses. When asked directly businesses denied cooperating with each other but closer examination showed a lot of informal cooperation took place between people who knew each other well. This included referrals between businesses offering different products as well as between those in the same sector and a lot of 'good neighbourliness'. Major factors underlying the interaction between businesses were being friends or former colleagues, being neighbours and family ties. Factors emerging from the examination of the milieu dimension included being related to other entrepreneurs in the area, having worked for another tourism firm in the area and belonging to a local family with a long involvement in tourism.

Quantitative techniques have also been used to analyse and map the spatial association or agglomeration dimensions of tourism clusters and industrial districts and to explore the nature and relative performance of the 'district' or 'cluster' effect (Capone and Boix, 2008; Fernández Cuevas *et al.*, 2008; Estevão and Ferreira, 2012; Segarra-Oña *et al.*, 2012; Majewska, 2015). The emphasis here is on establishing overall patterns of clusters and districts within countries rather than on establishing relationships between firms or other actors within a particular cluster or district as is the case with the qualitative research just outlined. The focus is on the relative distribution, composition and co-location of firms at a range of territorial units and the nature and extent of aggregate inter-firm relationships is determined on the strength of statistical tests of association rather than direct information on the ties between firms. Approaches differ but in general involve multi-step procedures whereby cluster analysis is used to classify then map spatial units based on the composition of tourism-related firms (Table 4.3).

The differences shown in Table 4.3 highlight methodological issues raised earlier and are apparent on each of the characteristics listed. Variations that occur in terms of the areal units used reflect both differences in the administrative structure of each country, data availability and methodological decisions. The two Spanish examples illustrate this last point well. Fernández Cuevas *et al.* (2008) analysed lower level municipal units which they then classify and attempt to aggregate up to larger districts. In contrast, Segarra-Oña *et al.* (2012, p. 1323) opted for provinces as their aim was 'not to define the geographical scope of the cluster but rather to compare territories whose boundaries have already been determined'. They acknowledged that provinces had a high degree of internal heterogeneity but argued that they were related to administrative areas responsible for tourism promotion and corresponded with destination brands (e.g. Costa del Sol, Malaga; Costa Blanca, Alicante).

Even allowing for differences in aims and availability, the challenges of deciding on what tourism variables to use to characterize a cluster or district are clearly illustrated by the wide range in number and type employed in the five studies, particularly those relating to the attractions sector. Majewska's (2015) is the only one which includes measures of demand. In three of the studies tourism units are designated using LQs. Estevão and Ferreira (2012) classified all the territorial units in Portugal without first differentiating them into tourism and non-tourism categories. Different types of typologies result from the different aims and procedures used. The clusters referred to by the authors result from cluster analysis and are statistical and not necessarily spatial in nature; they show similarities in the make-up of the units rather than their spatial connectedness. The spatial aggregation of units, one of the key elements of clusters and districts, is dealt with in varying degrees. Capone and Boix (2008) map all the local tourist production systems (LTPSs) identified without any geographical grouping: the 'art cities' (e.g. Rome, Florence, Venice), for instance, are not contiguous. In the case of Portugal the tourism regions are discussed in terms of whether or not they are homogenous with regard to performance of the lower level territorial units (Fig. 5.6). Segarra-Oña *et al.* (2012) effectively treat each of the provinces reaching the cut-off point of the LQs for tourism-related employment as a cluster. Fernández Cuevas *et al.* (2008) map all tourism municipalities and subsequently incorporate a spatial element into their identification of tourism districts in Spain on the basis of geographic functional and proximity; that is, tourism municipalities are close both physically and in terms of the dominant form of tourism. However, the actual procedure for doing this is not outlined. Majewska (2015) developed a spatially weighted location quotient (SWLQ) procedure to explore the interregional agglomeration effects of tourism in Poland. In this the geographic spillover effects are first captured by weighting LQs with the local Moran's I statistic which enables interregional tourism clusters to be identified and mapped. Majewska then distinguished between localization-driven and urbanization-driven clusters using a modified Hirschman–Herfindhal Index (HHI) and the share of tourism in the regional economy. In the third step, tree cluster analysis was undertaken combining these two measures of supply using data from the Polish Division of Activities together with demand variables (number of tourists staying overnight, number of overnight stays) to identify patterns of similarity in interregional tourism clusters. This resulted in eight groups of similar districts. Analysis of the performance of the clusters or districts identified by various economic measures and examination of the composition of clusters and the nature and extent of agglomeration provide very useful attempts to combine and explore elements of both spatial and organizational structure. The results of these studies are presented and discussed in Chapter 5.

Social constructs and assemblages

Research on destinations viewed through the lenses of social constructivism and assemblage thinking share a concern for the use of multiple forms of qualitative research and the inclusion of a wide range of actors to ensure the voice of those who may be neglected by other approaches, especially local communities, is heard (Hughes, 1998; Saarinen, 1998; Squire, 1998; Briassoulis, 2017; Bruttomesso, 2018; Cohen, 2018). Discourse analysis is favoured by some social constructivist researchers. Drawing on imagery and documents from a range of archival sources, Saarinen (1998) discussed the transformation of the Saariselkä tourism region in Lapland in terms of the discourses of region and of development. Hughes (1998) analysed how tourist places are produced in four different genres: guidebooks; advertisements; tourist brochures; and tourist trails. Squire (1998) suggested the experiences of different actors can be explored by such means as open-ended interviews and focus groups. She then goes on to examine gender in the history of tourism in the Canadian Rockies based on the textual analysis of women's diaries, letters, interviews and railway advertising material.

Cohen (2018, p. 113) proposes that urban backpacker enclaves can be seen as 'co-constituted by multiple co-functioning assemblages of young travellers, local service providers, landowners, guest houses, hostels, coffee-shop and bar owners, and more remote suppliers, tourism entrepreneurs, politicians, officials, planners, and the backpackers' families and friends'. Bruttomesso (2018) primarily collected her data on urban tourism conflict in Barcelona through 'engaged participation', taking an active part in local protest action while at the same time gathering information and recording events by means of photography, field notes and participant observation rather than by structured interviews. Given the range of actors being considered and the means used to collect the information required, considerable expertise and effort would be needed to provide an overall picture of a city by extending assemblage methodologies across the 'multiplicities' said by Briassoulis (2017) to make up a destination.

Systems

Comprehensive empirical studies of destinations as systems have proven difficult to operationalize and have been hampered by the range of data and techniques needed to

Table 4.3. Characteristics of quantitative national studies of tourism districts and clusters in Europe (author's summary).

Article	Fernández Cuevas *et al.* (2008)	Capone and Boix (2008)	Segarra-Oña *et al.* (2012)	Estevão and Ferreira (2012)	Majewska (2015)
Country	Spain	Italy	Spain	Portugal	Poland
Framework	Industrial district	Industrial district	Cluster	Cluster	Agglomeration
Aims	To identify types of specialization in tourist resorts and to establish tourist districts To measure relation between tourism and second home development and tourism and well-being	To establish a geography of LTPSs in Italy and to explore their sources of growth and competitiveness	To investigate whether hotels located inside clusters perform better economically than those outside the clusters	To evaluate tourism Clusters in Portugal and map them through multi-variate analysis	To identify and measure the 'interregional' effects of spatial agglomeration in tourism considering the occurrence and strength of the geographic spillover effects in both localization-driven and urbanization-driven clusters in Poland
Areal unit	Municipality	Local labour markets/ municipality	Province	Territorial units/tourism regions	District (NUTS-4)
Tourism variables	Demographic, housing and accommodation coefficients; indices of tourism and economic activity, registered items of cultural interest; national or regional parks; inland or coast location	19 activities in a 'broad tourism *filière*'	'Whole tourism industry' 3 NACE sectors Intermediaries, hotels and other accommodation: restaurants, bars, etc.	WTO tourism satellite activities disaggregated to 5 digits to give 53 activities	Supply (4 classes), structure and relative position of tourism market (modified HHI; tourism entities as % of all entities); demand (overnight tourists, overnight stays)
Designation of tourism unit	Positive values on dichotomous variables (e.g. national park) or greater than average score on other variables	LQ ≥ 1 in tourism employment	LQ > 1.25 in tourism employment	All units included	All units then those identified by significant local Moran's I statistics and by SWLQs.

Tourism typology	1) Non-specialized 2) Rural and nature 3) Second home 4) Traditional tourism (urban–cultural and 'vacational') 5) Coastal	– Large 'art cities' and some medium cities have lowest LQ – Specialized medium-small cities – Most probable 'tourist districts' have highest LQ. Further sub-divided by make-up into 3 clusters	All characterized as mature 'S S S'	3 classes labelled according to average turnover: 1) low-performance 2) medium-performance 3) high-performance	Classified by agglomeration types and tourism demand: 1) localization-driven 2) urbanization-driven 3) mixed (localization + urbanization 4) spillovers of urban type 5) spillovers of localization type
Cartographic depictions	Comprehensive map of tourism municipalities by type and separate maps of each typology type	Map of distribution of local production systems in Italy	Map of Spanish tourism clusters (6 non-contiguous provinces).	Clusters mapped by territorial unit and region, collectively and then by performance type showing compos-ition by economic activities	Maps showing: distribution of tourism firms in absolute numbers distribution of tourism firms by LQs Moran's I statistics of overnight stays (HHI) tourism firms by spatially weighted (by overnight stays) LQs
Other analysis	Grouping of tourism municipalities into tourist districts Regression analysis of performance variables for tourism and non tourism municipalities	Non-spatial and spatial modelling of growth in tourism employment	Analysis of variance of performance variables of hotels inside and outside of tourism clusters	Analysis of composition of cluster types by economic activities	Cluster analysis of most specialized districts within interregional clusters by degree of localization and urbanization economies

NACE, Nomenclature for Economic Activities in the European Union; NUTS, Nomenclature of Territorial Units for Statistics.

examine the multiple elements and relationships among them. Manente and Cerato (2000b) present and illustrate a methodology for studying a destination system based around four main tasks: analysis of demand; local actors; tour operators and other intermediaries; tourism brochures and communication tools. Baggio (2008) outlines a range of techniques for how destinations viewed as complex systems might be examined, but in general empirical work in this area is limited. The empirical aspect of Prats *et al.*'s (2008) study, for example, is confined to network analysis of actors using well-established techniques.

In a more comprehensive study Rognant (1990) used a systems approach incorporating flows, nodes, networks and surfaces to first develop a general model of the economic spatial structure of Italy which incorporated notions of the North–South imbalance of contrasts and gradients. He then progressively built up a national tourism model based on an analysis of domestic and international flows, networks of different types of destinations (coastal, thermal and alpine resorts, historic cities) and the dispersion of other forms of tourism (agri-tourism and second homes). Again a systems terminology was employed: he refers to sources, gradients, axes, poles, a core, peripheries and pioneer zones. Domestic flows are shown to radiate out of sources, the major urban centres, decreasing in intensity with distance. A tourism core is defined consisting of a mix of densely populated coastal zones and four major crossroads, with Rome constituting a secondary southern core. The southern periphery of the peninsula and islands is complemented by 'interior peripheries' in the north and areas such as the Po Valley, transit zones which are passed through quickly. These zones are then linked to the country's broader economic structures and various regional models are developed, ranging from the 'complete' model of Tuscany (including 'metropolitan', 'coastal' and 'mountain' components).

Market structure

As the spatial structure studies show, there is a wide range of demand-side data available but this has rarely been used to examine questions of buyer concentration and linked explicitly to other aspects of market structure (Viken and Aarsaether, 2013). The lack of readily available comprehensive data on ownership and the total number of firms in a destination has limited analysis of seller concentration. Some work has been done on concentration and competition in the hotel sector (Pan, 2005) and a more comprehensive cross-sectional approach has been taken in small island states where wider and more complete coverage is more practicable due to the size of these destinations (Britton, 1990). Product differentiation in the hotel sector has generally been considered in terms of standard variables such as room size, price and star ratings (Baum and Haveman, 1997; Urtasun and Gutiérrez, 2006; Cirer-Costa, 2013a). Differentiation of attractions, on the other hand, often requires a more subjective assessment. This is also the case with assessing conditions of market entry where much of the work is done by experienced consultants (MarketLine, 2018).

Discussion and Conclusions

As this chapter has shown, many different approaches to the analysis of the spatial and organizational structure of destinations are possible. This arises in part out of the varied assortments of complementary and competing products and services which tourists seek that are made available through the participation of multiple, interdependent and largely co-located actors. It is also a consequence of the various ways in which research on aspects of destination structure has been framed. Comprehensive studies of destination structure require multiple data sources, especially if destinations are to be studied as clusters, systems or places where the total visitor experience occurs. Difficulties in obtaining a full coverage of all destination elements and dimensions often lead to studies which take either a broad, generalized approach covering a range of sectors or which focus in more detail on specific sectors such as accommodation, on parts of a destination (e.g. the historical core or waterfront) or on a limited range of actors, most commonly firms. The type of data available and the uses to which they can be put often vary from one scale to another with the result that different issues may be studied at each scale. The paucity of relevant time series data results in many studies that portray patterns and relationships at a point in time rather than dynamic analyses showing how destinations evolve.

To date, there has been little convergence in the streams of literature on spatial structure and organizational structure. These studies are generally undertaken through different lenses, addressing separate problems, using different techniques and different sources of data. The quantitative analyses of tourism clusters and industrial districts start to bridge the gap by exploring the relationships between location and sectoral composition, even if the links between firms are determined indirectly at an area unit level rather than between firms directly (Capone and Boix, 2008; Estevão and Ferreira, 2012; Majewska, 2015). As discussed in Chapter 9, another area where the two streams come together more explicitly is in some of the work on the structure of ski resorts where the particular location and form of these destinations are directly associated with organizational aspects underlying their development and management. Early descriptive accounts of ski resorts (Préau, 1968; Cumin, 1970) have been followed by more analytical research using notions of geographical and organized proximity (Bocquet, 2008) or viewing ski resorts as networks (Favre-Bonté *et al.*, 2016; Gajdošík *et al.*, 2017). Such research suggests the two dimensions of destination structure are closely intertwined and that there is much to be gained by drawing the various stands of research together. The following chapters endeavour to provide such a synthesis by considering in turn structural aspects of destinations at a national, regional and local level, with separate chapters of the latter being devoted to urban destinations, coastal resorts, ski resorts and rural and natural area destinations. In each case spatial structure is considered first from both a supply and demand perspective before attention is given to organizational matters and the links between the two are explored.

5 Structure of National Destinations

As Chapter 2 outlined, countries are one scale at which destinations are considered. It is at the national level where many policies regarding the development and marketing of tourism are made. Understanding the structure of tourism at this level is crucial for decisions about what might be developed and where and how a country is to be marketed internationally. National level analyses also reveal where the impacts of tourism are being felt and shed light on destinations at a lower level by assisting in regional and local differentiation and identifying the various functions which smaller-scale destinations perform.

Analysis of destination structure at a country-wide scale raises some interesting conceptual and methodological issues. In terms of international tourism, a country might be considered a national destination *to which* and *within which* tourists from other countries travel. The analysis of international tourists' travel between countries is usefully complemented by research on their intra-national movements once they have arrived in a country and patterns of international demand there might be broken down to the level of regional and local destinations (Pearce, 1995a). International tourists talk about visiting a country but the extent to which this actually happens is unclear. Are they really visiting a country in the sense that they are travelling widely throughout it or are their visits confined to a limited number of regions or localities? What geographical coverage is needed to speak of truly visiting a country? Is it sufficient just to go to a single city, resort or region? Domestic tourism involves travel *within* one's home country between sets of what are effectively regional and/or local origins and destinations. The interregional flows of domestic tourists might be considered as the dynamic element under-pinning overall domestic demand. A comprehensive country-wide study of the structure of destinations at a national level would integrate these two conceptually different phenomena, a challenging task in view of the different types of data and analytical techniques needed to carry this out.

The following sections discuss the distribution of demand; the travel patterns of international and domestic tourists; the distribution of supply, including consideration of chain hotels and airline networks; and institutional frameworks for tourism. Two examples are then provided of how syntheses of the structure of destinations at a national level might be developed, first of Spain and then of New Zealand. Patterns and issues which arise throughout the chapter are then discussed and drawn together.

Distribution of Demand

Tourism demand is not dispersed evenly throughout countries but concentrated in particular regions or areas. Figure 5.1 depicts the Lorenz curves for measures of demand for international and domestic tourism in France (hotel bednights, 2016), the Republic of Ireland (expenditure, 2017) and New Zealand (expenditure, year ended August 2018). The three graphs show that on these measures the demand for tourism for all three countries is geographically concentrated and that the demand for international tourism is more concentrated than that for domestic tourism. In each case demand for both forms of tourism, but especially for international tourism, is heavily concentrated in the leading metropolitan area: the Île de France accounted for 48.5% of international hotel bednights in metropolitan France and 21.3%

of domestic hotel bednights; Dublin recorded 40.2% of international tourist expenditure and 16.3% of domestic tourist expenditure in Ireland; and in New Zealand Auckland received 37.1% of expenditure from international tourists and 23.3% from domestic tourists. The concentration of demand in these metropolitan areas is a function of their limited areal extent (which accentuates their LQs) and the accessibility and pulling power of the countries' largest urban areas, which are also their dominant gateways. The difference between the shares of international and domestic demand in large part reflects the fact that large urban areas are also the leading sources of domestic tourists who by definition are visiting other regions or counties. Though the measures are different, the shares of the top five of 13 regions in France and of 26 counties in Ireland (combined values were reported in some cases where demand was limited) were virtually identical: 77.1% for international demand in each case and 51.7% and 51.2% respectively for domestic demand. In New Zealand the top five (of 16) regions accounted for 68.1% of international expenditure and 53.8% of domestic expenditure. In France and New Zealand the rankings for four of the top five regions were the same for both international and domestic demand; greater variation occurred in Ireland.

Figure 5.1 reinforces an earlier study based on Lorenz curves for 11 European countries which also showed that the both international and domestic tourism were spatially concentrated and that, with a couple of marginal exceptions, international demand was more concentrated than that from domestic tourists (Pearce, 1995a). Various interrelated supply and demand factors

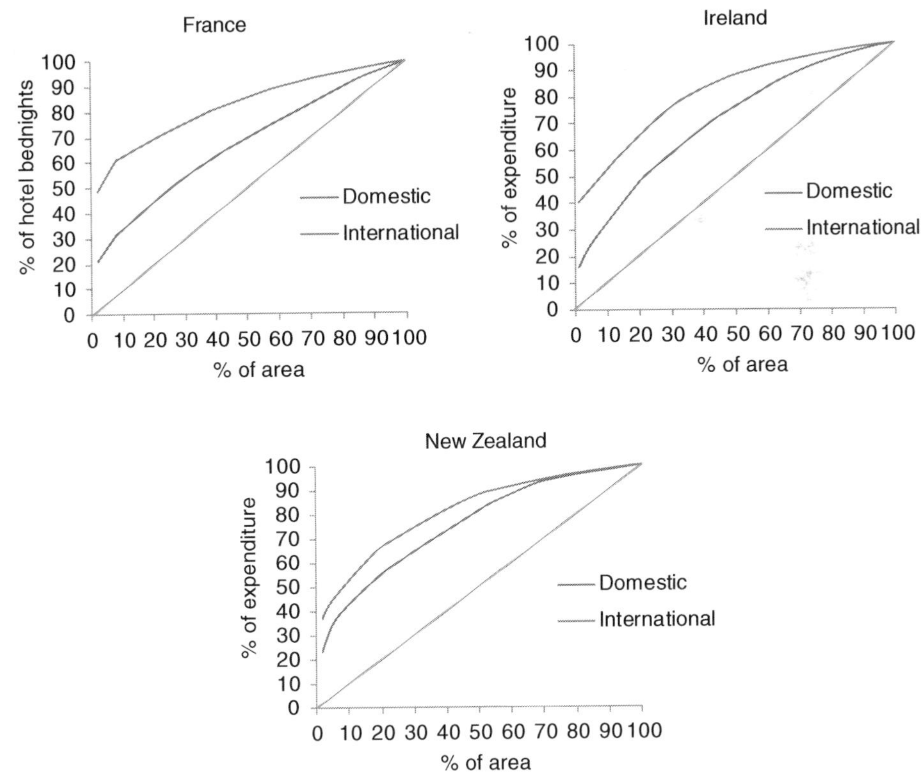

Fig. 5.1. Lorenz curves of measures of demand for international and domestic tourism in France (2016), Ireland (2017) and New Zealand (year ending August 2018). Data from Direction Générale des Entreprises, Failte Ireland and Ministry of Business, Innovation and Employment.

contributed to those patterns, particularly regional variations in resource endowment. The quality of the resources and experiences sought by international tourists is normally higher and perhaps more specific than that of many domestic tourists and thus the features they seek are likely to be more localized than those sought by, or at least able to be afforded by, many domestic tourists in their home country. The dispersion of domestic tourism is also influenced by the nature of the urban hierarchy and outward flows which decrease in distance from multiple, large urban markets.

Lorenz curves also show variations in the extent to which international tourism is concentrated or dispersed occur by market. In the late 1980s/early 1990s Japanese travel in Europe was consistently more heavily concentrated than the overall demand from international tourism there (Pearce, 1995b). At that time, Japanese travellers overwhelmingly favoured the major metropolitan regions containing the capitals, historic centres and other leading urban areas.

Koo *et al.* (2017) analysed the spatial distribution of international tourism in Australia seeking to examine the conjecture that the geographic dispersal of visitors follows the power law, that is, the number of tourist arrivals in each region could be determined from the rank of that region. They also explored variations in the way the law applies to different inbound markets. Their results showed that there was strong evidence for the power law and that 'the growth of destinations occurs in a way that preserves the log-linear relationship between size and rank' (Koo *et al.*, 2017, p. 116). The more established Western markets, such as the UK and New Zealand, were more dispersed than those from more recent markets in Asia (e.g. China and Korea). There was also evidence that over the period 1999–2012 the dispersal of the Western markets had reached a level of stability while the Asian markets were continuing to disperse, though at different rates.

The National Travel and Tourism Office (2015) reports data on visitation by state for overseas visitors (excluding those from Canada and Mexico) to the USA. On this measure the top five states in 2014 were: New York (29.0%); Florida (22.7%); California (20.8%); Hawaii (8.9%); and Nevada (8.9%). These figures indicate that the majority of the demand in the USA is concentrated in the five leading states and that the position of the leader, New York, is less dominant than in the other examples given in Fig. 5.1, a function in part of the large size of the country.

Table 5.1 depicts the places visited by tourists from Chile's three priority markets: Brazil, Argentina and the USA. It shows visits by tourists from all three markets are heavily concentrated in and around Santiago and in the nearby coastal cities of Valparaiso and Viña del Mar. This concentration reflects the highly centralized nature of Chilean society and economy in general, the dominant gateway role of the capital's airport and the long distances to be covered to reach destinations in the north (the Atacama Desert) and the south (national parks in southern Patagonia). Winter sports resorts are also popular with Brazilians. Although there are a number of overland gateways into Chile from neighbouring Argentina, demand from that market is not particularly dispersed, especially when compared to Americans who travel more widely throughout the country, including to the south (Puerto Montt, Torres del Paine, Puerto Natales and Punta Arenas). This concentration of international visitation contrasts markedly with the spread of officially designated destinations throughout the length of the nation (Chapter 2).

Travel Patterns

Research on patterns of tourist mobility and on the economic flows that accompany their travel complements studies on distribution by showing the links between regions and places. Researchers have investigated regions and places visited, which provides an indication of the extent of travel, and analysed patterns of tourist flows which offers a more direct means of exploring interregional links.

Table 5.1. Main places visited in Chile by visitors from Brazil, Argentina and USA, 2014. From Sernatur.

Main places visited	Brazil %	Argentina %	USA %
Santiago and environs	95.7	91.5	85.3
Valparaiso – Viña del Mar	57.3	35.3	36.7
Other coastal areas in the Valparaiso Region	7.1	9.7	8.0
Winter sports centres (e.g. Farellones, Portillo)	42.9	3.6	4.6
Valle de la Luna, San Pedro de Atacama, Geiser del Tatio	4.8		7.5
Puerto Montt, Puerto Varas, Pérez Rosales N. Park, Lake Llanquihue	4.3	1.2	9.3
Pucón, Villarica, Lican Ray, Caburga	3.5	1.4	5.2
La Serena, Coquimbo		3.5	5.4
Concepción, BioBio coast		1.7	3.7
Punta Arenas			10.8
Torres del Paine, Puerto Natales			10.2
Easter Island			3.0
Arica			1.1
Antofagasta			3.9
Valle del Elqui, Vicuna			2.3
Valdivia			3.1
Chiloé			3.4

International tourists

International visitors commonly restrict their travel to a single or a few regions. On average only 12% of first-time and 13% of repeat international holiday visitors to the UK in 2015 visited more than one region (VisitBritain Research, 2017). Multiple-region visits were more likely to be made by visitors from longer-haul markets such as New Zealand, Australia, China and South Africa than from nearby European markets (Table 5.2). Visitors from the USA and Canada constituted an intermediate group. London was by far the most visited region, being visited by 68% of first-time and 66% of repeat visitors. In a survey of visitors to London who had not travelled outside the capital the main reasons cited for not going beyond London were: there were other places higher up the list to visit; nervousness about driving in the UK; there was so much to do in London that there was no time to go elsewhere; and a lack of knowledge about what there was to do and see in other parts of the country (VisitBritain, 2013).

An analysis of US inflight data by Hwang and Fesenmaier (2005) showed 20.3% of Japanese travellers to the USA in 1997 made multi-state visits, 32.3% visited multiple destinations within a single state and the remaining 47.4% went to only one destination. The amount of multi-state and multiple-destination travel varied by demographic and trip characteristics. Free and independent travellers, for example, were more likely to visit several destinations within a state than those on a prepaid package trip.

About 40% of all tourists visiting Brazil in the period 2004–2010 visited more than one destination within the country (Santos *et al.*, 2012). However, the average number of destinations visited was just 1.7. Just under a fifth visited three or more destinations. Santos *et al.* (2012, p. 1347) concluded that:

> A tourist's decision to take a single or multiple destination trip is different and somehow detached from the decision on how many destinations to visit. Moreover, adding a second destination to the travel itinerary is relatively more difficult than expected.

More destinations were visited by long-haul tourists than those from neighbouring countries and by those travelling for leisure rather than for business or to visit friends and relatives.

The USA and Brazil are big countries and states within them are large but even in very

Table 5.2. Percentage of first-time and repeat visitors to the UK who stayed in more than one region on their trip, 2015. From International Passenger Survey.

Country	First-time visitors % staying in ≥1 region	Repeat visitors % staying in ≥1 region
Total	12	13
Europe		
Irish Republic	16	19
Switzerland	14	11
Germany	11	18
Netherlands	10	15
France	7	7
Italy	7	7
Austria	6	14
Spain	6	7
Sweden	5	6
Poland	4	2
Belgium	2	8
Americas		
Canada	23	28
USA	19	21
Brazil	12	11
Long-haul		
China	47	44
South Africa	44	24
Australia	29	36
New Zealand	28	50
United Arab Emirates	21	8
India	18	13
Japan	10	18
South Korea	8	6

small countries such as Estonia international demand may also be very concentrated. An analysis of the mobile phone calls of foreign tourists revealed nearly half (48.3%) of the visits to that country were made to only one county while 35.8% were made to two (Raun *et al.*, 2016). Only 4.4% visited four or more counties. Harju County, where the capital, Tallinn, is located, was the leading county, being visited by 40.7% of foreign tourists.

Due to its size and geography (it borders 15 countries and regions), China has some 180 points of entry. Most international arrivals are by air in the three major gateways of Shanghai, Beijing and Guangzhou while many Chinese visitors from Macau, Hong Kong and Taiwan arrive overland. Bowden (2003) used a variety of measures to analyse the dispersal of tourists after their arrival and showed variations occur in their travel patterns. Both Beijing and Shanghai, for example, are popular points of arrival for interregional (from outside Asia) countries but Guangzhou attracts only a small share. In terms of intraregional visitors (from within Asia), the Japanese favoured Shanghai over Beijing while the converse occurred with the Koreans. Relatively few tourists from both markets went to Guangzhou, which attracted mainly 'compatriots'. As might be expected, most of the border regions, except Tibet and Gansu, draw the majority of their visitors from neighbouring or nearby countries. Bowden cites a CNTA survey which showed that in 2000 about one-third of international visitors continued their travel within China, primarily within the province or region which they entered or by going on to Beijing and Shanghai. Emerging centres such as Yunnan and Shannxi also attracted a small share. Bowden (2003, p. 267) noted that international tourist flows 'after entry are increasingly scattered year after year. The destination choices of international tourists have extended from a few coastal regions gradually toward the Interior regions, owing to the rapid tourism development in these regions.' The direction of the flows after the point of entry also varied. Those who entered at Beijing went first to the Middle region before going on to the Southeast region whereas tourists who entered at Shanghai then travelled to the Northeast region, after which they stayed in its nearby provinces.

Zillinger (2007) drew on Hägerstrand (1973) and took a time–geographical approach based on travel diaries to analyse the routes of German car-tourists in Sweden. Two-thirds of her respondents travelled through Sweden on a round tour and one-third were defined as base tourists (those who made day trips from a single place or made a round trip but stayed at some places for several days before moving on). Zillinger divided the long narrow country into four regions and found visitors to each exhibited different travel rhythms. Those travelling to northern Sweden, for instance, exhibited a circular pattern. The relative location of sites to other sites, not just their own attractiveness, was found to be an influence on travel patterns along with distance from the tourists' home and their length of stay in Sweden. Those on shorter stays compensated by travelling further each day. On average, Zillinger's respondents travelled 500 km for every week spent in Sweden.

Domestic tourists

Insight into the dynamics of domestic tourism in France is provided by Figs 5.2a and 5.2b, which portray dominant interregional flows and patterns of intraregional travel. Figure 5.2a (departures) shows a significant amount of domestic travel occurs within the home region. Intraregional travel accounts for 30% or more of the bednights in the four southern regions, which have an appealing climate and attractive coastlines or mountains. Lower rates of intraregional travel are experienced in the northern and north-eastern regions, including the Île-de-France. The pattern of interregional departures is characterized by travel between neighbouring regions, with the Auvergne-Rhône-Alpes and Île-de-France also being the destination of travellers from further afield.

In terms of arrivals, the pattern of intraregional travel (the share of bednights originating in the home region) is more diverse (Fig. 5.2b). Over 28% of the bednights in the regions of the Auvergne-Rhône-Alpes, Hauts-de-France and Grand Est are generated by intraregional travel. In the former this might be attributed to regional residents holidaying in their own attractive region; in the latter two, it is perhaps more a case of residents from the rest of the country finding them less appealing. The pattern of arrivals for interregional travel reveals the dominance of the wealthy, heavily populated Île-de-France region which generated the largest flow into ten of the 13 regions and the second-largest flow into two others. Other inbound flows were largely from neighbouring regions. The Île-de-France's dominant position as an origin or source is further demonstrated by it being the only region to record a net negative balance. Corsica was the only region to register a net positive balance with all other regions; Occitanie had only one net negative balance and Nouvelle Aquitaine had two.

Similar patterns are found in Spain. Figure 5.3a, which is based on data from the national Familitur survey, highlights the importance of intraregional destinations. In 2012 the largest destination flows for all regions except Madrid, the small but densely populated capital, were intraregional. Travel to places within the home region constituted over half of all trips in ten out of 17 regions, reaching in excess of 80% in the case of the Canary Islands and Andalusia. Intraregional travel rates were lowest in Madrid, in small regions in the north of Spain, Murcia and Castilla–La Mancha. The second–Largest destination flows in most cases were to neighbouring regions. Four regions accounted for more than half of the trips received: Andalusia (18.2%), Catalonia (15.2%), the Valencian Community (11.5%) and Castile and Léon (10.0%). Intraregional travel constituted the single-largest source of demand in the majority of the regions (Fig. 5.3b). Madrid dominated the remaining flows, emphasizing its role as a source region. It had a net negative balance. Other net negative balances were recorded by Catalonia, the small industrial regions in the north (the Basque Country, Navarra, Asturias) and Murcia, although the difference between the number of trips generated and received was minimal in the last two cases (Instituto de Turismo de España, 2013).

Drawing on data from the 2016 Familitur survey, Álvarez-Díaz *et al.* (2017) analysed the determinants of domestic tourism in Spain using a gravity model and a spatial autoregressive (SAR) model. Their findings reinforce the importance of distance and location. They found distance and relative prices between the regions of origin and destination exerted a negative effect on flows and identified a significant spatial spillover effect, whereby regions benefited from the inflows of domestic tourists received by neighbouring regions. Other factors positively associated with the domestic tourist flows were levels

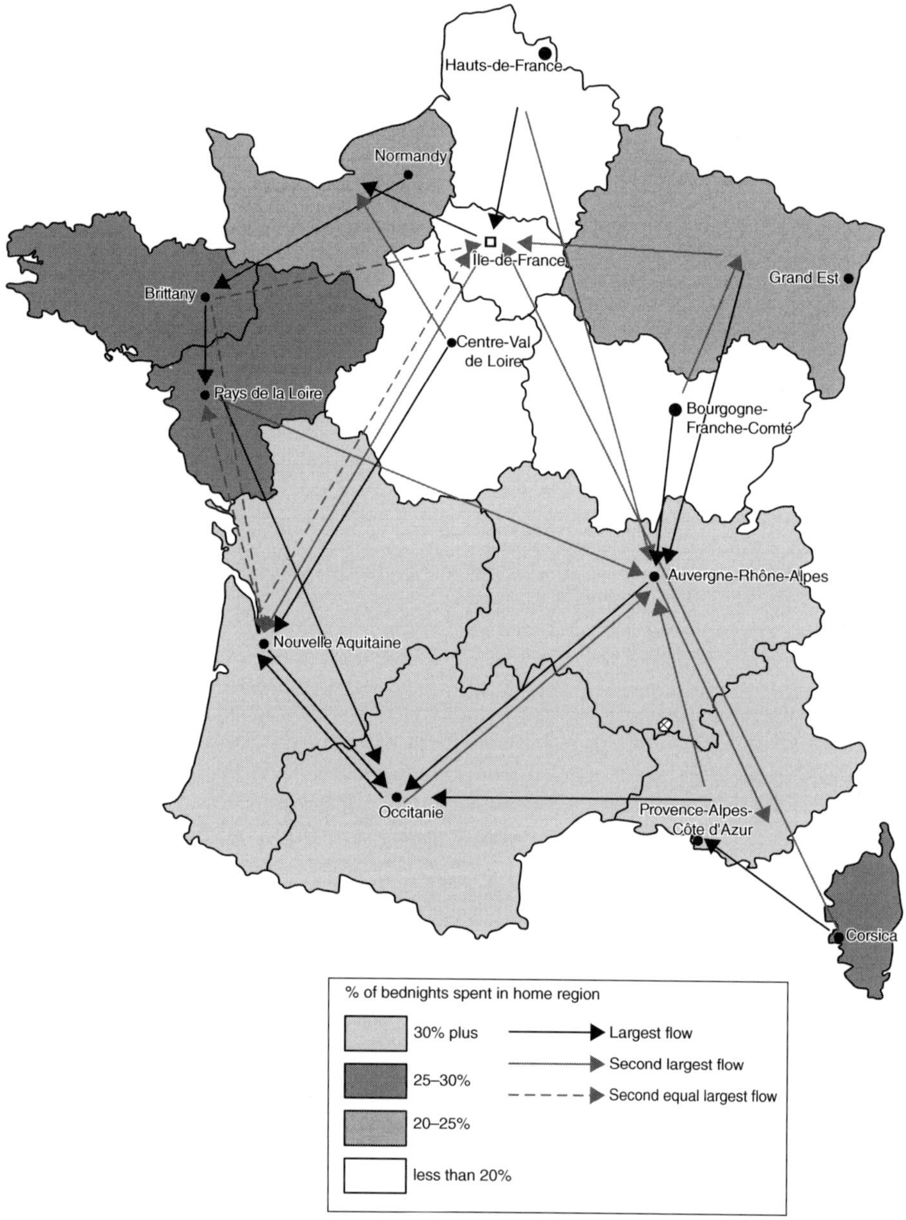

Fig. 5.2a. Patterns of domestic departures in France (2014–2016 average). Data from Khiati (2018).

of income in both origin and destination regions and destination characteristics such as the quality of beaches, level of accessibility, insularity and the number of museums, theme parks and natural parks. Álvarez-Díaz *et al.* (2017) pointed out that per person expenditure per day in Spain is much lower for intraregional trips (31 euros in 2016) compared to interregional trips (52 euros). This would reduce the relative contribution of intraregional travel when domestic tourism is considered in economic terms, again indicating the importance of the measures used in such analyses. De la Mata and Llano-Verduras

Fig. 5.2b. Patterns of domestic arrivals in France (2014–2016 average). Data from Khiati (2018).

(2012) undertook an econometric analysis of interregional monetary flows in Spain resulting from domestic tourism by linking travel and expenditure data and by distinguishing between overnight stays at tourist establishments and second homes. Their results showed that although in the former case there were strong flows connecting some distant regions, the largest flows in the latter appeared between adjacent regions, that is, a larger negative elasticity for distance is found with second homes.

Strong intraregional flows and flows to neighbouring regions or departments are also a feature of domestic tourism in Uruguay where the determinants of domestic travel are similar to those reported for Spain (Dominguez Perez, 2016; Brida *et al.*, 2017). Coastal departments and/or those with good quality accommodation have a significant comparative advantage over the others. Montevideo is a major source and net emitter of domestic tourists but also an attractive destination for tourists from other parts of the

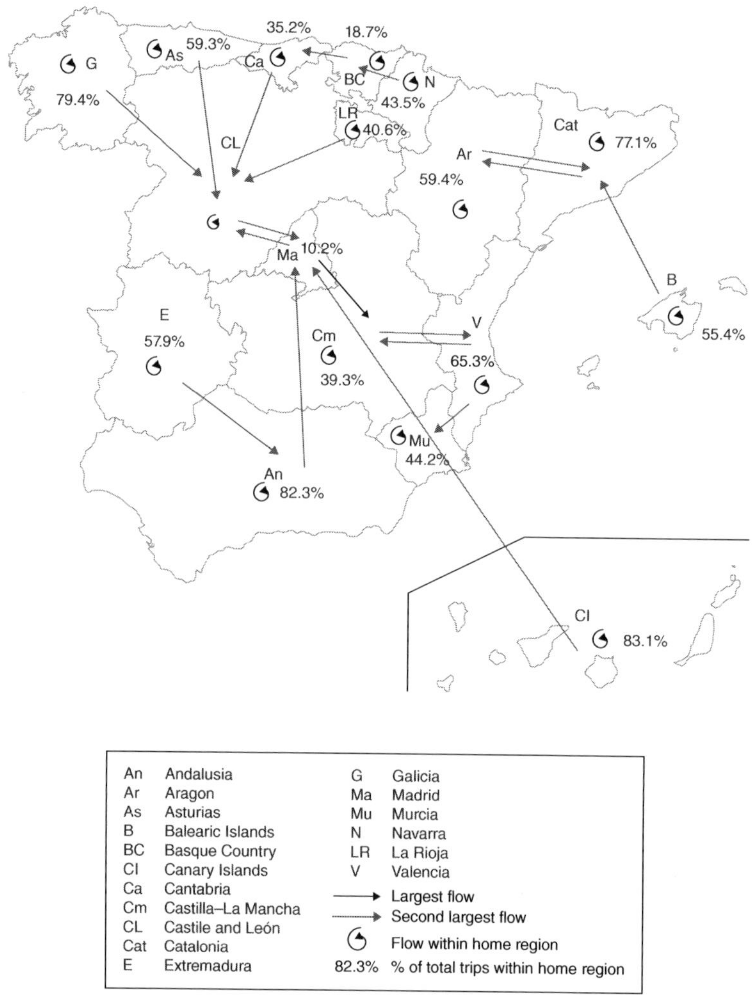

Fig. 5.3a. Patterns of domestic departures in Spain (2012). Data from Instituto de Turismo de España (2013).

country. In Brazil, Haddad *et al.* (2013) analysed the flows of domestic tourist expenditure between five macro-regions using an interregional input–output system and found that domestic tourism produced a redistributive effect of income from the richest region (the Southeast) to the poorest region (the Northeast).

The analysis of domestic travel in Sweden by three trip categories (visits to friends and relatives and stays in second homes (VFR/SH); leisure travel; and business travel) revealed marked differences in patterns of travel segmented by purpose of trip (Pearce, 1996a). The constraint of distance, as indicated by the prominence of intraregional flows and Relative Acceptance Index values, was shown to be important for each trip purpose but especially so with regard to VFR/SH travel which accounted for 60% of all bednights. The spatial pattern of net leisure travel in Sweden was characterized by a near neat east–west division running the length of the country, broken only by the southeastern destinations of Kalmar, Blekinge and the islands of Öland and Gotland. This pattern of eastern origins and western destinations is in large measure a function of the distribution of population and of recreational resources. Regional biases were evident in the dominant

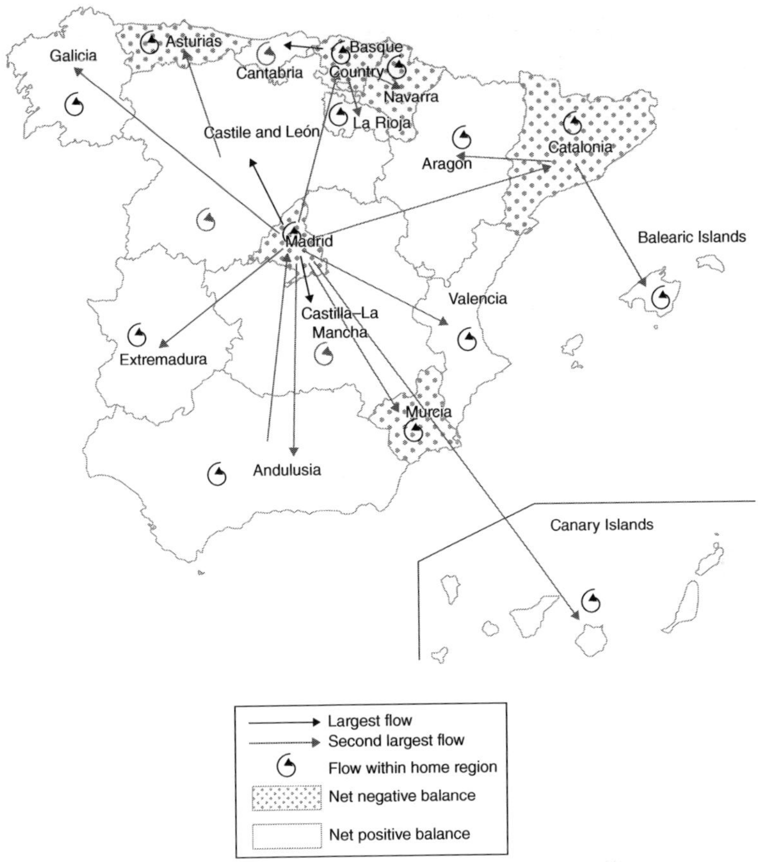

Fig. 5.3b. Patterns of domestic arrivals in Spain (2012). Data from Instituto de Turismo de España (2013).

destination flows for leisure and business travel but to markedly different regions: to Kopparberg and Jamtland in the former and to Stockholm in the latter. Seventeen (of 31) first-order destination flows and three second-order flows were directed into Stockholm, whose primacy in terms of business travel also stands out in terms of net demand and the absolute number of bednights received (over one-fifth of the national total). Much official business can only be transacted in the capital; the city also draws a large share of commercial traffic and is the venue for many international, national and regional conferences.

The flows derived from the origin–destination matrices represent aggregated travel between the generating and receiving regions but do not portray the actual itineraries of domestic tourists; that is, they do not show either desire lines or grounded flows. Very little research has been done on this topic. Hwang and Fesenmaier (2003) analysed data from the 1995 American Travel Survey which showed that only 5.4% of domestic pleasure trips by Americans involved multi-destination travel, the average number of destinations visited was 1.1 (including single-destination travel) and on average multi-destination travel consisted of visits to 3.6 destinations. Using the Lue *et al.* (1993) typology of travel patterns, Hwang and Fesenmaier found the vast majority (91.3%) of multi-destination trips consisted of 'en route' travel, 6.2% followed a 'base camp' pattern and 2.5% were classified as 'trip chaining'. Visitors to Yellowstone National Park exhibited more complex patterns with almost half of those surveyed by Mings and McHugh (1992, p. 92) taking a completely circular or 'full orbit' route 'visiting a very wide array of scenic attractions in their grand tour of the American West, returning home via a completely different route to that taken on the outward journey'. Such a pattern, however, would appear to be exceptional and not the norm for domestic travel within the USA.

Distribution of Supply

Good data on the distribution of accommodation are collected and published in France (Khiati, 2018). Maps depicting the distribution of hotel beds, camping places and second homes show the importance of considering a range of accommodation types as the patterns differ in each case. Hotels are concentrated more heavily in the main urban areas, along the Mediterranean coast and in parts of the French Alps. Camping grounds are distributed along the coasts, in upland and mountainous inland areas (the Alps, Pyrenees, Central Massif, Vosges), along the valleys of the Rhone and Loire and around the Île-de-France. The distribution of second homes follows a similar pattern with particularly dense concentrations along the coasts, in the northern French Alps, a greater dispersion across the Central Massif and the interior of Languedoc-Roussillon, around Paris and in Burgundy.

Other French studies have sought to quantify the impact of attractions on different aspects of the structure of tourism in France (Barros *et al.*, 2011; Truchet *et al.*, 2016). Limitations in obtaining comprehensive measures of attractions are acknowledged in each case. Although the results are somewhat predictable these studies do represent interesting attempts to go beyond the basic observation that accommodation in France is located on the coast, in the mountains and major urban centres. Barros *et al.* (2011) assessed efficiency in each of the country's 22 regions in terms of output measured by bed-nights and inputs by tourist arrivals and accommodation capacity. These scores were then related to a series of attractions for which data were available: monuments, museums, theme parks, beaches, ski resorts and natural parks. All of these variables contributed to regional efficiency but kilometres of beach appeared to have the greatest impact. The authors interpret their results in terms of E (Escape)-attractions and D (Discovery)-attractions (Botti *et al.*, 2008). Languedoc-Roussillon's good performance score is attributed to its numerous beaches (an E-attraction for tourists); that of the Île-de-France to a cluster of D-attractions (monuments, museums and theme parks). Conversely, low-ranked Lorraine possessed few of the attributes considered. On the premise that the appeal of tourist attractions is not constant in space but exhibits a distance decay function, Truchet *et al.* (2016) used municipal level data to analyse the spatial influence of different types of attractions – local, spread and diffuse – on the degree of tourism development as measured by the number of hotel rooms and employees in the tourism sector. Their econometric analysis using count data included four local (ski slopes, beaches, sites and facilities), two spread (lakes and watercourses) and two diffuse (ecology [at least one protected area] and green area) attractions. Again, most of the attractions had a positive effect on tourism development, particularly ski slopes, beaches, sites, facilities, lakes, and environmental quality. However 'access to green areas has a positive and significant effect on the presence of hotel capacity and tourist employment but has a negative and significant effect on their level' (Truchet *et al.*, 2016, p. 13). Differences were found between attractions with regard to their spatial influence. The positive effect of leisure and sports facilities was localized and limited to the municipalities in which they were located and negative in terms of tourism development in the neighbouring municipalities. In contrast, other attractions impacted tourism development over a greater distance: within a 25-km radius for beaches and tourist sites, and beyond 25 km for ski slopes.

The five studies introduced with reference to Table 4.3 combine multiple variables relating to aspects of the spatial and organizational structure of national destinations (Capone and Boix, 2008; Fernández Cuevas *et al.*, 2008; Estevão and Ferreira, 2012; Segarra-Oña *et al.*, 2012; Majewska, 2015). They also demonstrate the challenges of operationalizing the concepts of tourism clusters and districts and mapping, measuring and interpreting their structure and impact.

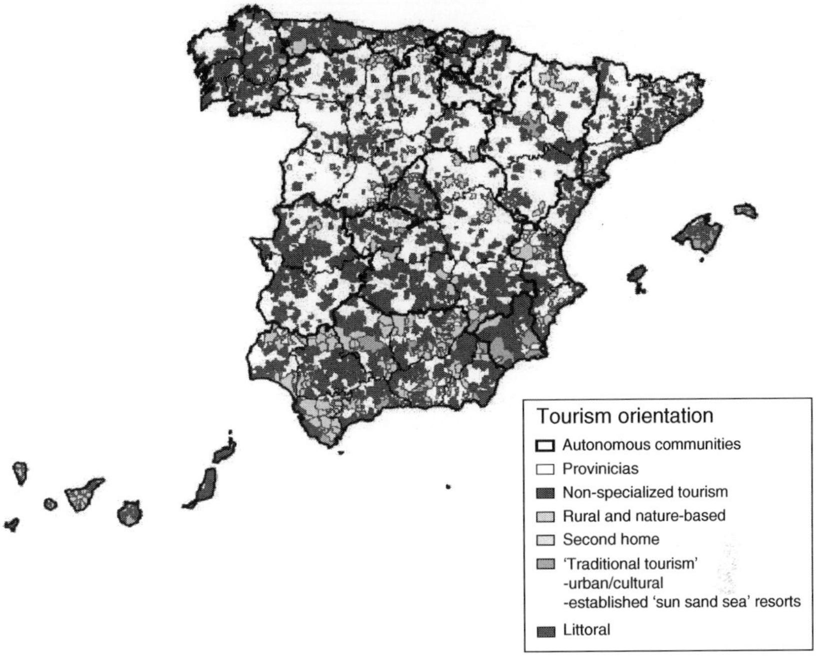

Fig. 5.4. Classification of tourism municipalities in Spain. From Fernández Cuevas *et al.* (2008), Mapa 1, with permission of Instituto de Estadística de Andalucía.

Figures 5.4 to 5.8 illustrate the spatial patterns which result from this work. The effects of the different approaches and the use of different areal units are readily apparent when the two maps of Spain are compared. Figs 5.4 and 5.5 present two rather different expressions of the distribution of tourism in Spain: the former presents a differentiated and more dispersed pattern with municipal level variations; the second, a highly concentrated one of homogenous mature clusters at the provincial scale. In Fig. 5.4 almost two-thirds of the municipalities were identified as having tourism functions. These are found throughout much of Spain, particularly along the coast. The most common and widespread are characterized as non-specialized 'tourism'. Large 'empty' areas occur in the interior. Figure 5.5 accentuates the degree of concentration: the five provincial level tourism clusters accounted for 30% of hotels in Spain but 50% of bookings received. The finer mesh of analysis and wider range of variables used by Fernández Cuevas *et al.* (2008) in Fig. 5.4 indicates that tourism in the five mature 'sun, sand, sea' clusters in Fig. 5.5 is not spread evenly throughout each of these provinces and that different forms of tourism are found there.

Fernández Cuevas *et al.* (2008) went on to identify 219 tourism districts based on the geographic and functional proximity of the specialized tourism municipalities. Of these, 114 consisted of a single municipality and 105 grouped together two or more nearby municipalities with similar characteristics. On this basis each one of the 'urban and cultural' municipalities constituted a separate district. Derivation of the other 'traditional' districts proved more problematic and resulted in municipalities from the 'littoral' category being grouped with dominant 'vacational' ones to form a cluster. For example, Benidorm, as the nucleus, was combined with 11 other nearby municipalities to form the tourist district of Alicante. In contrast, Segarra-Oña *et al.* (2012) effectively treat each of the provinces reaching the cut-off point of the LQs for tourism-related employment as a cluster.

Similar in some respects to the pattern in Fig. 5.4, Fig. 5.6 also shows a differentiated and dispersed pattern of LTPSs (districts) in Italy. Cluster analysis here, however, has produced different categories. According to

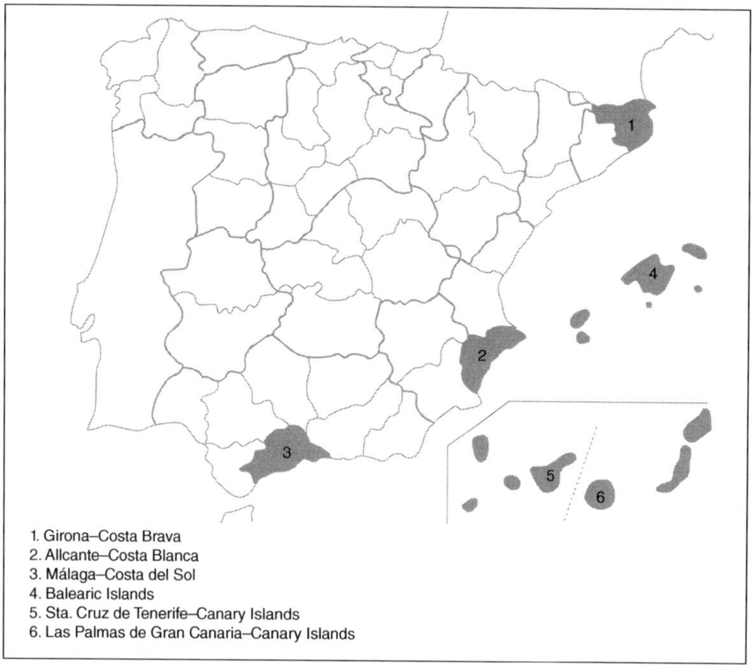

Fig. 5.5. Spanish tourism clusters. From Segarra-Oña *et al.* (2012), Figure 1.

Capone and Boix (2008, p. 216) the most specialized group, 55 LTPSs having the highest LQs, 'can be considered as the most probable "tourist districts' (even though the characteristics of the local community were not studied)'. This group could be further sub-divided into those with a dominance of transport and accommodation facilities and those with a concentration of food and recreational facilities. The authors do not comment on their location but Fig. 5.6 shows them to be concentrated in the Italian Alps, coastal Sardinia and Tuscany. The largest cluster (157 LTPSs) has the lowest LQs, diversified employment and rail, air and other infrastructure. It is described as being made up of large 'art cities' such as Rome, Florence and Venice as well as other medium-sized cities. The third cluster is an intermediate group of 56 LTPSs which are less well equipped, have a narrower *filière* (e.g. accommodation and recreational activities) and consist of specialized medium-sized cities and small rural systems.

Estevão and Ferreira (2012) did not distinguish tourism from non-tourism areas and as a result Fig. 5.7 depicts tourism as occurring throughout Portugal when mapped at the territorial area level. Discernible patterns, however, are apparent in terms of the distribution of the three performance clusters they identified. The low-performance cluster predominates in the north and in central inland parts of the country with bands of medium- and high-performance clusters further south, particularly on the coast. Performance is consistent across some tourism regions (e.g. Oporto and the North (low) and the Algarve (medium)) and variable in others (e.g. the Alentejo). Only the Alentejo Coast Tourism Region, the Madeira Islands and the Azores are shown to be high performance throughout.

Majewska (2015) puts greater emphasis on the spatial aspects of tourism clusters in Poland. Maps showing the absolute and relative distribution of tourism firms are followed by those depicting spatial autocorrelation statistics (local Moran's I statistics) in relation to the number of overnight stays and then interregional tourism clusters determined by SWLQs in which the values for

Fig. 5.6. Local tourist production systems in Italy. From Capone and Boix (2008), Figure 2, with permission from Springer, © 2008.

the local Moran's I statistics are used as the spatial weights of the number of tourism firms in each district (Fig. 5.8). This latter technique enables the effects of spatial proximity to be captured; that is, it shows whether or not the region is surrounded by neighbouring regions with similar or different values. 'Hotspots' occur where a centre with high values is surrounded by those with low values. Geographical spillover effects which give rise to 'interregional' clusters are identified where regions are surrounded by similar regions. As Fig. 5.8 shows, such clusters are concentrated along the coast in the north, in mountainous areas in the south and in the Warsaw agglomeration. The inclusion of measures of demand is a useful addition but as the number of overnight stays is not broken down by Polish travellers and foreign visitors no distinctions can be made between the patterns of domestic and international tourism.

These five studies also go beyond depicting the geography of the distribution of clusters and districts and start to provide insights into their functional structure and impact. Fernández Cuevas *et al.* (2008) assessed the 'district effect' of the tourist districts in Spain by comparing their performance against the country's other municipalities, first with regard to their relationship to the development of real estate (Spain's economy in the late 1990s having been driven by a real-estate boom, a boom which was to burst in the following decade) and, second, in terms of well-being as measured by the Index of Economic Activity. Overall, the level of economic activity was greater in the non-tourist municipalities than in the tourist districts. With regard to specializations,

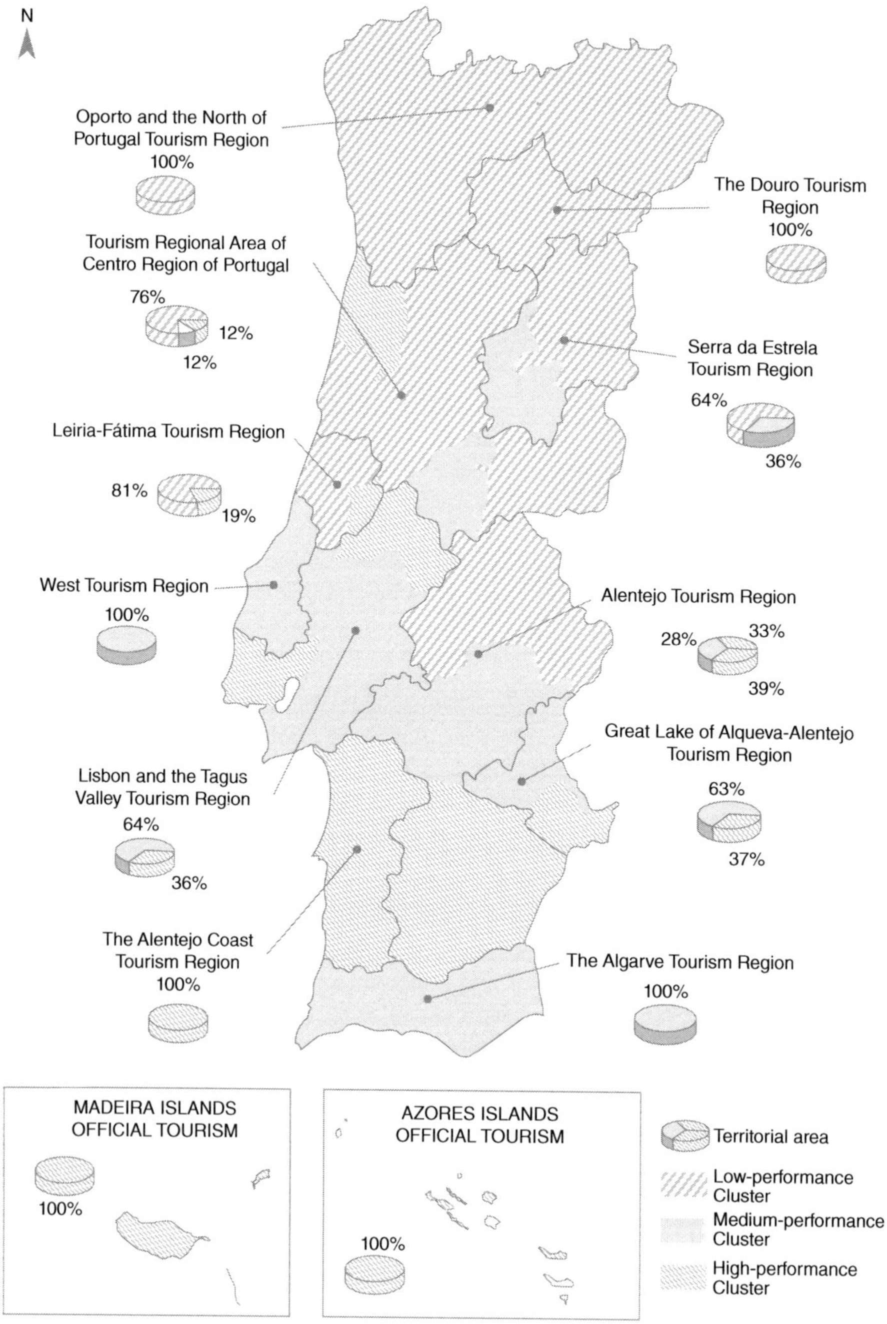

Fig. 5.7. Cluster performance distribution by tourism region in Portugal. Redrawn from Estevão and Ferreira (2012).

the 'traditional' tourist districts performed better than those having a large proportion of second homes. Only weak relationships between the levels of real-estate development and all tourist districts were found. In terms of the different types of tourist

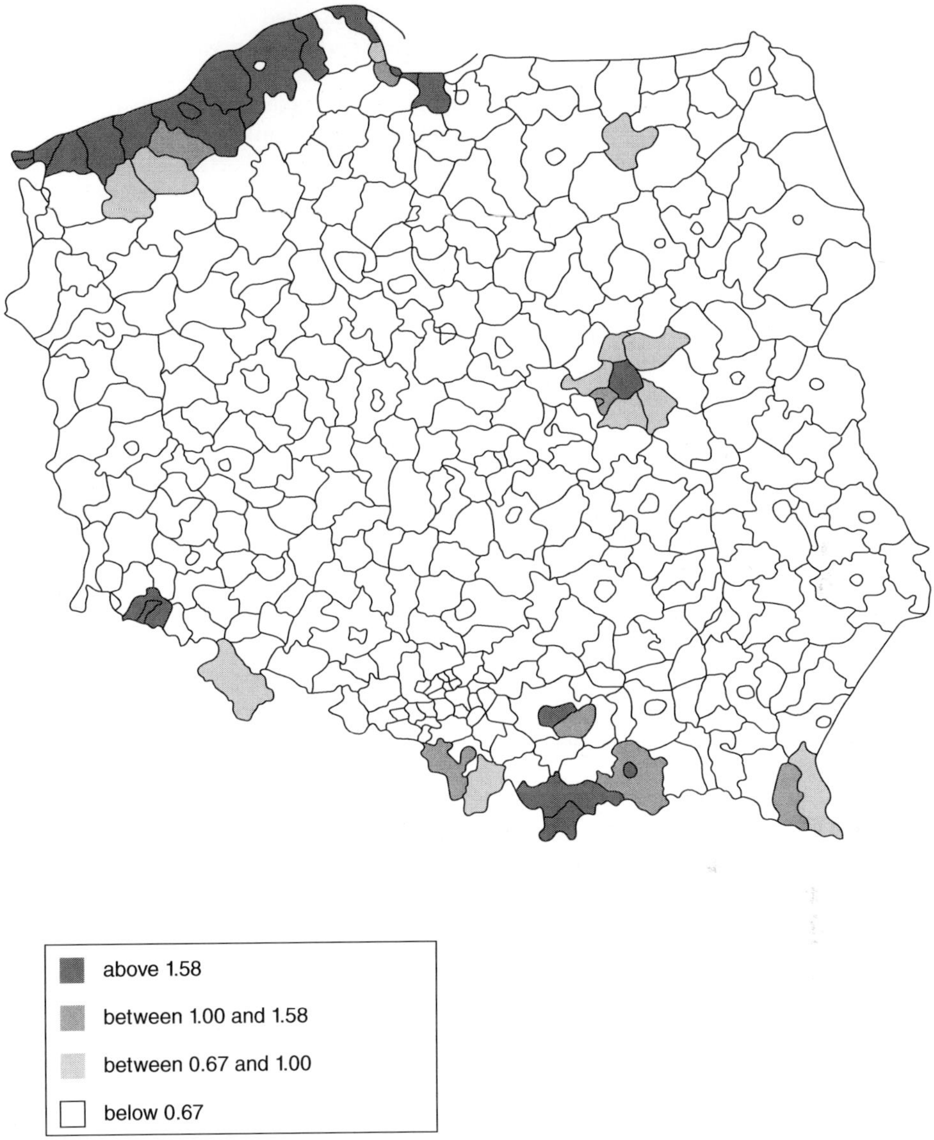

Fig. 5.8. Distribution of the spatially weighted LQ coefficients in tourism by value classes in Polish NUTS-4 regions in 2012. From Majewska (2015), Figure 5.

district, not surprisingly the relationship was strongest in those characterized as second-home districts and weakest in those specializing in rural nature-based tourism. In their more focused study on hotel performance in Spain, Segarra-Oña *et al.* (2012) found that overall hotels located within the tourist clusters were more profitable than those outside. Differences, however, occurred between the provinces leading these researchers to conclude (p. 1328) that 'each of these clusters should be considered as a different destination with its own features and trends'.

Majewska (2015), Estevão and Ferreira (2012) and Capone and Boix (2008) explored the composition of clusters and districts and their impact. Majewska (2015) evaluated the similarity of tourism clusters in the two highest

SWLQ categories by agglomeration type and tourism demand (number of tourists and overnight stays). Eight groups were identified depending on whether they were localization-driven (specialized), urbanization-driven (diversified), mixed or were spillovers of localization or urbanization types. Not surprisingly, the first two groups consisted of the urbanization-driven clusters of Warsaw and Krakow and the specialization-driven coastal and mountainous tourism regions. Others were characterized by a lesser intensity of tourism based on natural attractions. Spillovers of both the urbanization and localization types were also identified where districts benefited from the development of tourism in nearby cities. Majewska (2015, p. 426) attributed the pattern shown in Fig. 5.8 to the 'rather average degree of tourist attractiveness of many Polish regions' and (p. 427) to 'low levels of social capital and the willingness to take up the inter-personal or inter-regional cooperation' which limits spillover effects.

Capone and Boix (2008, p. 223) concluded from modelling growth in tourism employment in Italy that 'higher levels of growth of the tourism industry in local tourist production systems are based not on natural endowments but on agglomeration economies, with high relevance of localization economies'. Higher growth rates were especially associated with 'the presence of every phase of the tourist *filière* in the local production system', thereby providing support for the effects of complementarity and interdependence in the working of tourist districts. The results for urbanization effects were unclear. Evidence was also found for spillover among municipalities within the system. Their model performed well for two of the three groups in their typology but not for the third, the one they considered as the most probable 'tourist districts'. Estevão and Ferreira's (2012) analysis of the weighting of turnover by economic activity in the Portuguese territorial units revealed that the greatest variety occurred in the low-performance cluster followed by the high- and medium-performance ones. Certain activities were found exclusively in some clusters but the reasons for and impact of these patterns are far from clear. The high-performance cluster, for instance, included not only air transport and air transport renting, which might be indicators of accessibility, but also zoos, botanical gardens and aquariums, which would not appear to be significant factors in the context of tourism in Portugal. Other activities such as hotel establishments without restaurants were shared by both the high- and low-performance clusters.

Chain hotels and airline networks

The aggregate data used in the analyses of clusters and districts mask other structural supply-side characteristics which vary by sector. In terms of hotels, management type is a key factor, particularly the distinction between independent and chain hotels. A chain hotel is one whereby 'a professional organization owns, leases, manages or franchises the hotel and several other hotels, by means of centralized functions and under a central governance system, within its own strategic management guidelines and operational inputs' (Ribaudo and Domeniconi, 2013/2014, p. 9). Hotel chains may have several brands covering a range of hotel categories. In 2013 chain hotels accounted for about two-thirds of total hotel room supply in North America but only about one-quarter in the Asia Pacific region, Europe, the Middle East and Africa and a fifth in Latin America (Ivanova and Ivanov, 2015). A further distinction can be made between domestic and international chains, the nature and relative penetration of which vary from country to country (Ribaudo and Domeniconi, 2013/2014; Le *et al.*, 2018). Affiliation to a chain offers component hotels greater international and national market exposure through stronger brand recognition and access to larger and more sophisticated distribution channels; knowledge transfer relating to markets, operating procedures and service standards; and economies of scale. In return, the chain benefits from local market knowledge and from the additional revenue or increased management, franchise or lease fees which come from wider market and geographical coverage (Ingram and Baum, 1997; Ribaudo and Domeniconi, 2013/2014;

Ivanova and Ivanov, 2015; Le *et al.*, 2018). Affiliation to a chain, however, may bring about strategic constraints by limiting a component hotel's flexibility to respond to its local environment (Ingram and Baum, 1997).

The links between different facets of destination structure are clearly seen in the strategies of hotel corporations and the distribution of hotels by management type. In the USA, according to Ingram and Baum (1997, p. 73), the 'ideal local environment for most hotel chains is a medium-sized city on a busy highway. Chains' standard products thrive in markets where consumers are highly uncertain about the quality of independent hotels' rooms and service.' Much international expansion of hotel chains, particularly in the early stages, involved large, upscale hotels in the major gateways oriented to business travellers (Litteljohn, 1997; Johnson and Vanetti, 2005). Johnston and Vanetti (2005) found that the most important geographic variables for international hotels in East Central Europe were the size and nature of the city, the infrastructure within the region and perception of it as an attractive business location. In Italy hotel chains were 'strongly concentrated in solid and steady performance destinations such as Venice, Rome, Milan, and Florence' (Ribaudo and Domeniconi, 2013/2014, p. 19). The average chain hotel in the Asia Pacific region in 2017 had 164 rooms, more than three times the size of the average independent hotel which had just 52 rooms (Horwath HTL, 2018). In Japan, a country with a long history of small *ryokans*, chains made up just 5.3% of all hotels but provided one-third of all rooms.

The relationships between commercial and spatial structure at the national level are also very evident in the changing patterns of national airline networks. De-regulation of the airline industry in the USA in 1978 saw a reduction in the number of airports receiving jet services and a restructuring of airline networks into a more efficient hub-and-spoke structure whereby each of the country's major carriers centred its network on a small set of hub airports (Shaw, 1993). Through these hubs the mainline carriers have also strengthened the links between their domestic networks and international airports. Rubin and Joy (2005) characterized the structure of the industry as oligopolistic, with almost 70% of US passengers in 2001 being carried by the six major airlines and a further 12% by the largest low-fare airline. Subsequent consolidation within the industry has reduced the number of major carriers with the domestic market becoming dominated by four very large carriers during the second decade of the 21st century (Reynolds-Feighan, 2018). Considerable consolidation has also occurred among the regional carriers providing feeder services to the mainline carriers. At the same time, the number of airports receiving jet aircraft for these feeder services has increased with the result that some spatial dispersion occurred in the US air traffic although overall traffic distribution across the carrier networks remains highly concentrated. The use of multiple regional carriers with their different mixes of aircraft size gives the mainline carriers greater flexibility in handling variations in demand.

The hub-and-spoke American pattern is in part a function of the country's size. European scheduled carriers 'generally tend to focus their networks around two airports within the member state in which they are registered' (Reynolds-Feighan, 2010, p. 114), for example, Charles de Gaulle and Orly in the case of Air France. In the smaller countries of Europe, comparable flight times to those of American domestic flights commonly involve international connections which are particularly important to Mediterranean destinations.

Significant changes in the structure of air travel have also taken place in Europe. During the boom in charter flights in the 1980s the volume of passengers on non-scheduled flights to Spain was three times greater than those travelling by scheduled airlines (Pearce, 1987a). Almost three-quarters of scheduled passengers arrived in the two major metropolitan centres of Madrid (45%) and Barcelona (27%). The non-scheduled traffic was much less concentrated. Palma de Mallorca accounted for one-third of the arrivals, followed by a second tier of airports: Malaga (13.3%), Las Palmas (11.8%), Tenerife (10.6%), Ibiza (8.8%) and Alicante (7.6%). Significant regional biases occurred in the destinations favoured by the three major international markets. The UK traffic was more widely dispersed than that from Germany and Sweden: the first three Spanish airports accounted for 58% of British arrivals compared to 66% of the Germans and 73% of the Swedes. Similar trends were evident in other sunlust destinations in the Mediterranean such as Greece.

In contrast to the American experience, airline de-regulation in Europe in 1997 resulted in a growth in the number of airports offering commercial jet services. Their number reached 522 in 2008, an increase of 81 since 1996 (Reynolds-Feighan, 2010). Much of this expansion was due to the advent of low-cost carriers which 'preferably operate out of secondary uncongested airports where they can benefit from fast turnarounds and low airport charges' (Lieshout *et al.*, 2016, p. 69). The general pattern is one of short- and medium-haul flights on point-to-point north–south routes, competing with, if not superseding, charters and complementing the full service networks (Dobruszkes, 2006). The top five airports in Europe for Ryanair and EasyJet accounted for about one-third of their seating capacity compared with half or more for the other top ten carriers (Reynolds-Feighan, 2010). These developments may have added to the dispersion of gateways but as low-cost travellers are often taking short city breaks or sunlust holidays this would not necessarily lead to widespread travel once they arrived at their destination.

Institutional Frameworks for Tourism

Tourism in most countries is supported and shaped in varying degrees by some form of institutional framework. Figure 1.1 shows this is typically hierarchical in nature with different functions being undertaken by public or semi-public organizations or agencies at a national, regional and local level. From a national perspective, two main structural issues arise: what is the most appropriate hierarchical structure of tourism organizations in a country and what is the best institutional framework at the national level? The hierarchical structure and role of public-sector tourism organizations is generally mandated legislatively, that is, responsibility for tourism is determined by central government or the national constitution and tends to reflect the broader administrative structure of a country. National tourism organizations generally play a more critical role in strongly centralized nations and regional or state organizations tend to be more dominant in countries with a federal system of government (Pearce, 1992).

Key questions arise as to which functions are best carried out at which level and by which organizations. National policies can influence what happens at lower levels as the discussion in Chapter 2 relating to formally designated destinations in Chile showed. International marketing is a widely accepted, though not universal, function undertaken by a national body as the scale economies and efficiencies needed to develop a strong, coherent and effective image of the country and presence in international markets are generally best achieved by leadership and co-ordination at this level, with support from regional and local level organizations. International marketing is an activity at the forefront of promoting the growth of tourism which is commonly the prime goal of both the public and private sectors. It is also often seen as a public good which benefits the country as a whole and the many and varied businesses and communities associated in some way with tourism. Destination marketing, however, is also carried out at regional and local levels, though usually with a different focus. Provision of visitor information and visitor servicing has generally been carried out at a local or region level, for example through visitor information centres. Planning for tourism and destination management is done at a variety of levels.

National bodies for tourism can take a variety of forms: as a separate ministry or department of tourism; a division of some larger ministry; a semi-state organization or some more autonomous agency. Figure 1.1 depicts a single NTO (national tourism organization) heading a hierarchical structure. In some instances a distinction is made between NTAs (national tourism administrations) whose functions are primarily concerned with policy making and regulation and NTOs (national tourism organizations) whose main role is generally international marketing. These terms are often used interchangeably although in some countries there may be two different bodies with separate responsibilities. Even that may be an over-simplification as a variety of public and semi-public agencies may exist at the national level within a broader institutional framework. This may arise out of a concern for separating policy formulation and implementation. It may also result from the transversal nature of tourism where policies affecting the sector may also relate to or arise in such diverse areas as regional development, conservation, transport, public works and immigration and the need to have some structure to co-ordinate these overlapping responsibilities.

Interorganizational analysis shows change is a constant factor in terms of both hierarchical networks of tourism organizations and institutional frameworks for tourism at the national level (Pearce, 1992, 1996b, 1996c, 1997b; Guala Catalan and Pearce, 2019). Domain consensus is often fragile or contested, partial interorganizational conflict occurs and changes in the broader political and economic environment inevitably impact upon existing frameworks.

The impact which broader institutional developments may have on national and regional destinations is clearly demonstrated by the changes which occurred in Belgium as a consequence of its transformation from a unitary state to a federal one (Pearce, 1996c). This process saw the disappearance of the national tourism organization powers as responsibility for tourism was transferred from the central state to the country's three language communities (Flemish, French-speaking, German-speaking) and three regions (Flanders, Wallonia and the Region of Brussels-Capital). Organizational restructuring associated with processes of 'communitization' and regionalization in turn resulted in marketing strategies which saw a weakening of the national image and greater emphasis on the regions and new configurations of regional destinations. The Kempen became Green Flanders and in the south the 'Ardennes' rapidly shifted eastwards to encompass the whole of Wallonia, a geographical compromise to reflect both market awareness and the political need to be seen to promote all of the region.

Even where there is not the same far-reaching political restructuring as occurred in Belgium, significant changes in responsibility for tourism may result from new national policies. In England, for example, the introduction of Destination Management Plans as part of broader localization policies by the coalition government which came to power in 2010 reshaped DMOs, the tasks they were expected to undertake and the influence of the private sector (Hristov and Petrova, 2015).

Figure 5.9 depicts the changes which occurred in Chile in 2010 whereby the previous national structure for tourism which had been introduced in 1975 was replaced by a more wide-ranging institutional framework for tourism which redistributed the

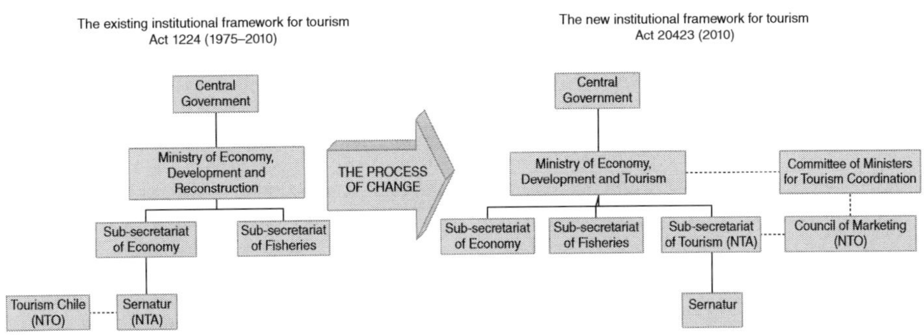

Fig. 5.9. Changes to the national institutional framework for tourism in Chile. From Guala Catalan and Pearce (2019), Figure 7.1.

roles of policy formulation, approval and implementation (Guala Catalan and Pearce, 2019). The 1975 act established Sernatur as the NTA (national tourism administration) in charge of policy making, quality assurance, coordination and marketing. Tourism Chile was created as a separate NTO (national tourism organization) to market inbound tourism. Under the 2010 act tourism's position within the public sector was significantly strengthened by the establishment of a separate Sub-secretariat of Tourism within the Ministry of Economy, Development and Tourism. Sernatur was restructured and a new NTO (Council of Marketing) and a Committee of Ministers for Tourism Coordination were included within the enlarged framework. In this the Sub-secretariat formulates and evaluates policies, the Committee of Ministers approves and Sernatur implements. In terms of marketing, the Council of Marketing provides the guidelines and evaluates while Sernatur implements the marketing strategies. Reflecting the very centralized nature of government in Chile, Sernatur has 15 regional offices throughout the country. These changes came about in a long and complex process in a decade characterized by rapid rates of growth in tourism, increased government support to the sector and the adoption of the New Public Management principles of governance, transparency and accountability in the Chilean public sector. With subsequent changes of government further modifications to the institutional framework were proposed but not implemented at the time of writing.

Towards a National Synthesis

As the preceding sections have shown, a wide variety of approaches and data sources have been used to examine aspects of the structure of destinations at a national scale but there has been little attempt to develop comprehensive and integrated syntheses of the type proposed by Rognant (1990) for Italy. Most studies have been selective: the majority draw on a single source of data and employ one technique; they concentrate either on demand or supply but only occasionally examine both dimensions; analyses of travel patterns rarely combine international and domestic travel; organizational studies are sector specific and deal either with hotel chains or airline networks; studies linking organization with other structural aspects are limited; in some cases a wealth of detail is provided but there is little attempt to integrate and interpret the patterns presented. Given the issues of data availability and the technical aspects of pursuing particular topics it is not surprising that specialization has outweighed synthesis. There is, however, much to be gained by drawing these various threads together. Two attempts at doing this are presented next to show how a national synthesis might be approached, though neither is exhaustive.

Spain

Pearce and Priestley (1998) systematically analysed a range of national data sets from the early 1990s and presented the results in a series of 14 maps. They began by establishing the spatial dynamics of domestic tourism through an interregional analysis of travel patterns derived from an origin–destination matrix (an earlier version of Figs 5.3a and 5.3b) before examining international tourism in terms of gateways and flows, both overland and by air. Pearce and Priestley then analysed patterns of distribution using a variety of different indicators – hotel and camping bednights, second homes, employment – and both absolute and relative measures. The key features of the patterns so identified were then combined into a composite map which provided the basis for a more integrated interpretation and explanation of the structure of tourism in Spain at a national level (Fig. 5.10). The emphasis here was on synthesizing often hitherto disparate but closely interrelated elements: the dynamic (flows) and the static (distributional), domestic and international, supply and demand. In particular, the flows of tourists link the origins, both international and domestic,

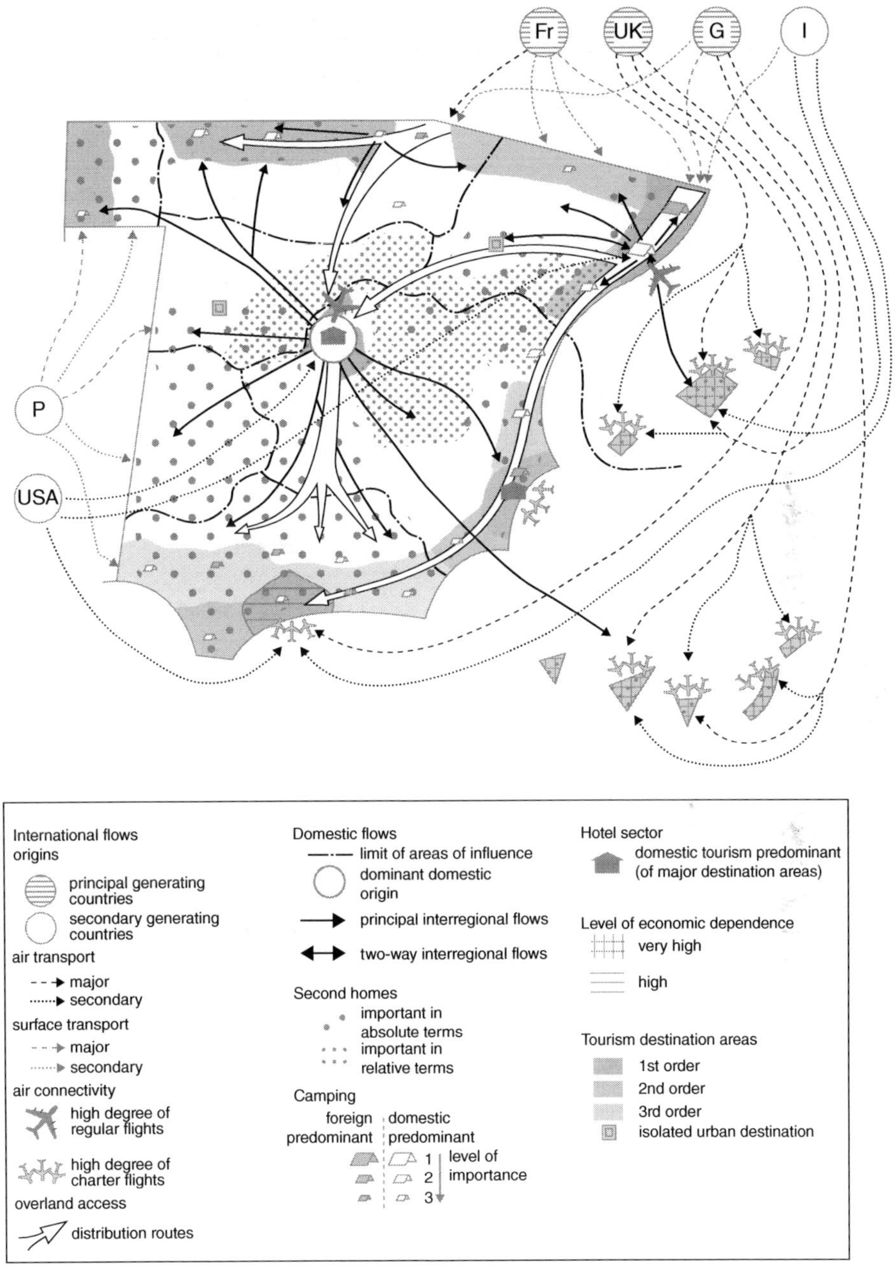

Fig. 5.10. The spatial structure of tourism in Spain: a synthesis. From Pearce and Priestley (1998), Figure 17, with permission from Cognizant Communication Corporation.

with a set of sub-national destinations whose characteristics vary. In the case of international visitors such an analysis goes beyond considering Spain as *the* destination to indicate their regional destinations within the country and the gateways through which they arrive.

The level of abstraction required to arrive at such a synthesis necessarily means only a generalized structure is presented in Fig. 5.10, with the detail being contained in the 14 supporting maps and their accompanying interpretation. Overall, four broad tourism areas were identified:

- the highly developed, tourism-dependent Balearic and Canary islands characterized by strong levels of international demand and a high degree of connectivity, a feature not usually associated with geographically peripheral areas;
- a very important Mediterranean coastal zone whose structure varies significantly as a function of the domestic/international mix and varying levels of accessibility, which in turn influence the volume and nature of the supply;
- a secondary area along the Atlantic and Bay of Biscay coasts of northern Spain, dependent largely on domestic markets, with some overland international visitors; and
- an informal interior dominated by Madrid, which acts as a major domestic source and destination as well as an international gateway and which exhibits a variety of capital city functions.

The national structure of tourism in Spain can also be set against broader structural features of the Spanish economy at that time. Ferrer Regales (1988) identified four dynamic growth areas, namely:

- the Mediterranean littoral from the French border to Murcia, with extensions to Malaga;
- the Ebro valley from the Catalan coast to the south of Alava;
- Madrid; and
- the Balearic and Canary Islands.

The development of tourism would appear to have been linked to these axes and areas in different ways. Along the Mediterranean littoral tourism has been both cause and consequence of the broader patterns of growth experienced, tourism being but one dimension of 'sunbelt' development contributing to further urbanization in the region while at the same time benefiting from more general growth through additional regionalized demand. Figure 5.10 shows tourism not to be very significant along the Ebro axis, although industrial and other development here would have generated additional demand for domestic tourist travel. Some of the continuing economic development of Madrid might have been attributable to its role as a domestic destination but the key feature here is the extent to which economic growth continued to foster the capital's role as a major origin. In contrast, the economic development of the islands is intimately intertwined with the expansion of tourism. In general, tourism had not yet had a major impact on the declining fortunes of much of interior Spain, the dominant tourism indicator there – second homes – being partly a consequence of this decline.

Figure 5.10 offers a wide-ranging synthesis of the distribution of supply and demand expressed through patterns of tourist travel but it does not incorporate institutional or organizational structures nor the public- and private-sector actors whose actions provide the supply and contribute to shaping the patterns of demand. However, other research contributes to filling these gaps. Pearce (1997b) examined change in institutional structures in post-Franco Spain as responsibility for tourism was devolved to the newly created autonomous communities for the matters that fell exclusively within their regional boundaries. Central government retained responsibility for the promotion of tourism abroad. In a process which might be interpreted in terms of exchange theory or power dependency, the community organizations initially did not function strongly as a network, neither vertically nor horizontally. Subsequently, greater coordination and cooperation developed between the different administrative layers. The institutional changes influenced marketing strategies at different levels. Products rather than places assumed greater emphasis in the activities of the national body, Turespaña, as these were to some extent geographically neutral. Faced with the new political reality the autonomous communities promoted their territories as a whole even though resources were not uniformly spread throughout them.

MarketLine (2018) provides a profile of the hotel and motel industry in Spain.

Nationwide there were more than 19,000 properties in 2018. Large chain hotels dominate, both those headquartered in Spain (e.g. Melia and NH Hotels) and those with head offices elsewhere (e.g. Marriott and Accor), but there are also numerous independently owned hotels. In 2014 ten foreign chains had some 328 establishments, 20% of which were in Madrid and 10% in Barcelona. The remainder were mainly in other large cities, both inland and on the coast (Marco-Lajara *et al.*, 2017). Chain-affiliated hotels (161 rooms) in Spain are on average over three times larger than those of independent hotels (Such Devesa and Mendieta Peñalver, 2013) (Fig. 5.11). The larger hotels are located in the major tourist resorts and cities while many of the smaller independent properties are situated in the countryside. MarketLine (2018) emphasized the importance of brand recognition in a mature and competitive market, noted that the major players have a wide portfolio of brands and reported that many of these were diversifying into such businesses as casinos, restaurants and shops. In terms of forces driving competition in the sector in 2018, they assessed rivalry as strong and buyer power, supplier power, the likelihood of new entrants and the threat of substitutes as moderate.

New Zealand

Figure 5.12 depicts the national pattern of demand for tourism in New Zealand expressed in terms of expenditure by domestic and international tourists in 2015. The greatest expenditure occurs in Auckland, the country's largest city and major international gateway, followed by Canterbury, Waikato and Wellington. In relative terms tourism makes a greater contribution to regional GDP in the more rural South Island, especially in Otago and on the West Coast. Domestic spending predominates in all but three regions: Auckland, Otago and the West Coast. It is especially important in regions close to Auckland and many of those destinations where total expenditure is relatively small such as Gisborne, Taranaki and Hawkes Bay and which might be thought of as primarily domestic destinations. The expenditure

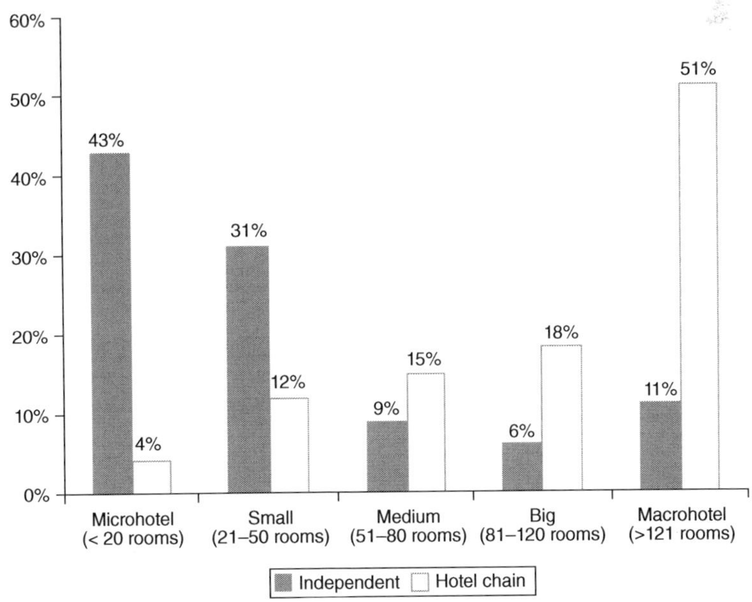

Fig. 5.11. Dual structure in the Spanish hotel industry: independent properties versus chain-affiliated hotels (2006). From Such Devesa and Mendieta Peñalver (2013), Figure 1.

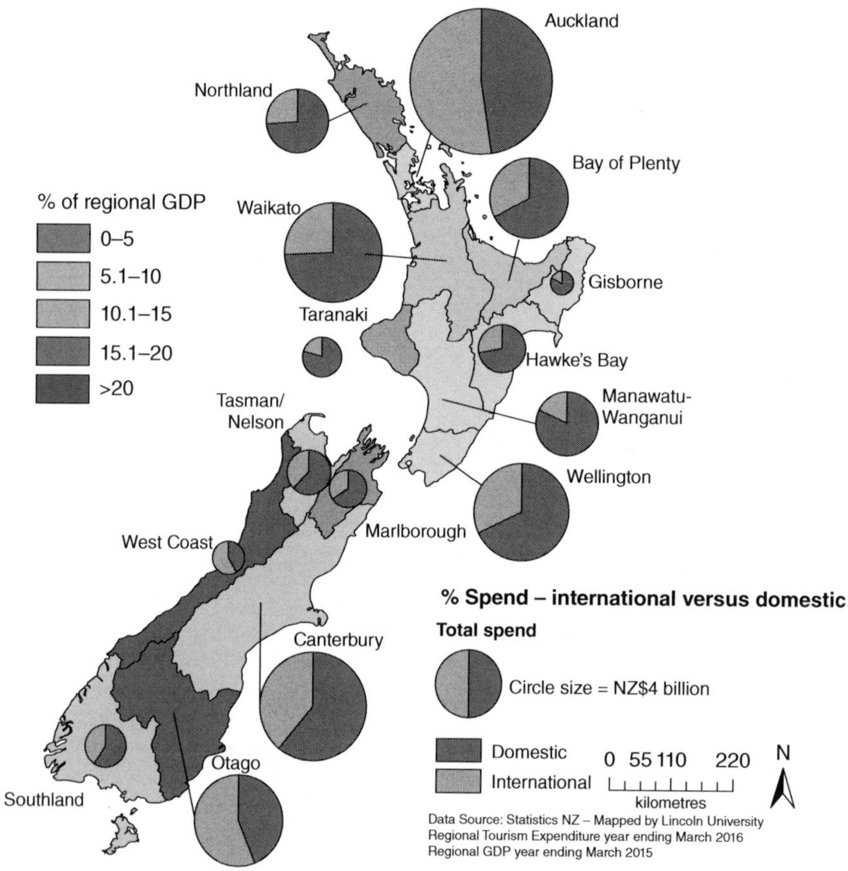

Fig. 5.12. Regional tourism expenditure as a percentage share of regional GDP in New Zealand, 2015. From Lincoln University, reprinted with permission.

patterns reflect the combination of the domestic and international tourist flows.

Domestic tourism involves a large degree of intraregional travel and distinct clusters of flows between neighbouring regions (Pearce, 1993b). Much international travel takes the form of circuit travel. Figs 5.13 and 5.14 depict the desire-line flows and grounded flows of Japanese and Australian visitors to New Zealand and show international travel within the country consists of a set of flows from the three principal international gateways (Auckland, Wellington and Christchurch) which are channelled along clearly defined corridors to other major destinations such as Rotorua and Queenstown as well as to a series of secondary centres. Travel throughout the North Island is more linear; that in the South Island more circular in nature. Trips by the longer stay and more numerous Australians were more dispersed than those by the shorter stay Japanese tourists, many of whose visits were confined to the Auckland–Christchurch–Queenstown triangle and a side visit to Rotorua. Figures 5.13 and 5.14, however, do not portray the centres visited by those going to a single destination, understating in particular the role of Auckland. These flows, it must be remembered, are aggregate patterns and do not necessarily reflect those of individual travellers.

Figure 5.15 shows how selected supply-side elements come together in a partial spatial and organizational synthesis of tourism in New Zealand by depicting the distribution

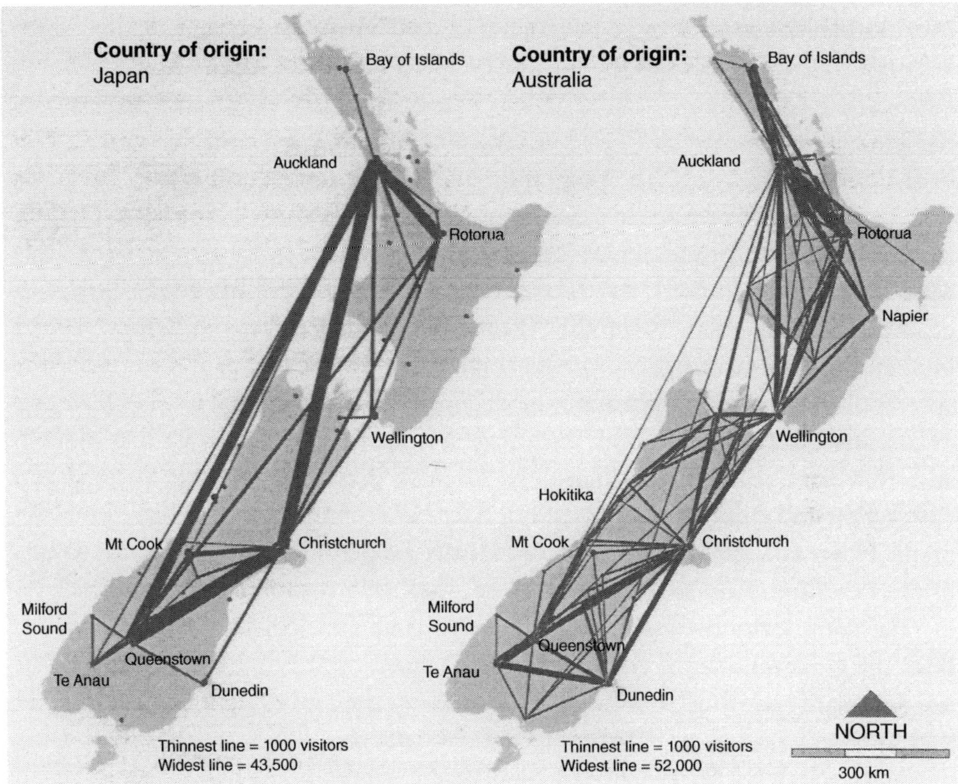

Fig. 5.13. Desire line flows of Japanese and Australian visitors to New Zealand (1992–1993). From Forer (2005), Figure 2.1, with permission of the editors.

of major chain hotels, together with the activities of major operators in the attractions sector and an example of a transport network. Chain hotels accounted for nearly half of all hotels in 2017 but 78% of hotel rooms (Horwath HTL, 2018). The 180 hotels of the international chains accounted for 55% of the country's 32,000 rooms while the 80 domestic chain hotels provided 23%. The three leading chains provided 26.5% of all rooms. Chain hotels averaged 96 rooms compared to the independent hotels' average of 25 rooms. AccorHotels was the leading international chain with 4366 rooms across its multiple brands in 31 properties. Scenic Hotels, the largest wholly New Zealand owned and operated domestic chain, ranked third overall with 16 hotels and 1571 rooms under its three brands. Each of the three leading chains has a different geographical spread, reflecting different strategies associated with their classes of hotel. Accor hotels are concentrated in Auckland (13), Wellington (6), the capital, and Queenstown (4), the South Island's leading resort. They are also present in some of the other urban centres as well as Rotorua, a major tourist destination in the North Island. The chain's range of hotels and their location reflects its orientation to business travellers and upmarket tourists. In contrast, the Christchurch-headquartered Scenic Hotel chain has five of its properties at smaller tourist destinations on the West Coast but is not represented in Queenstown; it is present in other provincial cities but has just two hotels in Auckland and none in the capital. The second-ranked Millenium and Copthorne chain (26.5%) constitutes an intermediate group with a dispersed presence in provincial cities and a range of tourist destinations as well as properties in Auckland and Wellington. As much

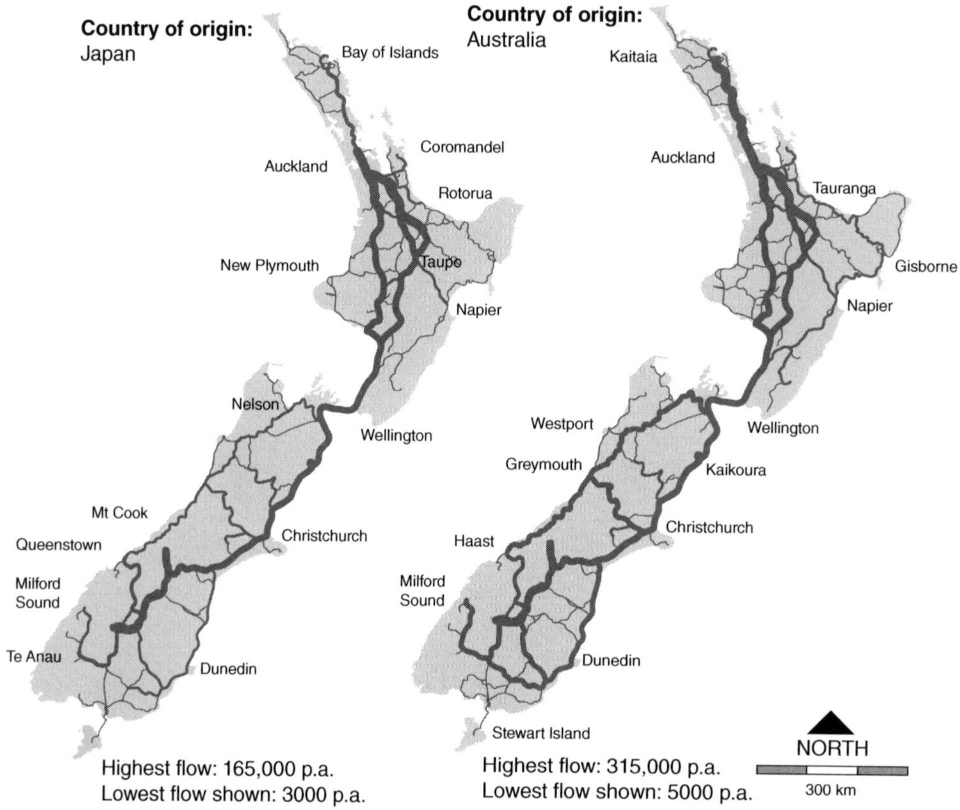

Fig. 5.14. Grounded flows of Japanese and Australian visitors to New Zealand (1999–2002 average). From Forer (2005), Figure 2.7, with permission of the editors.

international tourism involves a large degree of touring (Figs 5.13 and 5.14), a geographically diverse chain is able to cater to demand across its own properties.

At the same time it should be noted that in 2018 hotels accounted for just 36% of all overnights in commercial accommodation (excluding hosted accommodation and Airbnb) with the remainder being spent in motels (31%), holiday parks (21%) and backpacker hostels (13%). While some of these other forms of accommodation are affiliated to marketing groups or may be part of a chain, such as the YHA network of youth hostels, they are in the main small, independently owned and operated businesses with a much different spatial and commercial structure compared to the hotel sector.

Similar structural characteristics occur in other sectors. Many of the attractions and activities in New Zealand are small owner-operated enterprises or local government facilities such as museums, art galleries and parks and gardens. Three examples of the consolidation and geographical expansion which has occurred in this sector are shown on Fig. 5.15. Ngai Tahu Tourism is the tourism arm of Ngai Tahu Holdings, the investment company of the South Island's Maori tribal group. It has invested treaty settlement funds in a range of economic activities and acquired and built up a diversified portfolio of major experience-based tourism activities throughout the South Island and in Rotorua and Taupo in the North Island. Wayfare is a well-established privately owned family company which has its origins in operating sightseeing activities in Fiordland in the 1950s and has since diversified and expanded its operations and locations to include adventure activities in Queenstown, the Cardrona

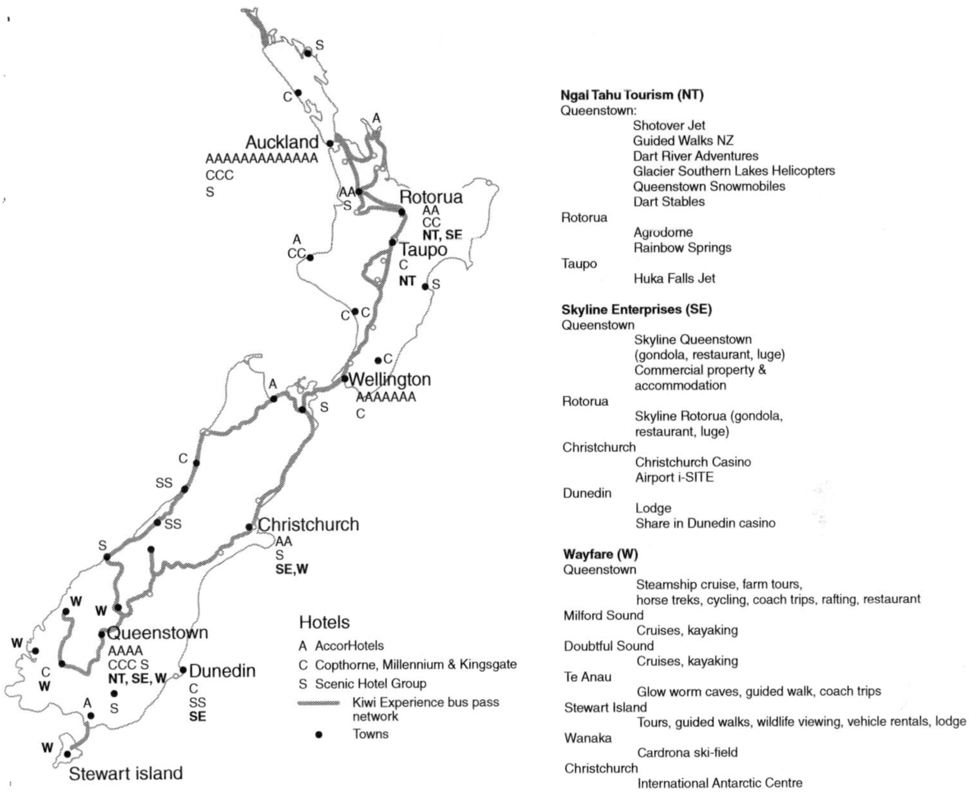

Fig. 5.15. A partial spatial and organizational synthesis of tourism in New Zealand: distribution of major chain hotels, selected attractions operators and a backpacker bus network. Data from company websites.

ski-field near Wanaka and the International Antarctic Centre in Christchurch. Skyline Enterprises has grown and diversified from a small Queenstown-based gondola operator by developing a gondola and luge in Rotorua, acquiring commercial property and accommodation in Queenstown and Dunedin, owning and operating the Christchurch casino, which also has shares in the Dunedin casino and the Christchurch airport i-SITE. The company also runs luge operations in Singapore, South Korea and Canada and until late 2019 offered a range of sightseeing, flightseeing and adventure tourism operations in the lower South Island.

The most extensive nationwide networks are those of the airline, coach services and major rental car companies. In 2019 Air New Zealand, the country's major carrier, served 20 domestic destinations; JetStar, the low-cost subsidiary of Qantas, had a more limited network of nine destinations. Figure 5.15 also depicts the routes of Kiwi Experience, a hop-on, hop-off backpacker bus network primarily serving the adventure tourism market. Kiwi Experience is part of Tourism Holdings Limited (THL in Fig 3.1), a publicly-listed company which is one of the country's largest operators of recreational vehicles in New Zealand and also in Australia, the UK and the USA. It also operates guided tours at the Waitomo Caves and black-water rafting nearby.

Horizontal integration occurs but there is only a limited amount of vertical integration or diagonal ownership and management of tourism businesses in New Zealand. Operationally, inbound tour operators play a major role in sourcing and bundling products together for wholesalers and other intermediaries in international markets, particularly more distant or emerging ones (Pearce *et al.*, 2007). In other instances, as in the Australian market, offshore intermediaries deal directly with New Zealand suppliers. The role of inbound operators is in large part a function of the complexity involved and the level of product knowledge required in making arrangements for packaged tours of the country, a prominent form of international travel in New Zealand. There is also, however, a large degree of independent travel, with visitors making their own arrangements while travelling between destinations within the country or once they have arrived there, 'at destination' distribution (Pearce and Schott, 2005). Much domestic travel is by private car and stays are often with friends and relatives and in holiday homes but when commercial services are required domestic tourists tend to book directly with providers.

New Zealand had the world's first designated national tourism organization when the Department of Tourism and Health Resorts was which established there in 1901. This was restructured from time to time (Pearce, 1992), most notably in 1991 when the international marketing function was devolved to a separate state-funded board, Tourism New Zealand, and in 2012 when the remaining tourism policy and planning functions were absorbed into a mega ministry, the Ministry of Business Innovation and Employment. Tourism Industry Aotearoa (TIA) is the peak organization for tourism businesses, large and small, and collectively represents around 85% of total tourism industry turnover. As well as advocating for the interests of its members TIA released its Tourism 2025 Growth Framework in 2014 to guide the development of the sector. Regional Tourism New Zealand (RTNZ) is the national organization which represents the interests of the country's 30 or so regional tourism organizations (Chapter 6).

Discussion and Conclusions

This chapter has illustrated the challenges of developing an integrated picture of national destinations resulting from the availability of different types of data and the range of analytical techniques used. The review of demand patterns has shown the importance of combining analysis of both domestic and international patterns as geographic variations in them influence the overall spatial structure. Differentiating between segments also deepens this understanding as the studies showing variations between segments illustrate, whether in terms of differences in long-haul or short-haul travellers, established or developing markets or purpose of visit (leisure, business or VFR). Likewise, conclusions drawn regarding the spatial patterns of supply depend very much on the range of data considered and the ways in which these have been analysed.

Despite these limitations some general patterns do emerge from the studies reviewed. The demand studies indicate that much international tourism is spatially concentrated and that intranational travel is rather limited. Although circuit tourism may be popular in countries such as New Zealand (Figs 5.13 and 5.14) or Sweden (Zillinger, 2007), many international tourists only visit one or a small number of regions (Tables 5.1 and 5.2; Hwang and Fesenmaier, 2005; Santos *et al.*, 2012). Thus while international tourists are by definition visiting a country most are only visiting a small part of it. Longer-haul tourists, however, generally tend to travel more extensively than those from neighbouring countries. The maturity of the markets also comes into play and visitation patterns may become more dispersed over time (Bowden, 2003; Koo *et al.*, 2017).

Figure 5.1 shows international demand to be more geographically concentrated than domestic demand, a pattern which was also evident in an earlier study based on Lorenz curves for 11 European countries (Pearce, 1995a). Figures 5.2 and 5.3 and the research reviewed indicate a pronounced distance decay effect in domestic travel with the result that much domestic tourism occurs in the home region or involves travel to neighbouring regions. Major urban centres are the sources of much domestic demand; that is, at a national level they may act as origins as much as or more so than destinations, even though this last function may be especially important with regard to international tourists. The nature of the urban hierarchy and the distribution of the major urban centres will thus have a significant impact on the overall pattern of demand and influence the degree to which international and domestic demand is concentrated. As might be expected given links between demand and supply, geographic concentration is also evident in the supply-side studies, though the degree to which this occurs in Figs 5.4 to 5.8 and in other studies (Khiati, 2018) is to some extent a function of the measures used. In general, the narrower the range of measures applied, the more spatially concentrated the distribution pattern appears. Use of a wider range also reveals spatial variations in the types of tourism to be found (Fig. 5.4).

The distribution of demand and supply is clearly also a function of the location of tourist attractions. The relative importance of these varies from country to country but most commonly is a function of the attractiveness and accessibility of the major urban centres (e.g. London, Paris, New York, Warsaw, Santiago); coastal regions (e.g. Florida and Hawaii, even the Baltic coast of Poland with its less desirable climate than that of Mediterranean Spain, France and Italy); mountainous areas; and, to a lesser extent, rural regions. Then too the basic products differ from country to country. It is New Zealand's varied and accessible landscape which appeals to many international visitors, whether for general sightseeing or the opportunities for adventure tourism it offers. This, coupled with the relatively small size of the country, fosters circuit tourism (Figs 5.13 and 5.14). But, unlike Spain which also has well developed touring patterns involving major urban centres and historic cities (Bote *et al.*, 2003), New Zealand attracts few international 'sun, sand, sea' visitors, although many domestic tourists head to the beaches in summer. Accessing and analysing comprehensive data relating to attractions in order to quantify spatial relationships with other sectors or indicators of demand is, however, particularly challenging in terms of data availability; comparable, relevant measures (length of coastline, number of ski slopes, presence of protected areas); assessment of quality; and areal units employed (Capone and Boix, 2008; Fernández Cuevas *et al.*, 2008; Barros *et al.* 2011; Majeska, 2015; Truchet *et al.*, 2016). Numbers of 'sites' or monuments in major urban centres, for example, do little to explain hotel demand from business travellers which accounts for a significant share of hotel rooms located there.

By drawing on a wide range of data and initially developing a series of more specific maps, Fig. 5.10 shows that it is possible to synthesize key elements of supply and demand and build up a more comprehensive and integrated picture of the spatial structure of tourism at a national scale. The inclusion here of domestic and international flows introduces a dynamic element and highlights the role of gateways and the generating role of Madrid, Barcelona and the industrial north. Aspects of the organizational structure of tourism in Spain, however, are only developed subsequently in the accompanying text above by reference to other studies, including various characteristics and assessments of market structure (Fig. 5.11; MarketLine, 2018).

In general, however, it is difficult to synthesize organizational structure at a national level as details on ownership and other market characteristics are not readily available. Data compiled on the hotel sector show differences in size between chain and independent hotels and reveal many of the chain hotels are located in key gateways and major urban centres and resorts (Litteljohn, 1997; Ribaudo and Domeniconi, 2013/2014). Distinctive spatial strategies of airlines are also apparent with differences occurring in the experience of the USA and Europe (Dobruszkes, 2006; Reynolds-Feighan, 2010, 2018; Lieshout *et al.*, 2016). Conditions of entry in Europe have seen the emergence of low-cost carriers serving a larger number of airports. Hotel and airline studies tend to be carried out separately, which also hinders synthesis. A more comprehensive approach can be taken more readily in smaller countries such as New Zealand. Figure 5.15 depicts the distribution there of the activities of leading organizations in three sectors – hotels, attractions and ground transport – whose presence at different locations throughout the county enables them to take advantage of and provide for circuit tourism as well as, in the case of hotels, serving other segments, especially business travellers. However, as noted, there is very little vertical integration and these larger companies are complemented by many SMEs.

Research framing destinations in terms of industrial districts, clusters and agglomerations is starting to produce some interesting insights into their spatial structure (Figs 5.4 to 5.8) and how they function in a country-wide context. The attempts to explore the performance of different types of tourist districts (Capone and Boix, 2008; Fernández Cuevas *et al.*, 2008), the varying composition of clusters (Estevão and Ferreira, 2012; Segarra-Oña *et al.*, 2012) and forms of agglomeration (Majewska, 2015) are particularly useful. The findings are mixed and far from conclusive with respect to how, or even if, the areas so identified function and perform in the ways discussed in Chapter 3 and the extent to which agglomerative forces outweigh natural endowments. Much more work remains to be done on these questions. Investigation at a national level provides the scope to assess performance across districts or clusters but complementary research into their internal structure and relationships is also needed at a regional or local scale.

Research at a national level has also highlighted the role of different places in terms of the various nodal functions discussed in Chapter 2, especially gateways, primary and secondary destinations and origins. This research has demonstrated how places may have multiple functions, acting in some instances as origins as well as destinations or gateways and destinations (Figs 5.2, 5.3 and 5.10), and shown how these functions influence and are influenced by the structure of airline networks and the spatial strategies of hotel chains.

6 Structure of Regional Destinations

Regional tourism destinations are a bridge and intermediate level between the country as a destination and multiple local level destinations. As discussed in Chapter 5, national travel surveys show a tendency for many visitors to limit their stay in any country to one or two regions. Important activities are carried out at this scale. Regional plans and strategies are frequently developed to take account of the differences which occur from one part of a country to another, to guide regional economic development strategies and to make provision for matters which are often dealt with most effectively at this level (e.g. major infrastructural works and environmental protection). Regional marketing activities complement those at a local and national level. Responsibility for carrying out these functions frequently falls to DMOs structured along sub-national administrative lines. For all these reasons it is important to analyse and consider the structure of regional destinations.

Much research on regional destinations focuses on specific questions within a given region rather than the pursuit of broader questions across multiple regions guided by common frameworks. Some of the research on regional travel employs network analysis and involves nodal regions but most of the studies deal with formal regions that follow administrative divisions which vary from country to country.

In this chapter a review of the regional distribution of supply is followed by a discussion of patterns of demand and regional travel. Consideration is then given to organizational structures and relationships. An attempt at a regional synthesis is presented before overall patterns are discussed and conclusions are drawn about the structure of regional destinations.

Supply-side Studies of Regional Destinations

Supply-side studies of regional destinations have a relatively long history, particularly in Europe where richly descriptive accounts of tourism regions have given way in more recent years to more quantitative analyses. Vera Rebollo (1987), for example, provided a detailed and insightful study of Spain's Costa Blanca. His synthesis of the different forms of tourist development along that coast included forms of land use and urban development processes: types of building; successful integration into the physical setting; infrastructure and recreational facilities. He also stressed that these physical attributes had to be seen in terms of the mechanisms which had transformed the littoral: planning, capital gain; demand and the road network. Vera Rebollo also delimited functional hinterlands based on commercial and service sector functions, derived a classification of the region's municipalities according to population size and economic base and distinguished between areas transformed by foreign, national and regional agents.

While generalizations must be tempered by contextual factors, some common structural characteristics can be identified in these early descriptive accounts of coastal regions in Europe (Pearce, 1995a):

- Dense tourism development is limited to a relatively narrow coastal strip but might be accompanied by the development inland of second homes and a more extensive recreational hinterland in which historic, cultural or scenic sites are incorporated into excursion circuits for day trippers.
- A hierarchy of nodes emerges in which the development of major resorts is

complemented by the growth of smaller ones.

- Major resorts have often grown up in relatively undeveloped areas some distance away from large urban and industrial centres. Smaller ones may be the coastal offshoot of inland hill-top villages or result from the transformation of fishing ports.
- Resorts throughout the region differ in terms of the mix of accommodation types: hotels, apartments, second homes, camping grounds and the manner in which these have developed. Many hotels and apartments are oriented towards international package tourists; the second homes and camping grounds cater more for domestic tourists and independent international tourists.
- The continuing growth in accommodation has been accompanied and made possible by the development of transport infrastructure which provides access to and throughout the region.

Later research has continued to analyse the spatial structure of tourism regions but has taken a more quantitative approach as illustrated by the three studies of Andalusia summarized in Table 4.1. Merinero-Rodríguez *et al.* (2014) identified four types of municipalities on the basis of their tourism potential:

1. Pure coastal: 16 municipalities with a predominance of coastal resources and a virtual absence of others.
2. Pure urban: ten municipalities marked by the presence of urban but few other resources, including Sevilla, Cordoba, Granada and some other cities over 100,000 inhabitants.
3. Mixed urban: 217 municipalities having a mix of resources including protected natural or heritage sites.
4. Rural: 527 essentially inland municipalities with few resources, either natural or heritage. Effectively, most of these are not tourist destinations.

Parrilla-González and Pulido-Fernández's (2017) latent class analysis of 760 municipalities characterized by the 30 variables selected for the years 1996 and 2011 produced three groups of municipalities in Andalusia in terms of their economic tourism dynamism:

1. Very dynamic tourism municipalities: 44 municipalities including a mix of traditional coastal destinations (e.g. Fuengirola, Marbella, Málaga); Seville and its wider metropolitan area; and small cities (20,000–40,000 inhabitants) with a rich architectural heritage and developing cultural attractions which were in the early stages of tourist development (e.g. Arcos de la Frontera and Baza).
2. Dynamic tourism municipalities: 56 varied municipalities experiencing lesser but still important development. These included a number of municipalities in the province of Almeria and some other middle-sized cities characterized by well-preserved resources, improvements in access, private investment in quality accommodation, attractions associated with golf courses and theme parks and a boom in second homes. Others in this category were provincial capitals and a group of municipalities which had received European rural development funds.
3. Non-dynamic municipalities: a large (660) and very heterogeneous category which included both traditional destinations which had witnessed little change in the measures over the period considered (e.g. Torremolinos) and those that had little or no orientation to tourism.

The outcomes of these two analyses are depicted and summarized in Figs 6.1 and 6.2. In contrast, Sarrión-Gavilán *et al.* (2015) present a total of 29 maps, that is, a lot of detail with very little synthesis. The maps represent the distribution of bed spaces, two indices of tourism pressure (DFTI (Defert's Tourist Function Index) and TIS (tourist bed spaces per square kilometre)) for the years 2000 and 2011, as well as the outcome of the analysis of measures of spatial association through a series of LISA maps. The latter reveal patterns of spatial clustering and the existence of outliers. The analysis of Sarrión-Gavilán *et al.* (2015) also differs from that of the other two studies in that it proceeds to examine several of these measures from an *a priori* division of Andalusia into the three areas used in the regional plan: coastal, inland urban and inland rural. The LISA maps reveal a trend towards an increase in the number and size of clusters, especially on

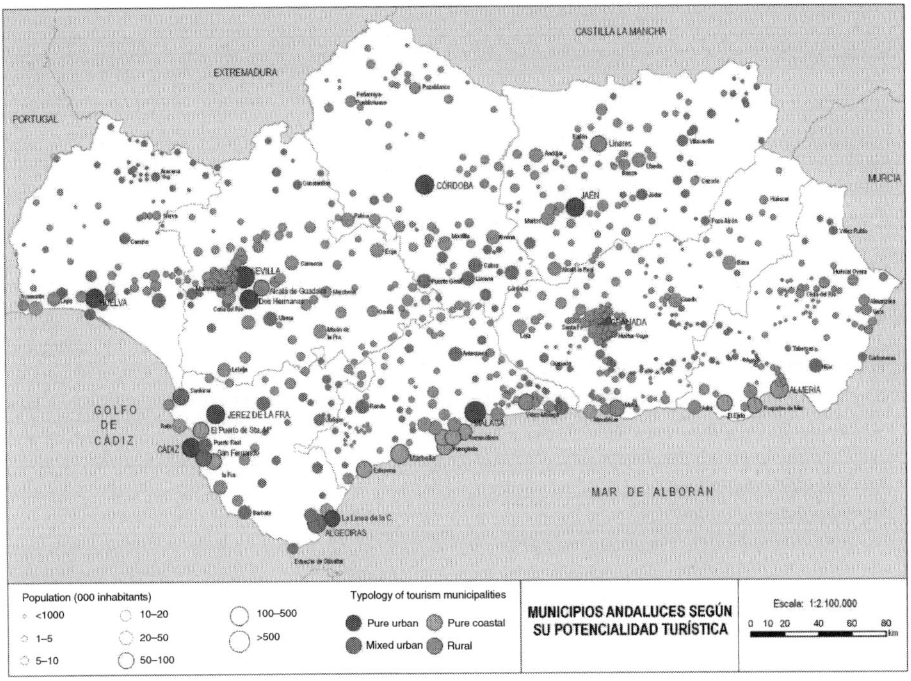

Fig. 6.1. Characterization of municipalities in Andalusia according to their tourism potential. From Merinero-Rodríguez *et al.* (2014), Mapa 1, with permission of the authors.

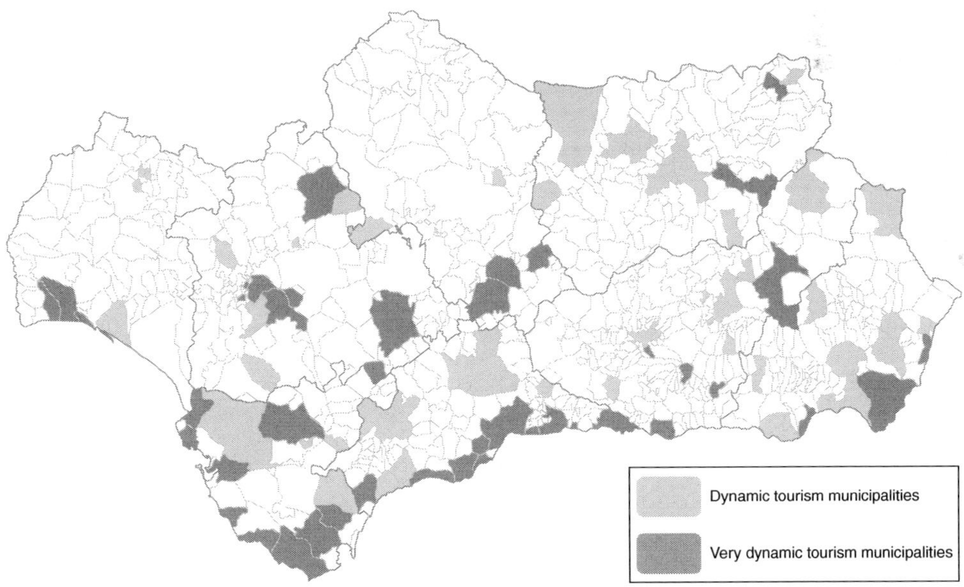

Fig. 6.2. Segmentation of municipalities in Andalusia as a function of their economic tourism dynamism. From Parrilla-González and Pulido-Fernández (2017), Figure 5, with permission of the editor.

the coast, with that from Malaga to Estepona being the most notable. Several local clusters of hostels and pensions and rural tourism establishments are also found in inland rural areas and around the cities of Seville and Cordoba. While variation occurs throughout the region there was evidence from the distribution of the DFTI values of a decrease in pressure on the local population in areas with intense tourism development. The TSI cluster map showed some growth in environmental pressure in mountain areas with high environmental and landscape quality and an increase on this indicator in already highly developed areas on the Costa del Sol.

At a general level it might be argued the sophisticated statistical analysis employed in these three studies does not greatly advance our understanding of the regional structure of tourism in Andalusia; that is, it is a region characterized by intensive development along much of its coastline, it is renowned for the historic and cultural resources of Granada, Seville and Cordoba and its inland rural areas have far less tourism potential and have experienced comparatively little tourism development. Conversely, as their authors contend, these studies and these approaches reveal intraregional differences and disparities more clearly and highlight the need for a much more nuanced approach to regional tourism planning, one which takes account of local needs and opportunities. Not all stretches of the coastline are experiencing the same issues; parts of the rural interior are developing; the mixed urban category draws attention to a particular set of destinations with diverse resources.

Moreover, the two studies which consider distributions at two different dates draw attention to the dynamic nature of destinations and show the need to go beyond a static account of regional tourism. Parrilla-González and Pulido-Fernández (2017) attribute the dynamism identified to the location of these municipalities, the tourism offer developed there and the employment generated in the sector. Furthermore, they interpret the dynamism of these municipalities as a manifestation of the networks, relationships and spatial associations portrayed in the theoretical literature of clusters and industrial districts. Though not embedded in that literature, the LISA maps of Sarrión-Gavilán *et al.* (2015) provide further empirical evidence of the spatial nature and distribution of destination clusters. Scope exists to draw these two approaches closer together.

However, in focusing on the selected supply-side variables of municipalities these is no explicit attempt in any of the three studies beyond analysing the patterns of distribution to explore how the different municipalities might come together functionally and to show how the different clusters or outliers function and might be linked together in a regional structure. In other words, the distribution of the supply-side elements is shown but not the functional relationships between them. The three major cultural-historical centres (Seville, Granada and Cordoba), for example, are often nodes on a common package tour circuit (Bote *et al.*, 2003). Malaga airport functions as the gateway for many of these tours as well as for stays on the Costa del Sol. Just over half of visitors to Ronda, one of the most important smaller centres in Andalusia, were on multi-destination trips (Bote Gómez *et al.*, 2003). It is also a popular destination for day trips from the coast. Of course it may be that few or no intraregional connections actually exist between some municipalities. While this lack of consideration of linkages is in part a consequence of the emphasis on validating methodological matters and the acknowledged lack of detail on demand data at the local level, it is also due to a failure to include infrastructural variables, particularly relating to transport, and to give a more considered account of factors underlying the patterns as is found in some of the more descriptive studies.

Other Spanish studies highlight the differences that may occur within regions at a smaller spatial scale. Sánchez Rivero *et al.* (2016) examined the distribution of the potential for rural tourism in the province of Caceres and identified four types of municipalities based on their assessed tourism potential as derived from a set of internal and external factors. Those with the highest potential were located in more upland parts of the province with water-based resources and complementary services and attractions. Cuadrado-Ciuraneta *et al.* (2017) show different forms of tourism developed at different rates and ways in different parts of the Catalan county of Alt Empordà, including the largest residential marina in Europe, Empuriabrava, and a more discontinuous urbanization occurs with the growth of second homes and various processes of suburbanization.

Attractions were the focus of Blasco *et al.*'s (2014) approach to delineating 'logical' tourism destinations or zones in the Pyrenees. Rather than relying on administrative boundaries, they took a more consumer-oriented perspective based on travel times between attractions. They obtained the number, type and distribution of attractions throughout the Pyrenees from guidebook listings and calculated the travel times between them so as to identify clusters of attractions that were within one to two hours' drive time of each other, travel times deemed to be representative of hub-and-spoke travel patterns typical of mountain recreation. In this way the attractions are seen to be linked by the trips taken by tourists. These linkages, however, are inferred from the travel times and not based on observed travel patterns. Cluster analysis of attractions and travel times resulted in the delineation of nine tourism zones. The new zones and the original destinations based on administrative boundaries were then assessed in terms of the number of borders in each area, number of beds, population and the intensity, specificity and variety of attractions. Compared to the existing regions, the new tourism zones were more uniform in extent and more similar to historical regions. Four types of tourism zones were then identified when hierarchical cluster analysis was conducted with these variables. These types varied in terms of the intensity, specificity and variety of attractions and the levels of the other variables: tourism zones found to have more first-class or specific attractions also had a higher number of total attractions; the specificity and intensity of attractions were also highly and significantly correlated with bed numbers and the number of administrative borders.

Employment data provide both a measure of the size distribution of tourism businesses and additional insights into the spatial structure of regional destinations. Judgements have to be made, however, about the scope of the tourism sector (Dean Runyan Associates, 2013; Chhetri *et al.*, 2017). The employment pattern for Queensland, Australia, is one that is often found in the tourism sector, one which is characterized by a majority of SMEs and a small proportion of large businesses (Table 6.1). The largest category is non-employing, essentially owner-operator businesses, followed by firms with one to four employees. Businesses with 20 or more

Table 6.1. Regional distribution of tourism businesses in Queensland by number of employees, 2015. From Tourism Research Australia.

			% of businesses by number of employees			
Region	Total	Total %	Non-employing	1 to 4	5 to 19	20+
Tropical North Queensland	3,658	7	39.6	32.8	21.3	6.3
Northern	2,093	4	38.6	29.7	24.7	7.1
Whitsundays	578	1	38.9	26.8	27.0	7.3
Mackay	1,170	2	37.9	26.1	26.3	9.7
Central Queensland	1912	4	38.9	27.2	25.5	8.4
Fraser Coast	895	2	39.6	33.0	24.1	3.4
Sunshine Coast	5,185	10	41.3	32.4	21.9	4.4
Brisbane	22,175	44	44.9	29.8	19.9	5.4
Gold Coast	9,067	18	46.6	30.6	17.7	5.1
Darling Downs	3,112	6	40.1	30.4	23.9	5.6
Outback	611	1	39.4	31.6	24.9	4.1
Queensland	50,456	100	43.3	30.3	20.8	5.6

employees account for less than 6% of the state total although their share of total employees would be much greater. Table 6.1 also shows there is relatively little variation in this pattern across Queensland. However, in terms of total employment in the sector there is a marked concentration of tourism jobs in Brisbane (44%), the state capital and major metropolitan centre, and in coastal areas (Gold Coast, 18%; Sunshine Coast, 10%).

California employment data highlight the importance of considering both absolute and relative measures. Figure 6.3 shows that half of all travel-generated employment in the state in 2012 was concentrated in the three metropolitan areas of the San Francisco Bay Area, Los Angeles County and San Diego County. However, in relative terms the travel sector accounts for over a quarter of all employment in the High Sierra (Yosemite National Park; Lake Tahoe, Mammoth Lakes), and more than 16% in the Deserts (Death Valley National Park, Palm Springs) but less than 5% in the San Francisco Bay Area and Los Angeles County and only slightly more in San Diego County. A similar pattern emerges with respect to the travel industry LQs. Again, these values are higher for the High Sierra and Deserts and areas in the north of the state and lower in the more diversified urban areas. Of the latter, San Diego has the highest LQ. Further differences occur within these areas: in the High Sierra travel-generated employment is spread amongst the counties; in the Gold Country over 70% of the sector's employment is concentrated in Sacramento, the state capital.

Distribution of Demand

As the patterns of supply would imply, demand too is not dispersed evenly throughout regions but concentrated in particular areas or at local destinations. However, explicit and detailed information on the distribution of demand within regions is generally less readily available than supply data and is generated most commonly by special purpose surveys. The extent to which this occurs is often a function of the relative importance of tourism to the region. In the USA, for example, good demand data are provided by the state tourism organizations of Florida and Hawaii.

In 2015 Florida recorded 91.3 million out-of-state visitors and 11.4 million overseas overnight visitors (Visit Florida, 2016). Figure 6.4 depicts the basic patterns of demand for the eight 'vacation regions' which Visit Florida, the state tourism organization, distinguishes within the state. In terms of domestic tourism, the three leading regions – Central, Southeast and Central West – accounted for 65% of visitors. The Central region is home to numerous theme and amusement parks, the Southeast is renowned for its climate and beaches. Domestic visitors coming from further afield and arriving by air are more concentrated geographically, particularly in the Central (40%) and Southeast (28%) regions, than those arriving by car. The Northwest (17%) is the second most popular region for auto travellers, attracting visitors from nearby states. Leisure visitors are more dispersed than those travelling for business. The Southeast (60%) is by far the most popular destination for international visitors, followed by Central (38%). Other profile data show variations in visitors to Florida's eight regions on a range of variables, including state or country of origin, age group, travelling party composition, purpose of visit, activities, length of stay, type of accommodation and expenditure.

Local Tourist Development Tax data show demand is even more concentrated: four counties – Orange and Osceola in Central and Broward and Palm Beach in the Southeast – accounted for just under half of the total collected state-wide. In 2017, 58.5% of visitors to Miami-Dade County who used commercial lodgings stayed at Miami Beach with a further 14.9% in downtown Miami (Greater Miami Convention & Visitors Bureau, 2018).

The domestic tourism data for Florida refer to out-of-state visitors and do not include residents visiting other parts of their respective states. To get a fuller picture of demand, account should also be taken of intrastate or intraregional travel, especially

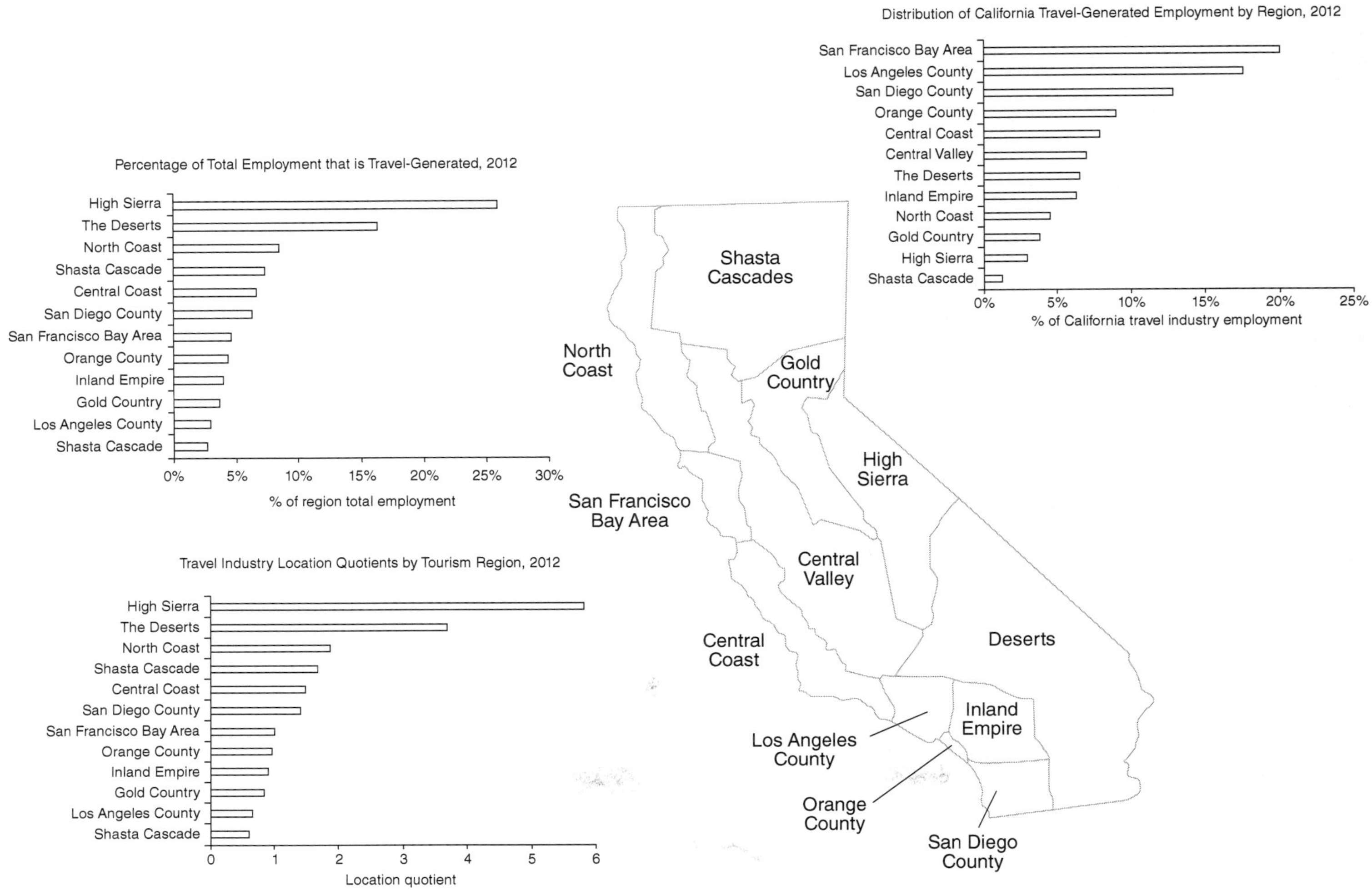

Fig. 6.3. Indicators of the distribution of travel industry employment in California, 2012. Redrawn from Dean Runyan Associates, Inc. (2013).

Region	Domestic					International[a]
	All	Air	Auto	Leisure	Business	
	%	%	%	%	%	%
Central	35	40	31	34	41	38
Southeast	18	28	13	18	20	60
Central West	12	13	11	11	15	16
Northwest	11	1	17	12	5	1
Southwest	9	9	8	9	5	incl. with Central West
Central East	7	5	9	8	4	incl. with Central
Northeast	7	3	9	6	9	4
North Central	2	1	2	2	1	incl. with North East
Top 3 regions	65	81	61	64	75	

[a]Do not sum to 100% since international visitors may travel to more than 1 region.

Fig. 6.4. Distribution of domestic and international demand within Florida, 2015. Data from Visit Florida (2016).

where states are large. In the case of Queensland intrastate visitors generate about one-third of all overnights and are particularly important in those regions where tourism is less well developed, such as Mackay, Central Queensland and the Outback (Table 6.2). Brisbane accounts for one-third of all overnights in the state and attracts by far the largest proportion of international overnights (58.6%). The highly developed tourism region of the Gold Coast, which adjoins New South Wales, is distinguished by its relatively high share of interstate (36.3%) and international (42.5%) overnights.

Important spatial variations also occur within Queensland in terms of purpose of visit (Table 6.3). Brisbane attracts a more diverse range of visitors than the Gold Coast. In the former, VFR (31.5%) overnights are marginally greater than for those on holiday (29.2%) while in the latter holiday overnights (56.1%) dominate but the VFR traffic (26%) is still important. Little difference in the share of overnights occurs in terms of purpose of visit between interstate and intrastate on the Gold Coast but VFR and business overnights are relatively more important for interstate visitors in the state capital.

These patterns of demand raise further issues of coverage, of what is being included or omitted in the data being analysed. As well as drawing attention to intrastate or

Table 6.2. Distribution of overnights in Queensland 2015. From Tourism Research Australia.

			% of overnights by origin		
Region	Total (000)	Total %	International	Interstate	Intrastate
Tropical North Queensland	16,930	13	39.6	29.0	31.4
Northern	5,741	4	29.3	25.2	45.5
Whitsundays	3,348	3	39.5	34.8	25.7
Mackay	3,363	3	19.3	14.0	66.7
Central Queensland	7,054	5	20.2	18.7	61.1
Fraser Coast	2,877	2	16.5	33.2	50.3
Sunshine Coast	13,600	11	19.8	28.5	51.7
Brisbane	42,420	33	58.6	19.4	22.0
Gold Coast	23,410	18	42.5	36.3	21.2
Darling Downs	7,196	6	24.6	16.6	58.8
Outback	2,799	2	11.4	22.3	66.3
Queensland	128,738	100	40.3	25.4	34.3

Table 6.3. Overnights in Brisbane and on the Gold Coast by purpose of visit, 2017. From Tourism Research Australia.

Purpose of visit	Holiday	VFR	Business	Other	Total
Brisbane					
Total overnights (000)	13,332	14,292	4,618	13,169	45,311
Total overnights %	29.2	31.5	10.2	29.1	100
International %	31.0	24.0	2.9	42.1	100
Interstate %	25.8	46.3	23.0	5.0	100
Intrastate %	28.0	36.4	16.3	19.3	100
Gold Coast					
Total overnights (000)	13,572	6,286	1,410	2,933	24,201
Total overnights %	56.1	26.0	5.8	12.1	100
International %	48.3	24.6	2.4	24.7	100
Interstate %	61.6	26.3	7.8	4.2	100
Intrastate %	60.7	27.9	8.5	2.9	100

intraregional travel, the purpose of visit data raise the question of the completeness of accommodation data sets, especially where these are limited to commercial accommodation. The majority of VFR demand, for example, is met by visitors staying with friends and relatives; business travellers generally use a narrower range of commercial accommodation, particularly hotels, than holiday makers. Consequently, spatial variations in the patterns of demand will mean that certain segments will be over- or under-represented by particular accommodation data sets. This will also occur to a greater extent with other indicators such as attractions or complementary services.

Regional Travel Patterns

Intraregional travel has been analysed by treating the regional destination as a network of local destinations (the nodes) which are linked by tourist movements (Shih, 2006; Liu *et al.*, 2012; Asero *et al.*, 2016; Piriou, 2016; Wang *et al.*, 2016). However, network analysis used in this context must be viewed cautiously as it is based on the assumption of multi-destination travel, that is, places in the regional network are only connected if tourists travel between them while places which are the focus of mono-destination travel are excluded from the analysis (Liu *et al.*, 2012; Asero *et al.*, 2016). The significance of this in terms of regional structures must not be discounted as mono-destination visitors may constitute a substantial proportion of regional demand. They made up 22% of Liu *et al.*'s (2012) sample in South Anhui (China). Asero *et al.* (2016) do not state the proportion of mono-destination visitors in their study of tourism networks in Sicily but research in the same region by Parroco *et al.* (2012) found over two-thirds of their respondents visited only a single destination. Moreover, the proportion of multi-destination visitors to Sicily varied across a range of profile variables: origin (international, 43.84%; domestic, 23.59%); first-time (48.23%) and repeat (21.84%) visits; and type of holiday (sun sand sea, 14.25%; partial sun sand sea, 38.15%; cultural/eco-tourism, 31.51%).

Asero *et al.* (2016) identified and mapped four clusters of destinations in Sicily based on their network characteristics. They also analysed the ego-networks (defined p. 752 as 'nets made of all routes linked to a destination') of the most attractive destinations and found differences occurred between them in terms of their size and shape while core and peripheral destinations existed within each ego-network. The destinations of the Catania ego-network, for example, were less linked than those of Palermo. Main and marginal routes were also able to be distinguished. The Ragusa ego-network was smaller than the others and mostly involved routes in the hinterland area. Associated profile data showed domestic visitors travelling with relatives were more represented in the Ragusa ego-network while international tourists were more prominent in the destinations included in the Agrigento and Palermo ego-networks. Although some order emerges from such analysis the overall picture is one of great complexity as in aggregate multi-destination visitors' major and minor routes overlap and intersect. However, in terms of the structure of Sicily as a regional destination, the dominance of the main nodes and a clearer hierarchy of places are likely to be even more apparent if single-destination visits were also included in the analysis. In other words, some places serve both as key nodes in multi-destination travel and at the same time attract large numbers of mono-destination visitors.

Shih (2006) and Liu *et al.* (2012) focused on identifying different types of nodal functions based on their network characteristics. For drive tourism in central Taiwan (Fig. 6.5), these included beginning and terminal nodes, core and peripheral nodes (Shih, 2006). In South Anhui, Liu *et al.* (2012) identified six types of destinations or nodal functions: tourist flow distribution centre; hub of tourist flows; passageway destinations; common touring destinations; attached touring destinations; and nearly isolated destinations. In both classifications each type of node was associated with or required a different set of facilities and services, such as transport, information and souvenir shops.

In a study which included the effects of changing accessibility, Wang *et al.* (2016)

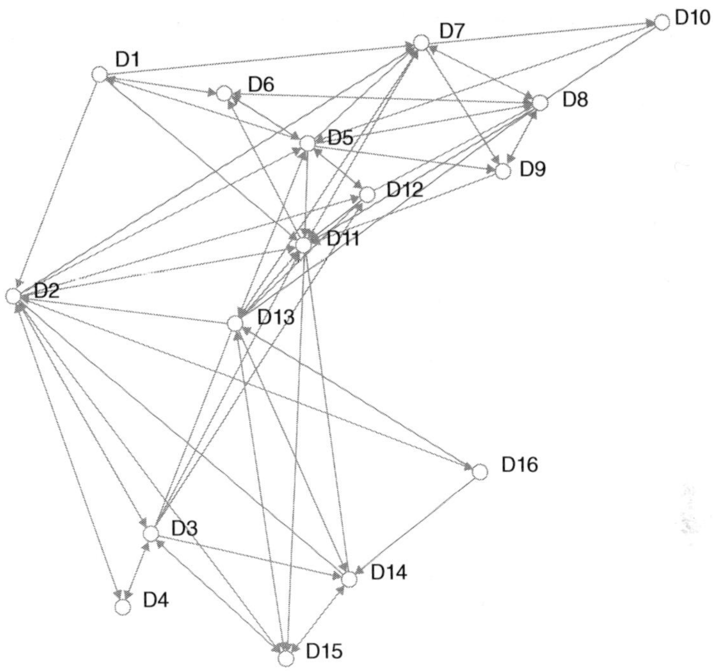

Fig. 6.5. Network graph of drive tourism destinations in Nantou, Taiwan. From Shih (2006), Figure 3.

examined the impact of the introduction of high-speed rail between Beijing and Shanghai on patterns of tourist travel from those two cities, notably the way in which faster travel – 'time–space compression' – affected the spatial structure of regional tourism. They went beyond a network analysis of survey data on before and after travel by tourists from Beijing and Shanghai by also incorporating in their analysis the role of four major destination attributes – tourism resource endowment, hospitality facilities, tourist transportation network density and location – for 17 destinations (including the two source cities). Their study identified different types of nodal functions and multi-functionality, including the roles of Beijing and Shanghai as both origins and destinations, and highlights the importance of transport infrastructure.

Four different effects were found, each of which was moderated by the destination attributes to produce different impacts on different destinations. First, a pronounced Matthew effect was observed whereby nodes which were already well-connected in the network benefited more from the reduced travel times. These were of three types: the two end-point destinations (Beijing and Shanghai) which ranked strongly on all the attributes; transfer nodes (Jinan and Nanjing) which had very strong locations benefiting from time–space compression but which also ranked highly on the other attributes; and resource-based nodes (Mt Tai or Suzhou) which were very well endowed with resources, were located near an end-point or transfer city and had strong reception capacity and traffic networks. A second set of destinations close to Beijing or Shanghai experienced a filtering effect, that is, traffic to them decreased as tourists took advantage of reduced travel times to go further afield. The degree of filtering (prominent or light) was influenced by the competitiveness of the nodes on the other attributes. Diffusion was a third effect observed, a function of more time being available to tourists who had gone to core nodes which could then be spent on visiting other destinations. Nodes affected in this way were classified into two groups: diffusing sources and diffusing destinations. The former were further sub-divided into three categories: the end-point cities; transition cities (e.g. Jinan and Nanjing) having superior locations and ranking highly on the other attributes; and resource nodes with very strong resource endowments and transport and reception advantages. Destinations benefiting from diffusion from these nodes were located nearby. Some diffusion between diffusion sources also occurred. Finally, an 'overlying' effect was identified whereby five destinations lying in an intermediate zone between Beijing and Shanghai experienced significant growth from both cities as time–space compression extended the range of travel from these two markets. Again, these were grouped into two categories depending on whether their advantages were based on superior location or resource endowment.

The effects of resource endowment and distance on network structure were analysed by Lee *et al.* (2013) who weighted links in the network of 43 rural villages in the southern part of South Korea by using a gravity model which combined a friction of distance measure (shortest path) with indices of rural amenity and population. Marked differences in the patterns of connectivity emerged using this refinement compared to measuring centrality and accessibility by geographic location alone. This led to the identification of a core village and sub-core villages around which a more integrated approach to rural tourism management could be developed.

Wu and Carson (2008) incorporated a temporal element in their analysis of multi-destination travel in South Australia. Adelaide, the state capital, was the primary destination for both domestic and international visitors to the state. The spatial patterns of domestic visitors showed little change from day one to day two except for a decrease in overall numbers as some of these were on short trips or travelled on to destinations outside the state. International travellers, on the other hand, gradually moved from outer regions to Adelaide, having travelled from neighbouring states and visited major regional destinations on the way.

In regions consisting of multiple islands the emphasis is on patterns of inter-island travel as the islands stand out more clearly as distinctive spatial units and because a change in mode of travel is usually needed. In the case of Hawaii, multi-island visits are relatively limited. In 2017 over 80% of all visitors arriving by air visited a single island within the state. The average number of islands visited was only 1.25 (Table 6.3). Oahu was the major destination for international visitors but domestic visitors making a single island visit were more likely to go to one of the Neighbour Islands, particularly Maui. Oahu, whose airport has the largest airline seat capacity, was also the island most visited in combination with one of the other islands. International visitors to Kauai and Hawaii were more likely to be making multi-island visits rather than just visiting those islands in isolation. These patterns indicate two major types of international visits to Hawaii: the majority restrict their stay to Oahu but a much smaller proportion engage in active multi-island travel, suggesting sightseeing is a key activity for this segment

Table 6.4. Patterns of demand in Hawaii by islands visited: arrivals by air, 2017. From Hawaii Tourism Authority.

	Domestic	International	All air arrivals
Visitor arrivals	6,239,748	3,037,865	9,277,613
Islands visited			
Average no. of islands visited	1.25	1.26	1.25
Any one island only %	81.9	80.8	81.5
Multiple islands %	18.1	19.2	18.5
Oahu only %	34.3	70.0	46.0
Neighbour Islands only %	51.8	11.7	38.7
Oahu & Neighbour Islands %	13.9	18.2	15.3

alongside having a beach holiday. Other data for 2017 show three-quarters of first-time visitors went to Oahu but just over half (54.8%) of repeat visitors did so.

In 2017, 82% of visitors to French Polynesia made it their sole destination and 18% visited there as part of a larger circuit (Institut de la Statistique de la Polynésie Française, 2018). On average, those going to French Polynesia visited just over two islands within the group. Tahiti (71.4%), the main centre of population and the site of Faa'a international airport, was the most popular island visited followed by Bora Bora (54%) and Moorea (50.3%). Differences between the travel patterns of destination and circuit travellers in terms of islands visited were only marginal (Table 6.5). This situation contrasts with visitor behaviour in the early 1980s when those having French Polynesia as their sole destination showed much greater mobility within the territory than circuit travellers (Pearce, 1995a). Greater differences appear on other profile variables. Those making a repeat visit to French Polynesia made up a larger proportion on the outer islands (Australes, 50%; Marquises, 31%) than on Bora Bora (13%). The latter is the most popular island for Americans, being visited by 77% of them, whereas Tahiti (89%) attracts the largest proportion of visitors from metropolitan France, followed by Moorea (65%) and Bora Bora (44%).

Table 6.5. Patterns of visit to French Polynesia by circuit and destination visitors, 2017. From Institut de la Statistique de la Polynésie Française.

	Circuit	Destination	All visitors
No. of visitors	36,182	162,777	198,959
% of visitors	18	82	100
Ave. no. of islands visited	2.14	2.20	2.18
% visiting each island			
Tahiti	73.7	70.9	71.4
Bora Bora	54.9	53.7	54.0
Moorea	47.5	50.9	50.3
Raiatea	13.7	16.6	16.1
Huahine	10.2	13.1	12.6
Rangiroa	9.6	10.4	10.2
Marquises	3.2	3.2	3.2
Australes	0.8	0.7	0.7

Organizational Structures and Relationships

Disparate studies have addressed a range of topics on the structure of the private sector and firm behaviour at a regional level, some of which have been framed in terms of networks or clusters. In France a small but growing body of research is emerging on structural matters relating to ski resorts in the French Alps. Much of the research on organizational matters at a regional scale has focused on the structure and functions of regional tourism organizations (RTOs) or regional DMOs.

Private-sector structures and firm behaviour

Networks

Research in the Surselva-Gotthard region in the Swiss Alps provides a rare example of SNA which takes a spatio-organizational approach, one which incorporates a geographic dimension in the analysis of tourism business links (Luthe *et al.*, 2012; Kelman *et al.*, 2016). The Surselva-Gotthard region is a relatively isolated region comprised of three communities: Andermatt, Disentis and Sedrun. Luthe *et al.*'s (2012) network graphic depicts the complete regional tourism supply chain in which the actors are differentiated by sector, community and centrality. The network is characterized by a low density, a centralized core of actors and a large number of peripheral actors whose ties are mainly towards the centre rather than to others on the rim. With further analysis Luthe *et al.* (2012, p. 848) concluded that 'the most central actors and gatekeepers are in the village of Sedrun from the public, the hotel, and the information sectors.' They also found (p. 849) that 'the main economic drivers of Alpine winter tourism destinations, the cableways, are often not sufficiently integrated with other actors such as the hotel and gastronomy sector, leading to ... inequalities in investments, risk management, and revenue generation.'

The correlation revealed by the SNA between propinquity and the strength of business links – the groups of businesses within each community tended to be linked more closely to each other than to those in the other communities – is not unexpected but additional enquiry using qualitative interviews provided insight into the behaviour underlying this pattern. Most of the businesses were small, either owner-operated or with few employees. This situation left little time for anything other than immediate, operational needs with most contact being with others close by. Other factors raised by the respondents were language differences (German and Romansh); location in different cantons; access; historic rivalry; and significant ongoing investment in Andermatt. Kelman *et al.* (2016, p. 7) advanced an embeddedness interpretation of the relationships between the communities: 'Sedrun and Andermatt's links are being enhanced by social embeddedness while links between Sedrun and Disentis are being limited due to lack of social embeddedness.' They also contrasted the Surselva-Gotthard network with that of Longyearbyen, the main but nevertheless small settlement on the remote Norwegian island of Svalbard. The quantitative SNA showed the Longyearbyen network was compartmentalized into five subgroups of businesses while the interview findings revealed 'informal day-to-day links and indirect business facilitations that characterise this small, isolated community' (Kelman *et al.*, 2016, p. 9).

Other network studies have also identified key actors in rural tourism regions. In South West Buenos Aires Province, Schroeder *et al.* (2016) differentiated between a network of excursion and accommodation providers and an institutional network and detected three main agents in each. Private and individual initiatives were supported later by the public bodies. Schroeder *et al.* (2016) also raise the issue of territorial scale, over what distances can working relationships develop? Strong inter-municipal links were observed between technical staff with common areas of work such as in tourism promotion. Romeiro and Costa (2010, p. 83) found that in the network of the Valle de Juerte, Spain, 'the positions of greater centrality are usually occupied by companies that focus on activities complementing the dominating activity in the network (leisure activities), rural tourism chains, or accommodation companies with a certain level of consolidation of the activity, whether it be due to the long time of activity, or the fact that entrepreneurs have fully dedicated themselves to the activity.'

Trust was found to be a key factor in the success of both formal and informal networks of rural tourism in the UK, Spain and Greece (Petrou *et al.*, 2007). Rural tourism businesses weighed up the time costs and commercial benefits of belonging to informal or formal networks; the former were often local and the latter wider. Risk reduction was a consideration in both cases, but especially with formal networks. In their study of inter-firm level relationships and core competencies in the province of Pavia, Italy, Denicolai *et al.* (2010) considered a set of network properties – knowledge sharing, trust, formal agreements and integration of services (expressed by the degree of packaging) – which they related to respondents' expressions of satisfaction with firm growth as a proxy for performance. Three out of the four properties – knowledge sharing, trust, formal agreements – were correlated with performance, with knowledge sharing being the main factor. In the second phase of the study they used cluster analysis to examine links between tourism core competencies and forms of networking. Three clusters were identified – specialized tourism firms, dynamic tourism firms and market-oriented tourism firms – whose different network approaches developed different core competencies. For example, the dynamic tourism firms, the best performers, focused on informal mechanisms such as trust and knowledge sharing to strengthen their tourism and marketing abilities.

Clusters

A variety of approaches have been employed to examine different facets of clusters within regions. For Erkuş-Öztürk (2009, p. 590) 'a "cluster" refers to a localized network of specialized organizations, which includes close local links between firms from different levels in the industrial chain.' She also recognized that global networks were important for clusters to be competitive and in her study in the Antalya region of Turkey sought to establish whether differences existed in the levels of local and global networking between agglomerated and non-agglomerated clusters and between specialized and non-specialized clusters. Agglomerated clusters were operationalized as those having more than a 10% share of the region's tourism firms; specialized clusters were those having a concentration of the same type of tourism firm. Her results showed that neither agglomeration nor specialization determined the level of connectedness. What was important was firm size: large firms were more aware of the need for global networks to attract customers and had more international connections; medium-sized firms had strong local connections; and those firms with no networking were generally small. Thus the size make-up of each cluster influenced the level of networking.

Jackson and Murphy (2006) applied Porter's theories to four regional towns along the Murray River in Australia and found that they had varying degrees of success in exploiting their factor conditions and that the most economically successful were those in which the presence of the prerequisites for cluster development were the strongest. Albury-Wodonga, the least successful, was characterized by a lack of leadership, the '"glue" needed to hold the cluster together' (p. 1032) and foster greater participation.

Chhetri *et al.*'s (2014) analysis of tourism employment in Victoria, Australia, is set squarely in a cluster-based theoretical framework. They drew on employment data as the basis for the co-location of tourism and hospitality firms and then identified and mapped the location of clusters using LQs to quantify spatial concentrations in the sector. The effects of location-specific attributes were then estimated using a range of spatial econometric techniques. Their results demonstrate the importance of differentiating between different types of employment: higher levels of employment in tourism operational services were found in regional Victoria whereas employment in hospitality services (predominantly accommodation and restaurants) was much greater in city-based areas. The latter had the most diversified employment structures as employment in entertainment services and tourism infrastructure were also concentrated there. Relative to total employment, however, jobs in the tourism and hospitality sector were proportionately greater in regional and rural tourism regions in Victoria. Drawing on the employment patterns identified, Chhetri *et al.* (2017) proposed a 'spatially nested and hierarchically structured cluster model' (Fig. 6.6). The model is schematic and not elaborated on in their text but portrays different types of tourism and hospitality clusters (T&H) linked to a CBD-based gateway-driven cluster (Melbourne) and to their associated attractions in a hub-and-spoke pattern of tourism circuits. These linkages, however, are inferred rather than based on observed data.

The clustering of tourism and hospitality throughout Victoria was shown to be largely driven by six location-specific factors:

1. the availability of tourism attractions;
2. proximity to the coast (but not accessibility to parks, open spaces, and water bodies);
3. the road density network (but not accessibility to the airport and train stations);
4. accessibility to employment within the Melbourne Central Business District (CBD);
5. the scale of the regional economy; and
6. the advantages and disadvantages associated with economic resources.

The importance of the distribution of attractions is highlighted by the resource-oriented clusters in Fig. 6.6. However, while the role of resources is acknowledged with regard to competitive advantages which particular regions have, Chhetri *et al.* also stress the impact which agglomerative processes and localization externalities have had in the creation of tourism and hospitality clusters in Victoria.

Resources are seen to be a critical component of nature-based tourism clusters.

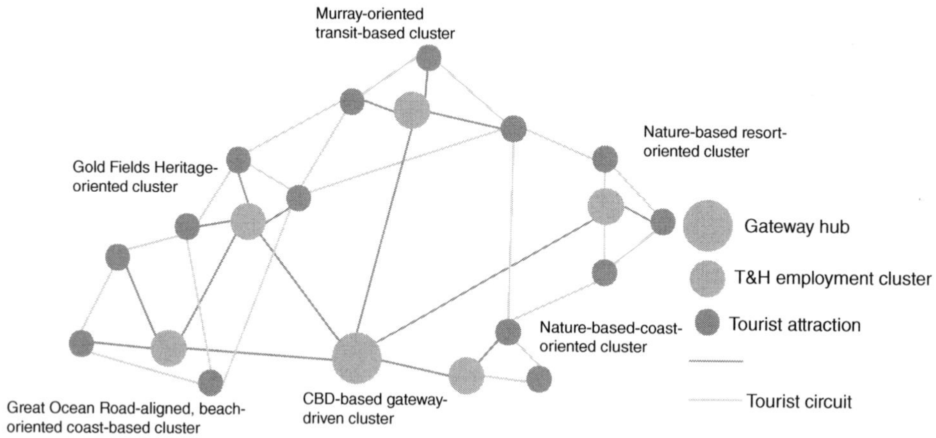

Fig. 6.6. A topological representation of spatially nested and hierarchical structured cluster model for Victoria, Australia. From Chhetri *et al.* (2017), Figure 5.

Huybers and Bennett (2003, p. 571) contend that 'inter-firm environmental cooperation is expected to materialise in nature-based tourism clusters because the environment is the major regional characteristic that determines the competitiveness of the cluster.' They develop this argument with reference to Tropical North Queensland, a large region but one where most of the long-established tourism operators are based in and around Cairns or Port Douglas. This proximity gives rise to frequent, informal contacts. More formal communication occurs through regional industry associations. These conditions have led to a range of cooperative activities, not just in joint marketing but also in intraregional cooperative environmental protection as businesses have a mutual interest in maintaining environmental quality. Both formal and unwritten codes of conduct exist. 'Peer pressure and social control are facilitated by the close proximity and the repeated interactions between the tourism businesses' (Huybers and Bennett, 2003, p. 584).

Industrial districts

Drawing on Lazerson and Lorenzoni (1999) and other industrial district studies, Cirer-Costa (2014a) argues that by the outbreak of the First World War tourism in Majorca exhibited the characteristics of a classic industrial district. The focal firms were the international range hotels, which monopolized contact with the tourists. They offered not only variously priced accommodation but also sub-contracted small firms to provide excursions and transport around the island. In subsequent years structural changes occurred (pp. 1254–1255):

> By 1935 the Majorcan tourist sector had adopted the form of an industrial district specifically focused on tourism, characterised by a high level of productive specialization, a complex network of relationships between firms and the intense flow of information. The grand hotels and the shipping firms occupied central positions in the district, but they did not reach the pre-eminence characteristic of the focal firms in the Italian model of the industrial district, since almost all the firms in the tourist cluster had direct contact with the customers and the capacity to provide them with all kinds of services in addition to lodging and transport from the mainland ... the Majorcan tourism cluster had a much more complex network of relations because it displayed not only the tacit agreements characteristic of all industrial districts but also the informal relations of recommendation and second-level flows of information that provided continuous feedback on the system and protected it from predatory or opportunistic behaviour from its members.

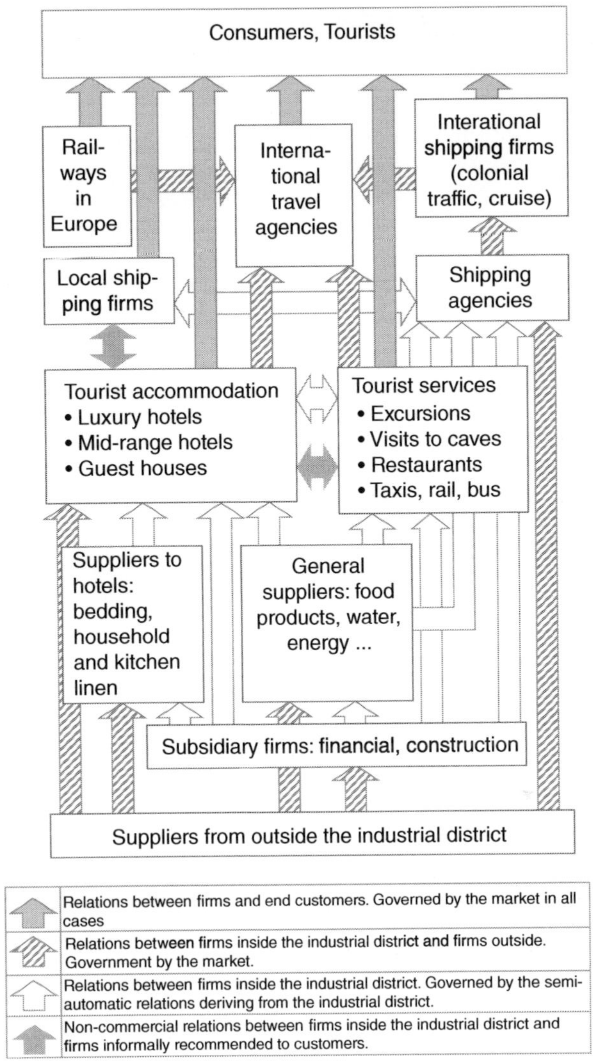

Fig. 6.7. Flows of goods and services in the Majorcan tourism cluster around 1935. Redrawn from Cirer-Costa (2014a), Figure 2.

> This diffuse and complex structure allowed the district to adapt rapidly to any change in demand or in the market conditions. If a hotel received tourists via a new foreign travel agency, for example, news of the result of this commercial innovation would quickly reach the other components of the district.

The flows of goods and services in the Majorcan tourism cluster around 1935 are depicted in Fig. 6.7. A high level of institutionalization also developed. The Fomento del Turismo de Mallorca, which was set up in the 1930s specifically to promote tourism, evolved into a strong pressure group that 'defended the interests of the tourist sector and acted as arbiter in cases of internal conflict' (Cirer-Costa, 2014a, p. 1255). Cirer-Costa concluded that the basis of Majorca's success as a tourist destination lies in the characteristics of an industrial district, particularly its capacity to adapt. At the same time, however, he refers here to inter-firm linkages as 'a complex *network* of relationships' and titles his paper

(Cirer-Costa, 2014a) 'Majorca's tourism *cluster*: The creation of an *industrial district*' (emphases added), suggesting some overlap in these concepts.

In an article dealing with the later development of the hotel sector in the Balearics over the period 1964–2010, Cirer-Costa (2014b) treats the island group as a whole as an industrial district and attributes the success of the sector to this structure and its accompanying competitiveness:

> Any small improvement spread rapidly through the islands thanks to the high degree of labour mobility, and also to the fact that the islands' travel agencies and transport firms and the tour operators themselves all worked with several hotel owners at the same time. This permeability spread to Ibiza and Minorca, as most of these service firms were present in the other islands as well and the Majorcan hoteliers also responded quickly to the opportunity to expand.
>
> The presence of a multitude of firms offering the same product, a flexible and mobile labour market, and the existence of widespread specialized knowledge constitute three of the main elements of the canonical Marshallian industrial district. The fourth – the presence of competitive providers of supplies and complementary services – has already been demonstrated in our description of the islands' business structure. If to these factors we add the presence of a clearly differentiated society directly involved in the productive structure of the region, then the concept of 'Italian-style' industrial district following the paradigm described by Becattini emerges in full force.

In a related paper Cirer-Costa (2013a) included not only the characteristics of hotel properties but also variables relating to firms (e.g. the number of units managed by the firm in the Balearics, its overall size and the location of its decision-making centre) and to each area (e.g. number of establishments and units, the distance from the airport, proximity to other tourist areas and the number of hotel firms present). The inclusion of the firm data enabled conclusions to be drawn not only about the location of tourism areas (primarily on the coast and separate from other economic centres) but also about the market structure of the sector and the strategies of the firms involved (Cirer-Costa, 2013a, pp. 20–23):

> The islands' firms display a high level of geographical dispersal, but also a high level of commercial specialization in specific market segments. In spite of sharing the same business concerns, and in spite of their evident geographical proximity, the different chains maintain the heterogeneity in capacities, strategies and behaviours that characterizes all clusters...
>
> ...the firms, in spite of their physical proximity, are not really competing for the same kind of client. In our case, this differentiation obliges firms to spread over a large number of geographical areas in order to reach the largest possible proportion of the potential demand.

Moreover (p. 22):

> Each tourist area is integrated in various ways in the industrial district, first through the hotel chains that spread their interests all over the islands, second through the inputs provided by the suppliers, since none of the chains are vertically integrated and all tend to buy their supplies in the marketplace.

Branding

The issue of consistency in destination branding has attracted some attention, particularly as it relates to relationships between the regional DMO brand and the brands of the multiple providers within the region over whom the DMO has no direct control. In these circumstances, as Cox *et al.* (2014, p. 85), observe: 'Delivering a consistent brand experience for visitors across multiple brand touch-points is a major challenge for destination brand managers.' This has led to an interest in processes for involving multiple internal stakeholders in the creation and delivery of the destination brand (Sartori *et al.*, 2012; Cox *et al.*, 2014). Cox *et al.* (2014) suggest DMO leaders need to actively promote buy-in, provide flexible leadership and empower stakeholders to deliver the brand promise. Sartori *et al.* (2012) found varying levels of commitment among stakeholders to the destination brand in South Tyrol and lower levels of consensus regarding planning communication consistent with the destination brand. Medium-sized hospitality enterprises were found to be more committed to the brand than smaller and larger properties. Pechlaner *et al.* (2007) examined how consciously and strongly hoteliers used the Alps brand and included it in their communication strategy. Their results revealed a large degree of consensus about how the destination was perceived and high levels of inclusion of listed destination values.

In their detailed study in Binh Thuan, one of the provinces in the South Central Coast region of Vietnam, Pham and Pearce (2019) operationalized brand consistency as the manner and extent to which tourism business brands are aligned with the destination brand in terms of brand elements and brand architecture is used to associate the business brands with the destination brand. Binh Thuan is branded very explicitly as a 'sea, sand, sun' destination. On the brand architecture dimension, the branded businesses split almost evenly with regard to whether or not their brands were explicitly linked to Binh Thuan's core values. The most consistent brand elements were the selling points and target markets; most variation with the destination brand occurred in terms of positioning. The elements of nearly two-thirds of the branded businesses aligned very consistently or consistently with those of the destination and a third were weakly or not at all consistent. The rather generic nature of the Binh Thuan brand means that the resorts and hotels located along the coast can be readily aligned with the destination brand. Factors contributing to a lack of consistency are location, ownership and management. The branding of many of the chain resorts and hotels, for example, is consistent with the corporate brand. This not only strengthens the corporate brand but also provides a means of differentiation within the coastal destination.

Ski resorts in the French Alps

The challenges of mapping the regional distribution of ski resorts are covered comprehensively by François *et al.* (2012) in their atlas of resorts in the French Alps. These include questions relating to the definition of the resorts (largely based on relationships between ski lifts and accommodation), issues of compiling comparable spatial data across a range of variables and derivation of indicators combining multiple variables (e.g. lift capacity and mean altitude or mean age of lifts). As discussed in Chapter 9, the relationships between ski lifts and accommodation are key factors. The atlas contains 33 maps, with a comparable number of figures plus tables. It is presented as a diagnostic tool to assist policy making and offers an abundance of detail. Particular patterns are discussed but there is little attempt to integrate the different variables and provide an overall synthesis of the structure of winter sports in the French Alps. In the present context, however, the atlas provides a valuable example of region-wide variations in patterns of management and the interplay of spatial and organizational features and public/private-sector relationships.

Under the *loi Montagne* French communes (local government) are responsible for lifts developed on their territory. This has been a determining factor underlying the pattern of ski-field development and management (Marcelpoil and François, 2008; François *et al.*, 2012). Communes may develop and operate the lifts themselves or delegate responsibility to a third party, by way of a concession or some other agreement. In the majority of cases the operation of the lifts has been delegated, often because the communes lack the resources to do so themselves. Considerable variation also occurs in terms of the extent to which ski areas coincide with communal boundaries. Less than one-third of the ski areas in the French Alps are located within a single commune; over half extend across parts of multiple communes and in 9% of the cases a commune had part of two or more ski areas. In addition, ski resorts have been developed at a range sites at different altitudes (Chapter 9).

As the atlas shows, despite this diversity certain broad patterns do appear in the spatial and organizational structure of ski resorts. A third of the ski areas were operated by the public sector, 38% were operated by the private sector and 19% by some para-public structure or public/private partnership. No data were available on the remaining 10%. There was a general correlation with size: the very large or large high-altitude resorts were primarily under private-sector management; smaller resorts lower down the mountain were mainly communal operations and the para-public or public/private partnerships constituted an intermediate category. The latter category is one means of enabling communes to work together in cases where the ski-field crosses multiple boundaries or of acquiring access to greater resources by joining with private-sector partners. Geographically, there is a north–south decrease in private-sector management as the larger resorts are located primarily in the northern French Alps.

The Compagnie des Alpes has developed into the largest ski resort operator in France. It was established in 1989 by the Caisse des Dépôts et Consignations, a French public-sector financial institution, to bolster the French ski industry. Tignes and La Plagne were the first of the 11 major ski resorts it acquired, with shares also being held in four others. The company has since expanded its activities to include a chain of leisure parks in France, Belgium, the Netherlands and Austria. The Compagnie des Alpes was floated on the Paris stock exchange in 1994 and privatized in 2004, with the Caisse des Dépôts et Consignations retaining a 40% share.

Another structural perspective is provided by Favre-Bonté *et al.* (2016) who adopted a network approach to the analysis of innovations in two extended ski areas: Paradiski and Portes du Soleil. Both ski areas combine multiple resorts linked by connecting lifts and common ski passes which provide skiers with greatly enhanced access to more pistes, effectively extending the destination beyond the individual resort and increasing levels of interdependence between resorts. Paradiski incorporates all the resorts of La Plagne with those of Les Arcs and Peisey-Vallandry, providing access to more than 140 lifts, 260 pistes and 425 kilometres of trails. The Portes du Soleil ski area includes seven French resorts (e.g. Avoriaz and Morzine) and five in Switzerland (e.g. Les Crozets and Morgins) with a network of some 200 lifts extending over 14 valleys. In a drive to remain competitive these ski areas have implemented a range of innovations relating to new products (e.g. audio guides, aquatic centres); front office innovations (e.g. training kits, the Vanoise Express (the world's largest cable car)); and back office innovations (a CRM, allocation of revenue based on distance travelled by skiers). The analysis by Favre-Bonté *et al.* (2016) shows the networks formed by the resorts vary according to the type of innovation developed and the network characteristic: relation type, regulation, architecture and geographical scope (Table 6.6). In particular,

Table 6.6. Network characteristics in French ski resorts according to the nature of innovations. From Favre-Bonté *et al.* (2016), Table 5.

		Network characteristics			
		Relational type (horizontal/ vertical/ inter-industry	Regulation mode (economic/ sociological)	Architecture (centralized/ decentralized)	Geographical scope (local/ national international
Type of innovations New what?	New offers	Inter-industry and horizontal	Sociological with local members Economic with other members	Centralized	National or international
	Front office innovations	Vertical	Economic	Centralized	National or international
	Back office innovations	Inter-industry	Priority to sociological	Centralized	Local

they suggest that ski resorts can no longer be treated as localized productions systems consisting of actors embedded in their own territory but rather they form part of much more geographically dispersed networks as in order to innovate they must seek creative partners who are often found outside the resort. In terms of network architecture, hub firms retain a critical central role regardless of the type of innovation.

Favre-Bonté *et al.* (2019) subsequently explored the influence of territory on the structure of innovation networks by examining the network characteristics of high-altitude and medium-altitude resorts in France. Differences between the two sets of resorts were found for three of the four characteristics (Table 6.7). Relationships in the medium-altitude resorts were localized, vertical and between inter-industry members, subject to sociological regulation (i.e. based primarily on trust). In contrast, in the high-altitude resorts the geographical scope of the network extended to national and international members so as to extend marketing reach and obtain access to partners capable of less

Table 6.7. Characteristics of innovation networks in French ski resorts according to the territory. From Favre-Bonté *et al.* (2019), Table 4.

			Characteristics of innovation networks			
		Innovation networks studied	Nature of the relation	Main regulation mode	Architecture	Geographical proximity with the main members
Territory	Medium-altitude resorts	Trail station	Inter-industry & vertical	Sociological	Centralized around the entrepreneur	High
		Natural Escape	Inter-industry & vertical	Sociological	Centralized around the entrepreneur	High
		Time investigation	Inter-industry & vertical	Sociological	Centralized around the entrepreneur	High
		Eco-Bivouac	Inter-industry & vertical	Sociological	Centralized around the entrepreneur	High
	High-altitude resorts	Aquariaz	Inter-industry	Economic	Centralized – private firm pivot of the resort	Low (international)
		You Can Ski	Inter-industry & horizontal	Economic	Centralized – Association des Portes du Soleil	Low (international)
		Multipass	Inter-industry & horizontal	Economic	Centralized Association des Portes du Soleil	High
		Rock the Slopes	Inter-industry & horizontal	Economic	Centralized – Association des Portes du Soleil	Low (national)
		Bike Park	Inter-industry & horizontal	Economic	Centralized – Tignes development (public organization)	Low (international)

readily imitated innovations; relations were inter-industry and horizontal; and the mode of regulation was essentially economic. In both cases, however, the networks were strongly centralized: in the one case they were centred on a local entrepreneur; in the other, around a private pivotal firm, a tourism office or a public development organization.

Elsewhere in France, inter-communality has been a response to the challenges faced by small, struggling ski resorts in the French Pyrenees developed by local municipalities whereby neighbouring communes come together to spread the financial burden of ski-field operations, diversify the tourism resource base beyond the communes where the ski-fields are located, and achieve economies of scale in terms of promotion and marketing (Vlès, 2012). Here again the destination is being extended. Vlès sees the setting-up of these inter-communal structures as a technocratic response led by the mayors, one which weakens local democracy as there is no direct representation of the locals and one which is not guaranteed of success.

Regional tourism organizations and destination management organizations

Similar issues to those experienced at the national level occur with regional frameworks, namely which functions should be undertaken by which organizations. A primary consideration is whether tourism-related functions can be carried out effectively by existing agencies or whether a separate regional tourism organization should be created. If the former, where should responsibility for tourism lie; if the latter, how should the new organization be structured and funded, what functions should it have and how should it operate? An organization's competencies might be thought of as elementary or operational, intermediate or functional (R&D, production, marketing), or upper level (inter-functional, e.g. quality control) or general (coordination, decision making) (Twyoniak, 1998). Functions might also be classified as statutory or discretionary, regulatory or enabling. Relationships between the units or organizations responsible for particular functions may be formal and tightly structured or informal and loose (Ruekert and Walker, 1987). These relationships may be dyadic, networked or take some other form.

Destination marketing is a commonly accepted regional function which complements that of the NTOs and local destination organizations. Promotion of regions contributes to national marketing activities by creating greater awareness of a country's range of attractions, diversifying demand and spreading the impacts of tourism developments. It is generally recognized that as distance to the market increases so the need to create a stronger image than that provided by multiple local destinations grows in order to avoid fragmentation in the market place and to obtain economies of scale. Destination management is another major function often carried out at a regional level. Debate occurs over whether destination marketing should come under the over-arching role of destination management or whether it is best regarded as a separate, stand-alone function (Pike and Page, 2014; Pearce, 2015a, 2016a). There is as yet no consensus on what destination management encompasses but commonly occurring functions include destination planning, product development, research, visitor management, capacity building, leadership and relationship building (Costa *et al.*, 2014; Pearce, 2016b). Debate occurs as to whether these are best undertaken by an autonomous or semi-autonomous RTO or by some regional/local government agency. Regional/local government is usually responsible for broader functions which may impact on regional destinations such as provision of infrastructure, spatial planning, environmental management, policy making, regulation and the management of public good assets such as parks, museums and art galleries. Moreover, these functions are interdependent (Fig. 6.8). Figure 6.8 presents destination management as an over-arching concept which embraces a range of interdependent functions and emphasizes the need to balance supply and demand and to manage the integration of the destination into the territory within which it is embedded through a set of interface functions.

As a result of these diverse organizational considerations and regional variations in

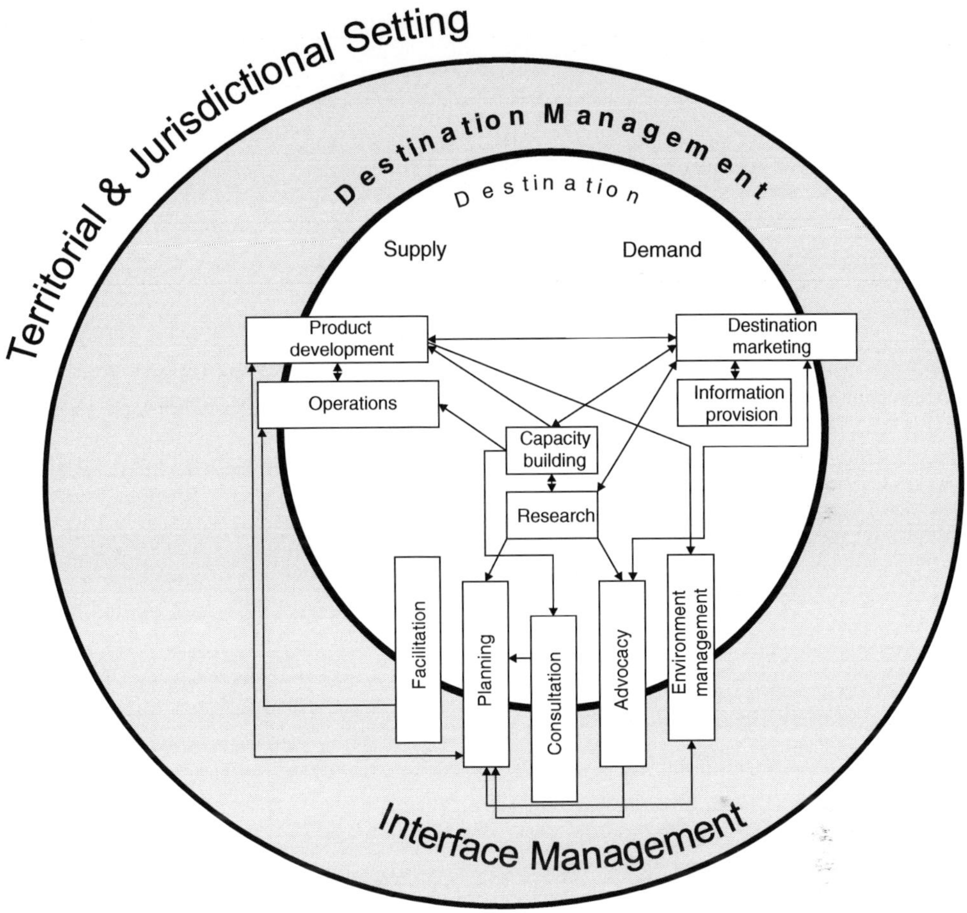

Fig. 6.8. A conceptual framework of interdependent destination management functions. From Pearce (2016b), Figure 1.

patterns of supply and demand, a range of institutional frameworks for regional destinations are to be found. Differences occur not only between countries but also from region to region within countries as each endeavours to find the most appropriate structure to suit its particular needs and circumstances (van der Borg, 2008; Paniagua, 2012; Costa *et al.*, 2014; Pearce, 2015a; Pearce *et al.*, 2017). In Spain, for example, the autonomous communities have adopted different policies towards rural tourism which reflect regional variations not only in tourism (Fig. 5.4) but also in the rural economy and countryside setting (Paniagua, 2012).

Figure 6.9 depicts the diversity in the structure of regional tourism organizations in New Zealand. New Zealand has some 30 regional tourism organizations, some of which are in effect representing local rather than regional destinations. RTOs vary in size and structure but their prime function is destination marketing. They may exist within a TLA (Territorial Local Authority or council) (Fig. 6.9a); be established at arm's length as a council-controlled organization or some form of not for profit organization (Fig. 6.9b); or form part of an economic development agency (EDA) which has a wider brief than just developing tourism and which itself might be located inside or outside the TLA (Figs 6.9c and 6.9d). Most RTOs serve a single LTA but some may serve several and in other instances a TLA may

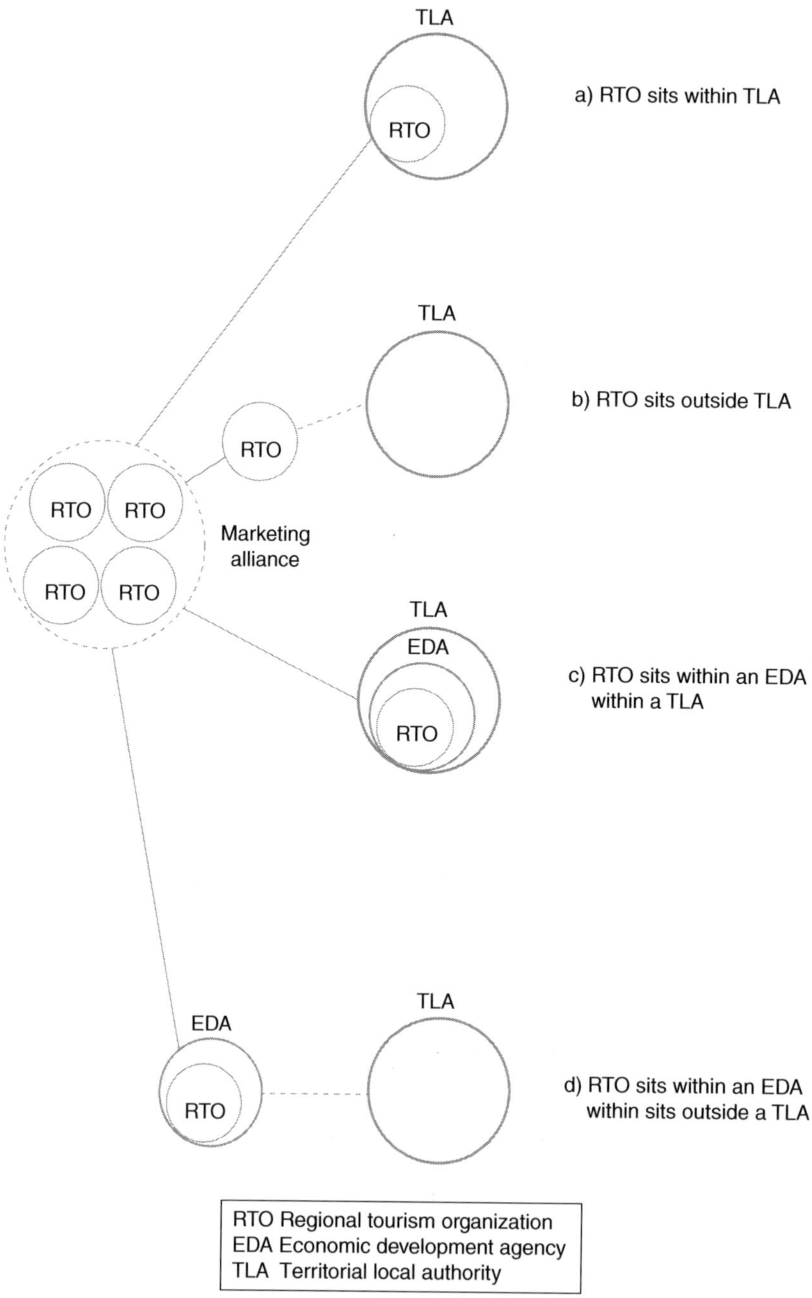

Fig. 6.9. Main regional destination management structures in New Zealand. From Pearce (2015a), Figure 1.

have more than one RTO. RTOs may also come together to form regional alliances to more effectively fund and carry out marketing campaigns with Tourism New Zealand, the NTO. In addition to regional marketing the RTOs, in varying degrees, also undertake such activities as provision of visitor information through i-Sites, building capacity, advocacy for the tourism sector and facilitation of product development. Local government has statutory responsibility for providing local infrastructure, spatial planning and environmental protection and may also foster economic and social development.

In Vietnam sub-national destination marketing is carried out at the provincial level; there are no tourism organizations at the larger regional scale. Nguyen and Pearce (2015) investigated the inter organizational relationships between the DMOs of the eight provinces comprising the South Central Coast region, the country's most dynamic tourism region, in terms of the nature and extent of their individual and joint destination marketing activities. Table 6.8 shows the four most developed provinces are the most active, especially in terms of international marketing, while the less developed provinces have a more modest programme of mainly domestic activities. The mix of individual and joint marketing is influenced by the type of activity and level of tourism development in each province. Activities such as hosting familiarization trips and organizing events around local festivals are often carried out by individual provinces. In some instances international marketing is carried out jointly, in others it is undertaken by the individual provinces. In varying degrees the DMOs also provide various linkages to the websites of other provinces. The joint destination marketing that does occur is essentially between small groups of provinces rather than across the region as a whole. The most established joint destination marketing partnership is that between Da Nang, Qang Nam and Hue in the neighbouring North Central Coast region. Other joint marketing also occurs with provinces outside the region. One reason for this is the location of the nearest gateway airport. Cooperative activities were thought likely to become more common and some move to more trans-regional marketing was foreseen.

The Vietnamese DMO representatives identified both pros and cons with joint marketing. Joining forces was thought to result in a stronger regional image based on the region's common coastal resource and to offer diversity to the visitor. At the same time it was recognized individual provinces might receive less attention. Joint marketing was seen to offer economic benefits (cost savings and a stronger budget) but less flexibility. In these respects, and despite structural differences in the Vietnamese economy, much of the joint marketing in the South Central Coast is motivated or conditioned by economic factors and might be explained in similar terms to those in the Western literature. In Central Florida, for example, Wang *et al.* (2013) found the motives among Convention and Visitor Bureau executives in Central Florida for collaboration in joint destination marketing fell into three broad categories: cost sharing and reduction, regional product enhancement and learning. The Vietnamese interviewees were also able to identify with the competition–collaboration continuum (von Friedrichs Grängsjö, 2003; Krakover and Wang, 2008). Representatives of six of the eight provinces identified their situation as being one of coopetition; those in Da Nang and Quang Ngai saw theirs as cooperation. Those identifying with coopetition said the combination of competition and cooperation enabled them to achieve their own targets as well as pursue shared interests.

The bridging role of regional tourist organizations between the local and national scales is even more complicated in countries with multi-layered sub-national administrative systems. Not only are there five different administrative levels in Spain, including three between the state and municipal levels, but there are multiple organizations at each level (Fig. 6.10). These organizations are independent and not formally related hierarchically but are nevertheless 'inextricably linked in various ways as they interact when implementing specific marketing actions' (Datzira-Masip and Poluzzi, 2014, p. 52). One consequence of this multilayered structure in Catalonia has been a proliferation of destination brands which are not clearly articulated and integrated into some overall scheme. In some cases brands overlap, in other instances multiple lower-level brands cloud those at a provincial or autonomous community level. Such a situation calls for the implementation of a carefully structured brand architecture model.

Table 6.8. Main individual and joint marketing activities undertaken by the provinces of the South Central Coast region of Vietnam. From Nguyen and Pearce (2015), Table 2.

Provinces	Individual destination marketing activities	Joint destination marketing activities
Da Nang	Twice yearly marketing activities for MICE tourism in Ha Noi and Ho Chi Minh City Host familiarization and press trips Organize fireworks festival	Attend trade fairs in Russia, Korea (KOTFA), Vietnam International Travel Mart (VITM) in Ha Noi[a] Organize regional tourism fair and conferences
Quang Nam	Host familiarization trips and press trips from Ho Chi Minh City, Ha Noi, Asean markets and Russia Organize events and festivals	Attend activities in Thailand and Korea 2013[a] Attend domestic tourism fairs and travel expos e.g. International Travel Expo (ITE) Ho Chi Minh, VITM in Ha Noi[a]
Quang Ngai	Organize promotional activities in Champasak, Laos Attend tourism fairs in Ho Chi Minh city Organize festival and photo contest	Attend International Sea Festival Nha Trang
Binh Dinh	Organize familiarization and press trips from Ha Noi, Ho Chi Minh City and Central Highlands	Organize joint promotional activities with Quang Ngai, Phu Yen, Gia Lai, Kon Tum Attend regional tourism conferences in Da Nang
Phu Yen	Organize promotion conferences in Gia Lai – Dak Lak, Kon Tum and Ho Chi Minh City	Attend tourism fairs in Da Nang and Sea Festival organized in Nha Trang Host national tourism year of South Central Coast region organized by the Ministry of Culture, Sports and Tourism
Khanh Hoa	Attend tourism fairs in Germany (ITB), Spain (FITUR), Russia (MITT and Intourmarket), Korea (KOTFA), Japan (JATA), Thailand (Travel Mart Plus); ITE in Ho Chi Minh, VITM in Ha Noi Advertise in magazines Host familiarization trips Organize sporting and cultural contests within the tourism industry of the province	Organize regional tourism fair and conferences
Ninh Thuan	Promote tourism through cultural and sports events Organize events in Ninh Thuan (on-site promotion)	Attend regional tourism events with its own booth in Nha Trang Sea festival in June 2013 for the first time
Binh Thuan	Attend International Travel Expo (ITE) in Ho Chi Minh City and Ho Chi Minh Tourism Day Attend promotional tourism activities in Nha Trang, Russia, Australia, Japan and South Korea Establish tourist support centre and training centre	Attend promotional tourism activities in Da Nang Attend regional tourism conferences in Da Nang, Khanh Hoa

[a]Joint marketing efforts by Hue – Da Nang – Quang Nam.

Other research has examined factors relating to DMOs and levels of cooperation. Elbe *et al.* (2009) found varying levels of cooperation were associated with perceptions of the DMO's moral and pragmatic legitimacy. Pechlaner *et al.* (2012) concluded that DMOs generally have a positive influence on the level of cooperation in destinations; that the level of acceptance of DMOs is positively correlated with aspects of their corporate

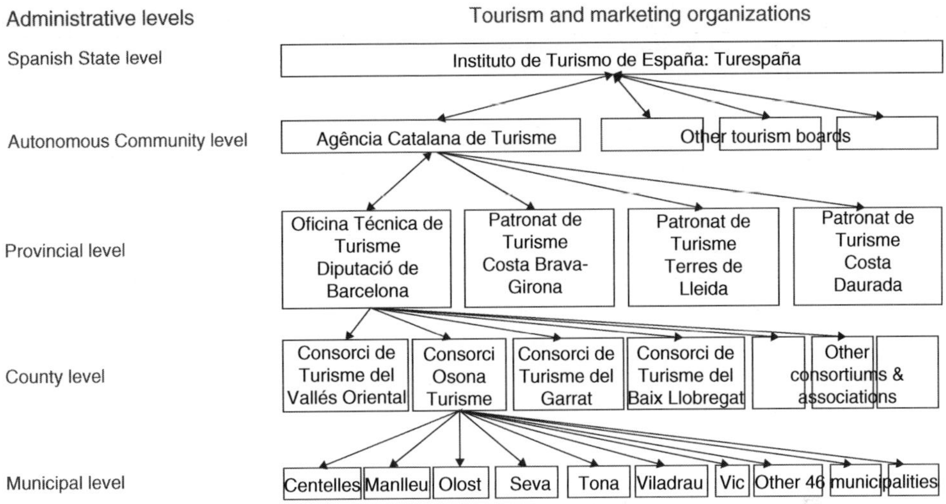

Fig. 6.10. Administrative levels and tourism organizations in Catalonia. From Datzira-Masip and Poluzzi (2014), Figure 3.

governance (the extent of stakeholder involvement, efficiency and evidence of performance); that their acceptance influences their cooperative effect; and that destination governance is associated with a high level of cooperation.

A variety of compulsory and voluntary associations exist in tourism and related sectors (e.g. environmental) in Antalya. Erkuş-Öztürk (2011) found richness in associational capacity contributed to the growth of tourism in the region through such activities as infrastructure development and advertising initiatives. In their comparative study of institutional networks for promoting integrated rural tourism in Ireland and France, Cawley *et al.* (2007) highlighted the need to take account not only of tourism organizations at the national, regional and local scales but also other institutions such as regional development agencies and local councils. They distinguished between horizontal networking actions aimed at promoting local embedding and vertical actions with non-local agencies and markets. Different institutional networks were active in the four domains they considered: (i) policy and planning; (ii) business support and development; (iii) regulation; and (iv) promotion and marketing. Variations also occurred between the two countries.

Towards a Regional Synthesis

Recent studies on the structure of regional destinations share many of the same characteristics as research at a national level in that they focus on specific structural features drawing on selective sets of data, rarely combine both supply and demand nor integrate spatial patterns with analyses of organizational configurations or industry structure. Some do analyse a large number of variables but provide little synthesis (François *et al.*, 2012; Sarrión-Gavilán *et al.*, 2015).

One attempt to take a more integrated approach is that by Marco-Lajara *et al.* (2014) who assessed the performance of hotels in six autonomous communities in Spain against their territorial characteristics using Porter's (1990) Diamond model and the concept of industrial districts as their theoretical framework. The study, however, analyses differences between regions rather than their internal spatial structure. It also demonstrates the challenges involved in obtaining and evaluating a range of appropriate indicators. In order to focus on comparable destinations Marco-Lajara *et al.* (2014, p. 780) studied 'only those [autonomous communities] mainly and directly related to sun-and-sand tourism', namely Catalonia, the Valencian Region, Murcia and Andalusia, and the Balearic and

Canary Islands. Just how comparable they are is debatable given the size and internal differences of the peninsular communities (including a significant degree of urban tourism) compared to the insular ones (Figs 5.4, 6.1 and 6.2). Four performance measures were selected: (i) ADR (average daily rate); (ii) RevPar (revenue per available room); (iii) occupancy; and (iv) length of stay. Application of the Diamond model involved the inclusion of measures of firm structure, strategy and rivalry alongside indicators of factor conditions, demand conditions and related and complementary industries. Most of the factor and demand conditions and related and complementary industry indicators were 'relativized' to take account of the population differences between the regions. Industry structure was analysed in terms of the number and size (number of beds) of hotels; the percentage of 4- and 5-star hotels; the percentage of hotels with a quality certificate; and employees per hotel bed. In addition, they included two other factors said to have an impact on all the other determinants: government and chance. The mix of government indicators consisted of the position of tourism within the regional government's organizational chart, the percentage of the total budget allocated to the tourism department, the existence of a strategic plan and the percentage of the tourism budget allocated to commercial promotion destination. Chance, seen as unforeseen international circumstances, was measured by the growth of foreign tourist arrivals in each region, a somewhat tenuous indicator.

The results show the best performing hotels are located in the Canary and Balearic Islands and Catalonia which outperform the other autonomous communities on at least three of the four profitability indicators examined. The values of the territorial variables are then compared in turn against the performance of each community. On this basis Marco-Lajara *et al.* (2014, p. 792) concluded that:

> Generally speaking the factors that seem to influence hotel profitability to a greater extent are: knowledge resources, demand by non-residents, some factors related to industry structure (... establishment size, and percentage of 4- and 5-star hotel establishments), and the political priority attached to this industry by the government of each autonomous region, along with the strategic view of tourism. What surprises us, though, is the fact that some indicators usually associated with high-quality tourism (such as conference venues, golf courses and marinas) do not seem to play a determining role in the efforts to increase hotel profitability.

The relative contribution of each factor to hotel performance was not calculated but the authors suggest that those relating to the firm characteristics are perhaps the reasons for the higher profitability of the establishments located in the Balearic and Canary Islands.

Discussion and Conclusions

Different approaches have been taken to examine the spatial structure of regional destinations but some common patterns do emerge. Tourism is concentrated geographically within regions, a hierarchy of local destinations exists and spatial variations occur in the types of local destinations, their structure, functions, levels of dynamism and types of visitors they attract (Figs 6.1–6.3; Tables 6.2 and 6.3). Major urban centres commonly form the apex of the hierarchy, supported by other primary and secondary destinations based predominantly on natural resources. Much of the supply-side distribution research is based heavily on resource endowment but other studies show this is only part of the explanation of regional structures alongside such factors as geographic positioning (Wang *et al.*, 2016) and firm characteristics (Marco-Lajara *et al.*, 2014). Blasco *et al.*'s (2014) analysis of attractions offers an alternative to administrative boundaries in the Pyrenees while the research on ski resorts in the French shows considerable variation in the geographical scope of ski areas and the extent to which these correspond to communal limits (François *et al.*, 2012; Tables 6.6 and 6.7).

Network analysis has proved a useful means of establishing regional structures through examining linkages between these local destinations, identifying the impact of positioning and associated nodal functions (Fig. 6.5). Network analysis has also revealed linkages of different sorts between businesses which contribute to the identification of organizational configurations at the regional level (Tables 6.6 and 6.7). Studies framed in terms of clusters and industrial districts have found support for these concepts although the extent to which the actual functional relationships between firms have been examined varies (Cirer-Costa, 2013a; Marco-Lajara *et al.*, 2014; Chhetri *et al.*, 2017; Parrilla-González and Pulido-Fernández, 2017). Other research has drawn attention to the degree to which joint activities occur within and between the private and public actors; factors such as trust which influence these relationships; and the structure and functions of RTOs and the role they play in uniting multiple stakeholders within a region.

A limited number of studies have also integrated aspects of the spatial and organizational structure of regional destinations. Work in the French Alps has revealed structural differences in ski resorts at high and medium altitudes (François *et al.*, 2012; Table 6.7); the spatio-organizational approach of Kelman *et al.* (2016) and Luthe *et al.* (2012) has demonstrated a geographic dimension to the tourism business links in the Surselva-Gotthard region; and the integrative study by Marco-Lajara *et al.* (2014) has indicated the importance of industry structure in influencing hotel performance between regions. Considerable scope exists to build on this research to develop a more comprehensive understanding of the structure of regional destinations but data limitations and methodological challenges in doing this should not be underestimated. In particular, quantitative studies might be fleshed out with more qualitative accounts of contextual factors and relationships.

7 Structure of Urban Destinations

Analysis of the structure of urban destinations is challenging due to the complex nature of cities and the diverse forms of tourism found there. Cities are characterized by high physical densities of structures, people and functions; social and cultural heterogeneity; economic multi-functionalism; and a physical centrality within regional and inter-urban networks. Considerable diversity is found among towns and cities; many are monocentric, others, especially large cities, are often polycentric; some have a very long history, others are more recent. In cities tourism is but one of several functions or activities. The relative importance of tourism will vary depending on the scale and scope of other functions, the cities' attractiveness to tourists and the emphasis given by developers and local authorities to promoting the sector. In varying degrees urban tourists and tourism providers share and/or compete with residents and other users for many services, spaces and amenities. Moreover, as the preceding chapters have shown, cities and towns may not only be important destinations in their own right in a larger regional or national context but they may also play one or more roles there, acting as gateways, hubs and distributing nodes as well as being significant sources of demand. As a result of these characteristics urban areas generally attract a wider range of visitors than other types of local destinations: in addition to visitors coming for a variety of leisure purposes, cities and towns also tend to have a larger share of VFR, business, meetings, incentives, conventions and exhibitions (MICE) and other visitors than coastal and ski resorts and rural areas.

These characteristics have also influenced the way research on the structure of tourism in urban areas has been carried out. Early descriptive studies provided a general overview of tourism in cities and towns whereas much of the more recent work has taken a narrower and more focused approach with regard to scale and sector(s) or some combination of the two. In terms of scale, the most common approach has been to take a city-wide perspective, for example by analysing patterns of tourist activity or hotel distribution across the city. Other researchers have focused on particular parts of cities presented as districts (Pearce, 1998a, 2001) and precincts (Hayllar *et al.*, 2008).

This chapter begins by reviewing the spatial structure of urban attractions and activities followed by accommodation. The focus switches next to research on tourist districts and precincts and then to consideration of broader network and relationship issues, a theme that has attracted relatively little attention to date in urban areas. The chapter concludes by drawing these different threads together.

Urban Attractions and Activities

Identifying the nature, relative importance and distribution of attractions and activities in cities is challenging. Due to the multi-functionality and heterogeneity of cities there are multiple reasons why tourists visit cities. Moreover, many tourists may have multiple motives. Many may be attracted by what Matley (1976, p. 30) referred to as 'the individual character and atmosphere which transcend the mere sum of buildings and other physical attractions. An obvious example is Paris.' In a similar vein, Karski (1990, p. 16) argued that the attractiveness of urban destinations lies in the 'rich variety of things to see and do in a reasonably compact, interesting and attractive environment, rather than in any one component. It is usually the totality of the overall tourism and town centre product that is important.' But establishing just what constitutes the totality of the overall product or tourist experience is not straightforward given that it may be made up of a mix of elements, some of which are more tangible and quantifiable than others (Pearce, 1997a).

Distribution of urban attractions and activities

Inventories and surveys at a city-wide level show attractions and activities are not distributed randomly across the city but are concentrated in particular locations or districts, most of which are found in the city centre, particularly in the historic centre (Burtenshaw *et al.*, 1981; Ashworth and de Haan, 1985; Pearce, 1998a; Jansen-Verbeke and Lievois, 1999; Shoval and Raveh, 2004; Duhamel and Knafou, 2007). Variations on this general pattern occur from city to city due to their individual characteristics, morphology and size. Figure 7.1, for example, shows that most heavily visited sites in Paris according to entry data are located in central Paris along the east–west monumental axis close to the Seine, with outliers in Montmartre (Sacré-Coeur) and in the redevelopment project at la Villette (Cité des Sciences).

Variations on the general pattern are illustrated well by the comparative study of García-Palomares *et al.* (2015) in which they use various quantitative measures (density of images, standard distance and Anselin Local Moran's I statistic) to analyse the density, dispersion and clustering of geotagged

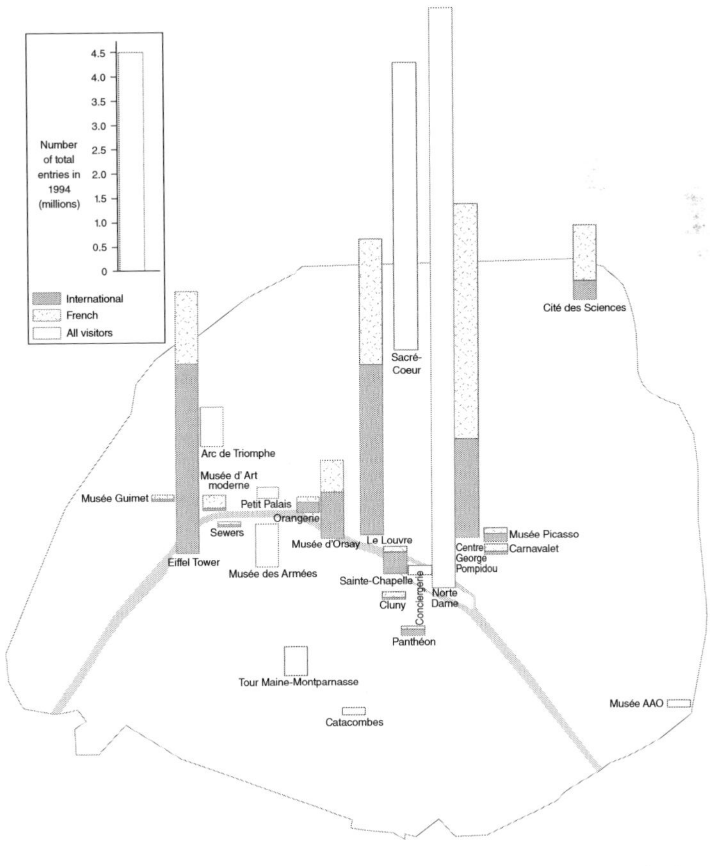

Fig. 7.1. Distribution of major formal attractions in Paris, 1994. From Pearce (1998a), Figure 2.

images in eight European cities. The highest concentrations were observed in Barcelona, Rome and Rotterdam; Paris and Madrid presented a medium degree of dispersion; London and Berlin exhibited a scattered distribution; and Athens showed the highest value for deviation from the centre. In London a large number of clusters were dispersed throughout the city whereas in Athens only three widely separated clusters emerged: the centre, port and Olympic area. Monuments and landmarks dominated the clusters but other features were also recorded by García-Palomares *et al.* (2015, pp. 414–415):

> Berlin ... presented a very scattered distribution of tourism clusters. Those in the centre corresponded to historical monuments (Museum Island, Brandenburg Gate), shopping areas (Kurfürstendamm) and renovated urban areas (Alexander Platz, Postdamer Platz). More peripheral locations presented clusters at Charlottenburg Palace and the East Side Gallery ...
>
> In Barcelona, tourists' photographs formed clusters around the historic centre, the works of Gaudí (Sagrada Familia, Casa Batlló, Casa Milà, Park Güell), the beach, the Forum of Cultures site, Nou Camp stadium (Barcelona FC) and green areas with scenic views (Montjuïc, Tibidabo).

Their description of the distribution of attractions in Paris derived from the images largely matches that shown in Fig. 7.1 but with more detail on some of the peripheral attractions:

> Clusters in the centre of Paris were primarily located along the Seine River from Notre Dame to the Eiffel Tower, but also occurred in other nearby areas such as the Champs Elysees, Montparnasse and Montmartre. On the periphery, they appeared mainly around historical monuments (Château de Vincennes), parks (Parc de Villette) and football stadiums (Stade de France).
>
> (Garcia-Palomares *et al.*, 2015, p. 414)

Connectivity and mobility

Clustering provides one perspective on how individual attractions are spatially associated. Other research focuses on the ways in which attractions are interconnected to form a more integrated picture of the activity spaces of urban tourists. Comparative research is again very useful in identifying commonalities and variations. Kadar (2014, 2015) analysed the distribution of tourist activity in Vienna, Prague and Budapest based on geotagged photography retrieved from Flickr and explored the links between the nodes identified. All three cities, Kadar (2014, pp. 94–95) writes:

> have an extended historic core with intensive tourist usage, but have also some main sites outside the centre. These isolated attractions are either not connected at all to the centre (Vienna, Schönbrunn) or connected with just one route (Prague, Vysehrad; Budapest, City Park). The most characteristic patterns in the central urban cores are axial ones related to some historical urban compositions, avenues or squares; in fact these urban places are attractions themselves, and tourists use representative historical routes instead of less appealing streets to move between other central attractions to maximize the foreseen experience in a given time.

In a second paper Kadar (2015) used network graph analysis to establish the degree of integration of the points of interest (nodes) in each city, with the pedestrian street infrastructure connecting them constituting the edges or links in the network.

Other researchers have used GPS tracking data to examine patterns of interconnectedness of attractions more directly. Aranburu *et al.* (2016) investigated the interaction between cultural points of interest (POIs) in Bilbao and found the most central places in network terms were located in the Old Town, that is, they were also central geographically (Fig. 7.3). Although further out, the Guggenheim Museum also had a reasonably high level of centrality due to its appeal. Measurement of average path length (1.92) emphasized the high concentration of the POIs in the city centre.

In other analyses of GPS tracking data, maps are prepared depicting the aggregated itineraries of visitors as they move around the city, going from one attraction or feature to the next. Bauder and Freitag's (2015) study

Fig. 7.2. Gaudi's works are the focus of many of Barcelona's tourist clusters. Here tourists take photos at the Parc Güell. Photo: the author.

in Freiburg showed most visitors followed a circular-shaped itinerary connecting the squares and main streets of the historic centre (Fig. 7.4). GPS tracking of visitors in Sydney and Melbourne highlighted 'areas of substantial intense use while contiguous areas of potential interest remain "underutilised", and the significance of iconic attractions and places as shapers of spatial activity' (Edwards and Griffin, 2013, p. 588). In Sydney, tourist movements were concentrated in the city core, along the 'spine' formed by the major thoroughfares of George Street and Elizabeth Street and at key nodes such as Circular Quay, Darling Harbour and The Rocks.

Tourist movements are more than just lines on a graph or map, however. Tourists' experience of the city is closely linked to their mobility, to how they move about the city (Simon, 2010; Lepan, 2013). In her study of tourist practices in Paris, Simon drew on Amar's (1993) concept of *adherence*, the notion that different forms of movement bring one into contact with the urban environment in varying degrees. A first dimension of this adherence is whether the movement is continuous, as in walking, when one has the possibility of changing destination or direction with each new step, or discontinuous, for example when taking the metro and one's route between stops is fixed. The Seine acts a structural axis for visits to Paris as walks along its banks provide pedestrians with a continuous space giving access on foot to many of the city's major attractions (Lepan, 2013). As Duhamel and Knafou (2007) observed, from Notre Dame to the Eiffel Tower, passing by the Place de l'Opéra, is only 4.5 km. A second and related dimension is the extent to which the mode of transport exposes one to the urban environment, again as in walking, or seals one off from it, as in taking the metro. In particular, it is argued that walking the streets and public open

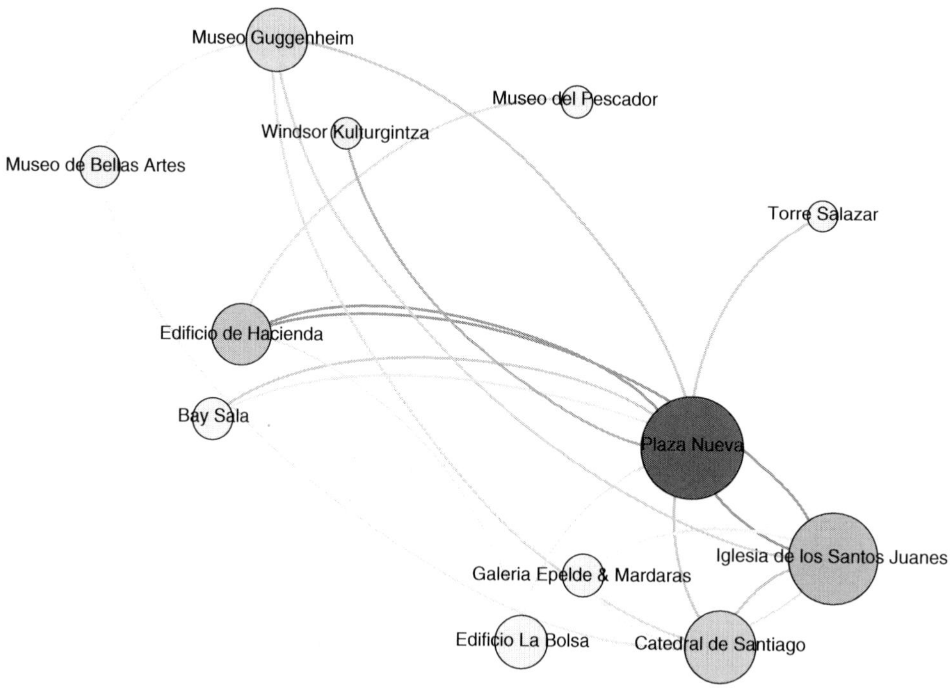

Fig. 7.3. Network of the most visited cultural points of interest in Bilbao. From Aranburu *et al.* (2016), Figure 3, with permission of the authors.

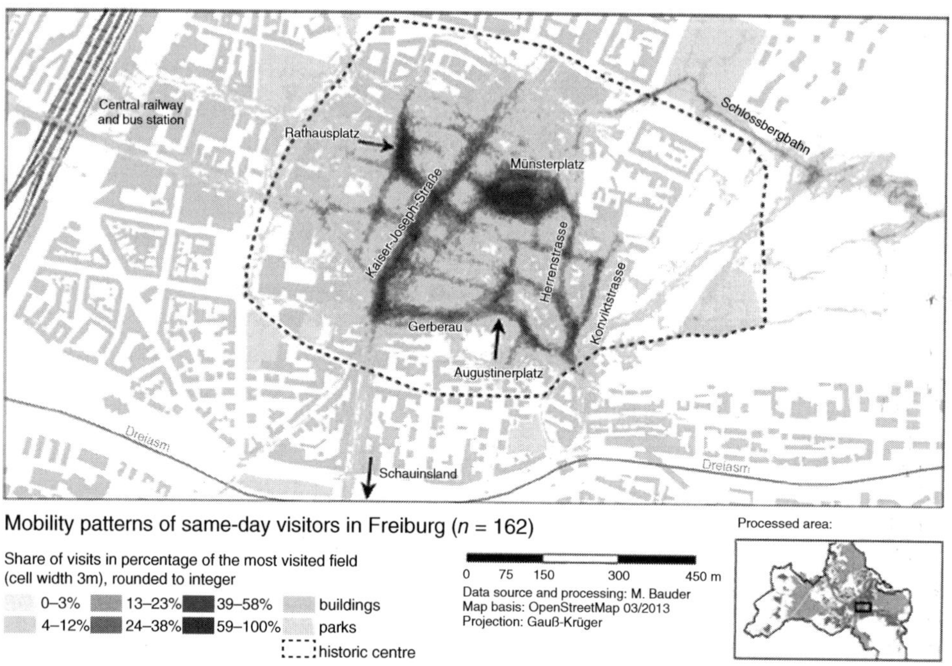

Fig. 7.4. Spatial mobility of same-day visitors in Freiburg. From Bauder and Freytag (2015), Figure 2.

spaces of Paris is not just a means of getting from A to B, as with the metro, but a way of experiencing the atmosphere and very essence of the city with the hustle and bustle of urban street-life providing visitors the anonymity to gaze at passers-by and observe and participate in the diversity of everyday activities. 'To walk is to appropriate public space and become part of it' (Lepan, 2013, p. 184). The Champs-Élysées, for example, is a place not only for walking but also for shopping, for discovering, for socializing and commonly for a combination of these activities.

Observation studies and surveys also shed light on tourist activities as they move about the city. Hartmann (1988) reported a diversity of behaviour among young Canadians and Americans in Munich. They had little contact with local residents and in addition to visiting the city they spent time on organizing travel for subsequent stages of their trip around Europe. Keul and Kühberger (1997, p. 1011) found stereo-typed patterns of visits to Salzburg, an '"ants-trail" structure of touristic pathways' and a highly non-random pattern of stops. Some 60% of the tourists' stopping activities involved gazing/looking around and 20% photography/videoing.

Salas-Olmedo *et al.* (2018) point out that tourists leave a digital footprint in ways other than geo-located photos and show how the analysis of attractions from such data can be complemented by Foursquare check-ins as a proxy for consumption and the analysis of tweets can be used as a proxy for accommodation as most such interaction on Twitter occurs in the evening. While subject to limitations, such an approach begins to provide a more complete picture of how tourists are using the city. Analysis of these three sources of data for Madrid enabled the identification of not only areas where sightseeing was concentrated but also those that were linked to tourist consumption, internet consumption or some combination thereof. All three activities were high in the historic centre. Sightseeing activity was also high along the Paseo de Castellana axis; at particular squares, museums and the like; and at good vantage points. High levels of consumption also occurred in areas surrounding the historic centre with luxury shopping or at specific places such as the Real Madrid Stadium museum. Tourist densities reflected in Twitter traffic were also dispersed into tracts surrounding the centre and along the Paseo de Castellana. Taken together, the three sets of data reveal the historic centre of Madrid as a multifunctional area which is surrounded by tracts which generally combine two activities whereas the most peripheral specialize in just one.

As noted with reference to Fig. 2.1, the activity space of urban tourists is not limited to the city in which they are staying overnight as some will also use the city as a base from which to make excursions to other places within or beyond the surrounding region or even to neighbouring countries. Half-day tours from Paris, for example, are offered to attractions outside of the city such as Versailles, Giverny, Fontainebleau and Chartres; full-day tours take visitors to the Loire Valley or even across the border to Bruges in Belgium (Duhamel and Knafou, 2007).

Segmentation of demand

Not all tourists, of course, behave in the same way. Differences occur in the patterns of intra-urban behaviour of particular market segments but these are often differences in degree more than in kind and many commonalities are to be found. The most common focus has been on the behaviour of first-time and repeat visitors (Shoval and Raveh, 2004; Freytag, 2010; Ipsos, 2017; Budović and Ratkaj, 2018; Caldeira and Kastenholz, 2018). Variations occur from study to study but there is a tendency for first-time visitors to focus more on iconic landmarks in a more intense manner and for repeaters to be more selective in terms of attractions and places visited and to seek out places offering different opportunities to experience a city's atmosphere. Major attractions, nevertheless, retain their appeal. Almost all domestic visitors and over half of foreign tourists to Paris in 2012 had already visited the city in the preceding five years (Observatoire Économique du Tourisme Parisien, 2014). In a large scale study of repeaters, Ipsos (2017, p. 11) observed:

> revisiting [places they have already been to] is ... synonymous with renewal, it is a means of (re)discovering places from a different angle. At the same time they also want to get off the beaten track and discover another Paris, as yet unknown and desirable: Repeaters do not consider themselves as ordinary tourists, their experience of Paris leads them to expect a more authentic Paris, an 'unknown Paris', and to live it less as a tourist and more as a Parisian.

Mapping and survey research by Dejbakhsh *et al.* (2011) revealed marked differences in the spatial behaviour of tourists of visitors to Melbourne from various national cultures. Individualistic visitors, for example, preferred to stay in central city accommodation and explore nearby areas on foot, their attention being captured by a number of attractions while going to or from their main destination. In contrast, collectivistic visitors tended to stay in suburban accommodation and travel to a single destination within the centre and then return home without any diversions. These and other differences draw attention to the links for certain segments between attractions in the central city and accommodation in the suburbs, a dimension that is neglected in most of the tracking and observation studies based on central city samples. Han *et al.* (2018) found both similarities and differences in the movements of Japanese and Chinese tourists in Seoul. While a few well-known attractions were visited by both groups, Chinese tourists ranged more widely across the capital, showed a greater preference for ancient temples and ended up at duty-free shops whereas the Japanese exhibited a more concentrated movement pattern, showed less tendency to explore the city and had different shopping behaviour. In Belgrade European tourists visited more locations than those from further afield (Budović and Ratkaj, 2018). Jaakson (2004) identified four segments of cruise-ship tourists in Zihuatanejo, Mexico, based on their behaviour: (i) the Shopping Browsers; (ii) the Cafe Crowd; (iii) the Pack (characterized by their 'cruise wear' and group behaviour); and (iv) the Explorers, a minority who ventured well beyond the tourist 'bubble.'

The focus in these studies, however, is essentially on different segments of leisure tourists and there is a general neglect of research on the intra-urban behaviour of others such as event attendees (Pettersson and Zillinger, 2011), pilgrims (Chadefaud, 1981), business travellers, conference goers and those visiting friends and relatives.

Activity spaces

Analysis of tourist attractions and patterns of tourist activity in cities highlight the importance of public open spaces: iconic monuments, pedestrian malls, well-known streets and squares, parks and gardens. To a greater or lesser degree these are spaces shared by residents and visitors alike. Researchers have distinguished between tourist and non-tourist spaces and offered typologies of tourist spaces based on changes made for or by tourists, the location of different types of attractions and the functions that places perform. Building on Goffman's (1959) front–back dichotomy of social space, MacCannell (1973, 1976) contended touristic space exists as a continuum along which six stages or regions might be identified depending on the modifications made expressly for tourists. Schmidt (1979) proposed a four-category classification of sites based on two variables: (i) whether the site exists purely to serve tourists or whether it has some ongoing (non-tourist) purpose in itself; and (ii) whether the setting is internally highly structurally differentiated or whether it is mostly comprised of public space, with little structural differentiation. Later writers have sought to differentiate between tourist and non-tourist spaces in particular cities.

Simon (2010) distinguished between 'monumental' Paris and 'banal' Paris, between the Paris of striking, grandiose, iconic landmarks and the ordinary, everyday lived Paris of its residents. But if the former is readily recognized and mapped, the commonplace is less easily articulated and specifically located. Effectively, 'banal' Paris is that which is subjectively experienced, it is 'the rest of the city', its street-life, restaurants, cafés and bars, its shops (e.g. specialized retailers such as *fromageries* and *boucheries* that may be common to locals but exotic to many international visitors) and also particular places thought to be typical of the city, such as its flea markets. 'But it is because [these features] are common', Simon (2010, p. 320) argues,

> that they constitute an attraction for tourists, as a departure from (or an aside to) the 'classic' tourist monuments, where one heads to more dispersed urban spaces that are less readily classifiable, at least for non-tourists. Observing unexpected situations or participating in everyday activities (going into a bar or restaurant, going out in the evening) leads to out of the ordinary experiences.

Such spaces would contribute significantly to experiencing what Matley (1976, p. 30) referred to as 'the individual character and atmosphere' of cities. Lepan's (2013) interviews revealed that many of her respondents made multiple visits to such places as Montmartre and the Latin Quarter to experience the ambience of these districts. The diversity and intensity of urban activities and populations are characteristics of large urban areas which foster unexpected meetings, sights and experiences but it is doubtful whether the everyday in every city is as attractive and interesting as that of Paris or indeed whether all parts of the French capital are equally as appealing in this regard.

Both Simon (2010) and Lepan (2013) draw attention to differences between places visited and activities carried out during the day and those that occur during the evening, that is, there are diurnal fluctuations in tourist activity spaces, a feature also observed by Vu *et al.* (2015) in Hong Kong. Although Paris is known for its night-time attractions – Burtenshaw *et al.* (1981) identified distinct nightclub areas off the Champs-Élysées, in Montmartre and in the Latin Quarter – evening activities among Simon and Lepan's respondents were rather limited. Some did take evening cruises on the Seine or went to see the lights of the Eiffel Tower to get a different experience from seeing it in daylight but often they just went out for a meal or drink or a short walk near their accommodation. Simon (2010, p. 580) concluded that '"monumentality" and the "day" dominate "banality" and the "evening"'. She attributed the restricted range of nocturnal activity spaces of visitors to Paris to factors such as fatigue, security issues and limited budgets (many of her respondents were backpackers on a grand tour).

Maitland (2013) writes of the search for the 'real London', the 'allure of the ordinary' and the 'appeal of the uncontrived.' The 'real city', he concludes, is a complex concept which can be found in the presence of local people and buildings and spaces which characterize distinctive, hard-to-reproduce areas which provide an opportunity to experience everyday life and in which tourists construct their own narratives.

Kotus *et al.* (2015, p. 108) suggested that Poznań, Poland, 'can be clearly separated into two different worlds: areas in practically sovereign use by residents, and areas in which tourists appear.' The latter is made up of different types of places based on their functions and spatial arrangement:

- 'induction spots' leading to the city or simply known attractions that enable itineraries to be planned even before the visit;
- 'gates to the city', entrance hubs such as the railway and coach stations, the airport and accommodation facilities;
- 'anchor spots', or attractive places one keeps coming back to during a visit; and
- 'bridges' or pathways between attractions which offer the possibility of chance discoveries.

Pedestrian counts in Zihuatanejo on days when cruise ships were in port enabled Jaakson (2004) to delineate a tourist 'bubble' with a core where most of the tourism shops and other retail outlets are located along a beachfront promenade and two streets going inland for a couple of blocks and a periphery extending another block or two having a mixture of tourist and non-touristic features.

In Barcelona, López Palomeque (2015, p. 494) distinguishes between a traditional tourist space encompassing the iconic landmarks of the historic centre and the Ensanche and a new poly-nuclear tourist space that is intermediate in structure between a concentrated and a dispersed distribution pattern which now extends beyond the city in a strict sense to form 'a functional tourist space with imprecise boundaries.' This newer area includes such places as the zone around the Forum; the Olympic Port and City; the waterfront; and Montjuic and the Diagonal *montaña*, including FC Barcelona. Such areas are given over to a range of leisure activities beyond sightseeing of historic or iconic landmarks.

Shoval and Raveh (2004) generalized from the Jerusalem case and the literature to propose a model describing different types of tourist spaces in large multi-functional cities based on visitors' length of stay and number of visits, factors shown to have a strong impact on consumption patterns. In this, the main attractions are concentrated in the 'tourist city', though some may lie outside (e.g. a main observation point). Additional sights are located within walking distance. Together, these two clusters constitute the Central Tourist District (Burtenshaw *et al*., 1981). Concentrations of such attractions as museums, artistic quarters or historic monuments in more distant locations create special tourist spaces. A fourth cluster is said to consist of 'sights, which, by definition, do not belong to the urban tourism product and are visited mainly by the local population and tourists who have visited the city previously and who stay for longer periods of time' (Shoval and Raveh, 2004, p. 748). These appear to be located in parts of the city where non-tourism functions dominate.

Temporary tourist spaces may be created when events or cultural activities are staged over a few days or even weeks, for example in parks or sports stadia. Though of short duration, these may have a medium/longer-term impact and contribute to a greater dispersal of urban activities (Duhamel and Knafou, 2007; Pasquinelli, 2017).

Factors influencing the spatial patterns of attractions and tourist behaviour

Three major sets of interrelated factors influencing the spatial patterns of attractions and tourist behaviour in urban areas are the underlying morphology of the cities and the time/space budgets and mental maps of urban tourists.

Urban morphologies

Tourism is a relatively recent urban function which has developed in a range of cities with an already established morphology. The spatial pattern of attractions and tourist behaviour reflect the size and character of each city, its history and underlying spatial structure but generalities are also found. The balance between the general and the specific is especially evident in the comparative studies of García-Palomares *et al.* (2015) and Kadar (2014, 2015). The central city concentration of attractions is accentuated in smaller cities with an historic core (Poznań, Freiburg, Norwich), while in larger polycentric cities central clusters are complemented by more dispersed secondary clusters and outliers (London, Paris, Jerusalem).

Recent studies have emphasized the impact that different street patterns, public open spaces and buildings have on the movement and activities of urban tourists (Gospodini, 2001; Edwards and Griffin, 2013; Kadar, 2015). Gospodini (2001) combines Boerwinkel's (1995) types of spatial order with Hillier's (1996) concept of space syntax to propose a typology of spatial patterns which are more or less attractive to tourists (Fig. 7.5). According to Boerwinkel (1995), a successive arrangement occurs when visitors uncover spaces by sight and movement in a step-by-step fashion whereas a simultaneous arrangement provides visitors in any particular space with multiple choices. In a similar way, Hillier (2005) argues that the spatial configuration of predominantly linear urban space features (streets, boulevards, avenues, alleys) shapes social patterns such as the movement and the co-presence of people and activities. Spatial systems with high syntactic depth are tree-like in form and offer tourists limited choices in their movements as they must move through many spaces to reach other ones. In contrast, spatial systems with low syntactic depth are more integrated; they have a semi-lattice or matrix configuration offering many choices for direct movements between places affording multiple experiences.

Kadar (2015) examined the syntactic spaces of Vienna, Prague and Budapest by using network graph analysis to establish the degree of integration of the points of interest (nodes) in each city, with the pedestrian street infrastructure connecting them constituting the edges or links in the network. Following Hillier (1996), the integration value *i* describes the level of integration of a node into the system. The fewer the connections needed to go from one node to all the others, the higher the *i* value and the more integrated that node is. Some nodes were individual attractions; others consisted of 'continuous public spaces where tourists can visit and photograph attractions without moving further' (Kadar, 2015, p. 18). The highest *i* values were measured for Vienna, a city with a semi-lattice-like graph where a few branches extend outwards from an integrated core. The lowest *i* values were recorded for Prague which has a tree-like graph with only a few linear elements that intersect at a small number of nodes and many tree-structured sub-graphs. In other words, Prague has a much less integrated system than Vienna. Budapest

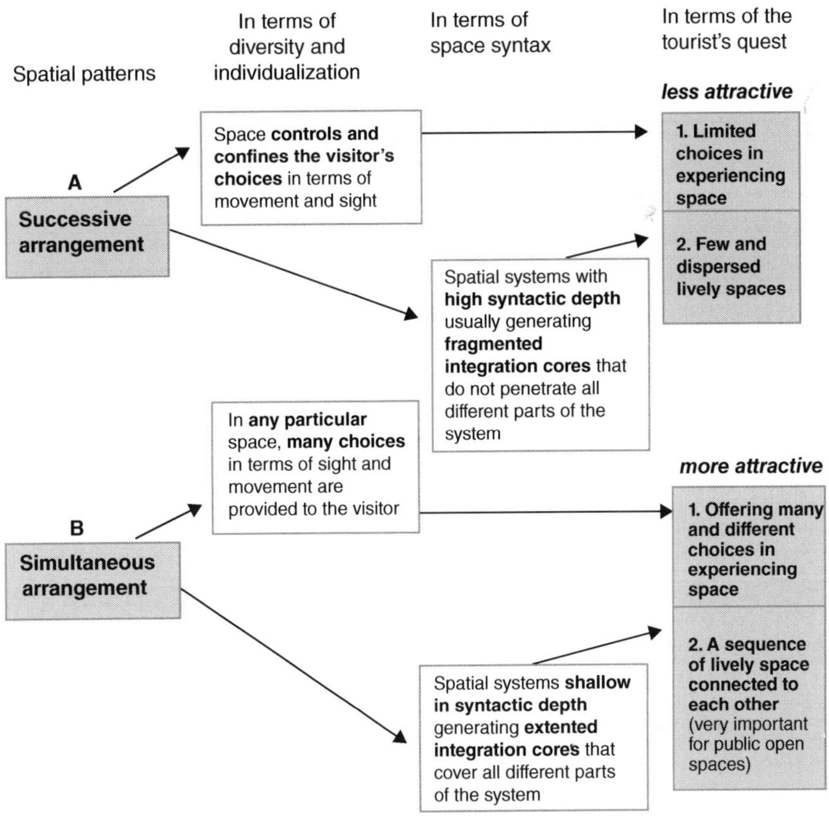

Fig. 7.5. Basic spatial patterns (public open spaces and buildings as tourism resources): a description-evaluation in terms of diversity, individualization and space syntax. From Gospodini (2001), Figure 3.

represents an intermediate case, having a tree-like structure on the Buda side and a semi-lattice form on the Pest side. Kadar demonstrates how the morphology of each city together with recent planning measures have resulted in different patterns of tourist behaviour: Vienna has a well-balanced, pedestrian-friendly core in which the main attractions are well-connected whereas movements in central Prague are channelled along three often congested axes that link the main historical attractions.

Edwards and Griffin (2013, p. 590) interpreted the pattern of visitor trails in Sydney in terms of the successive arrangement of the city's street pattern which induces 'a high degree of repetitiveness, with participants returning to the same places and using the same routes throughout the day.' In contrast 'tourists' dominant patterns of use in Melbourne represent a simultaneous arrangement in which its spatial order shapes movement, facilitating a pattern which maximises co-presence.' Similarly, Kotus *et al.* (2015, p. 107) found that the 'rich structure in the downtown zone [of Poznań] does not channel tourist traffic into distinct paths; just the opposite, this area is full of places of shallow syntactic depth.'

Time/space budgets

Many urban tourists spend only a few days in a city, either as a 'short break' holiday or as part of a longer tour (Hartmann, 1988; Pearce, 1997a; Simon, 2010). Consequently, as time is a limited resource they must allocate it accordingly if they are to achieve what they have come to see and do. Patterns of tourist behaviour can thus be interpreted from a time-budget perspective, that is, how much time do visitors spend on what activities and where? (Pearce, 1988; Lew and McKercher, 2006; McKercher *et al.*, 2012; Grinberger *et al.*, 2014). The spatial dimension is important as time (as well as effort and money) is spent on moving from one attraction or activity to the next. Trade-offs may therefore be required between spending more time at particular attractions and visiting additional or more distant attractions. Such trade-offs may also involve decisions regarding the mode of transport to be used if additional and more distant attractions are to be visited, for example, taking the metro or a city bus tour, rather than exploring the city on foot (Simon, 2010). Time-budget considerations thus play a key role in explaining the popularity of visits to landmark attractions in the central city and the clustering of activities which results in greater accessibility and less travel time.

GPS studies in Hong Kong provide further insights into tourist time–space consumption. Grinberger *et al.* (2014, p.119) found different patterns of time–space use but concluded that activity time was valued more than space. However, most tourists 'still selected behaviour strategies that traded some loss of time for the consumption of more space.' Differences in time–space usage also emerged between first-time and repeat visitors suggesting:

> first-time visitors tend to sample the destination, by wandering throughout it widely, often by foot. They also show a clear preference to visit icon attractions, while the repeat visitor tends to concentrate his or her activities in fewer places ... disproportionately in shopping or dining districts.
>
> (McKercher *et al.*, 2012, p. 157)

Time spent is also a critical factor used to identify the most important attractions or clusters. In Hong Kong McKercher *et al.* (2012) found a strong correlation between the share of tourists within specific cells and the amount of time spent there; that is, both movements and time were concentrated in specific areas. However, in Bilbao the cultural points of interest where visitors spent the most time were not always those that attracted a larger share of visitors (Aranburu *et al.*, 2016). Nor were the attractions most visited by independent cruise-ship visitors in Palermo, those attractions nearest the port, the ones where the average length of visit was the longest (de Cantis *et al.*, 2016). Likewise, in Poznań the areas most photographed were not always the ones where tourists spent the most time. Kotus *et al.* (2015, p. 107) concluded that:

> the places most attractive in tourist terms among the ones indicated can be considered those that both, stand out enough from the urban structure to attract visitors for a longer time and that are interesting enough to leave a trace in cyberspace ... Five of those places are situated in the central part of the city.

Mental maps of urban tourists

Time is not the only constraint influencing visitor behaviour in cities. Visitors' knowledge of a city will almost inevitably be incomplete, particularly in the case of first-time visitors. What they do and where they go will also be a function of what they know or perceive a city has to offer and their mental map of the place they are visiting (Lynch, 1960). Mental maps are formed by prior knowledge derived from a variety of sources and markers (e.g. publicity material, guide books, television programmes, news reports, information and advice from travel intermediaries), as well as by previous experience. According to Lynch, mental maps are based on five main elements: (i) paths (lines of movement); (ii) edges (barriers to movement); (iii) districts (distinctive areas); (iv) nodes (strategic meeting points); and (v) landmarks. Visitors, he argued, orient themselves initially by landmarks, further underlining the popularity of the iconic attractions of cities. Jansen-Verbeke and Lievois (1999, p. 103) concluded that:

> the appreciation of tourists of heritage resources in a historic city tends to be based on areas (squares, streets) rather than on single buildings. Apparently, the mental map before the visit was strongly based on markers (single buildings), whereas the appreciation after the visit is mostly structured by clusters and a more general appreciation of the urban tourism opportunity spectrum.

Elements of a city that contribute to a strong and clear mental map will facilitate tourist movements. Aranburu *et al.* (2016, p. 12) argued that due to its street layout and the location of its main attractions, central Bilbao has a simple legibility that facilitates the 'rapid consumption of the tourist destination.' The downside to this for the tourism sector is that visits are brief. Both well-prepared and less-prepared tourists to Freiburg visited the city's major attractions in the historic centre but some of the former also visited the Schlossberg with its observation tower and the railway station (Fig. 7.4) while the rather unstructured visits of the less-prepared were confined to the centre (Bauder and Freytag, 2015). Fuller preparation would have resulted in a more detailed mental map of the city and a more directed visit. As a structural axis in Paris the Seine as acts an edge for tourists and footpaths flanking it form paths (Fig. 7.6). Jaakson (2004) suggested cruise-ship passengers are aware of when they are approaching the outer limits or edge of the tourist bubble in Zihuatanejo through such 'boundary cues' as there being fewer other passengers but more different looking people present, fewer signs in English, the general state of the zone and the reaction of other tourists.

Tourist Accommodation in Urban Areas

A large body of research has been undertaken on tourist accommodation in urban areas. This reflects the tangible nature of accommodation and availability of data, the substantial investment in properties that are fixed in space and the weight of accommodation as a component of visitor spending. Linkages between demand- and supply-side studies are not well developed. Much of the work on hotel demand has developed as a separate research thread in which location has been identified as a key factor but one which is treated in general rather than site-specific terms. There are two broad strands to the supply-side research. The first is set in an urban morphological tradition and examines the distribution of accommodation in the city relative to other sectors and forms of land use. As such it provides a broader picture of accommodation as one part of the spatial structure of urban destinations. The second approach focuses on the impact of agglomeration or co-location of hotels and emphasizes spatial relationships within the sector.

Demand for urban accommodation

Identifying the importance of location in studies which examine the process by which guests select their accommodation and analyse the attributes that they take into account is not straightforward. Comparing the results of these studies is complicated by the different dimensions of location (Bull, 1994; Yang *et al.*, 2018), the complexity of the selection process (Lockyer, 2005; Jones and Chen, 2011; Sohrabi *et al.*, 2012), the number and variety of attributes considered (McCleary *et al.*, 1993; Jones and Chen, 2011), the sample used and the diversity of methodologies applied (Pearce, 1987b; Lockyer, 2005; Jones and Chen, 2011; Sohrabhi *et al.*, 2012; Yang *et al.*, 2018). Despite this heterogeneity it is clear that location is one of the key factors along with others such as price, quality, comfort and services that guests weigh up in deciding which hotel or motel they will stay at and where.

The importance of location and of its different dimensions (e.g. closeness to attractions, distance from city centre, proximity to public transport) varies from study to study. In some, location or some dimension of it tops the list (Lewis, 1985; Pearce, 1987b; Rivers *et al.*, 1991); in others it ranks in the leading two or three attributes (Zaman *et al.*, 2016; Njite and Schaffer, 2017). In factor analytical analyses locational attributes may be bundled with others among the leading factor scores (Chu and Choi, 2000; Sohrabhi *et al.*, 2012). However, location does not feature at all in Jones and Chen's (2011) experimental study.

In one of the seminal studies of hotel demand, Arbel and Pizam (1977) found tourists in Tel Aviv showed relatively little sensitivity to distance and that many would

Fig. 7.6. 'As a structural axis in Paris the Seine acts as an edge for tourists and footpaths flanking it form paths.' Copyright: iStock.

tolerate travelling further from the centre with little or no reduction in rates providing adequate public transport was available. Later, Bull (1994) used hedonic pricing to assess the value of location and other attributes in an Australian coastal town characterized by the strip development of motels along the road leading into the town centre and amenities and attractions. The implicit price of distance away from the town centre declined by about AUS $3 to AUS $6 per kilometre, but this effect diminished 3–4 km away. As expected, prices also increased with quality (number of stars) and possession of a restaurant.

The importance of location may also vary by market segment. In an early study by Lewis (1985) location was shown to be more important for business than leisure travellers, though the leading attribute for both. Little difference occurred in Yavas and Babakus's (2003) study between these two segments in terms of the importance of this attribute, ranked high in both cases but not top of the list. Likewise, Chu and Choi's (2000) research in Hong Kong showed that business and leisure travellers held the same perceptions towards hotel selection factors. McCleary *et al.* (1993) found location was the single most important factor influencing hotel selection across all types of American business travellers. In McCleary *et al.*'s (1998) comparative study of business travellers the Koreans ranked convenience to business fifth and to downtown sixth compared with respective rankings of tenth and 12th by the Americans. Convenience to tourist attractions was the most important attribute for Mainland Chinese visitors to Hong Kong, ahead of safety and security, with that order being reversed for foreign travellers (Tsai *et al.*, 2011). For short-stay visitors location ranked a narrow second behind cleanliness in Zaman *et al.*'s (2016) survey of hotel guests in Paris but sixth for those staying more than three nights. Those on a higher income gave greater weight to location than those with an annual income of less than US$25,000 for whom value for money was by far the most important attribute.

While some of the above research specifies location with respect to attractions, to downtown or to the airport, in general location is dealt with in the abstract; that is, specific decision attributes are not related to the choice of particular hotels or motels in particular parts of a given city. In contrast, the sample frame for Pearce's (1987b) survey of motel choice in Christchurch provided for guests to be sampled in ten motels in each of three locations in the city: in the centre, along major access routes and in the suburbs. Moreover, instead of using a preselected list of attributes the question of motel choice was set as an open one: 'What was your main reason for choosing this motel?' Respondents expressed three types of location: (i) simply 'location'; (ii) a central location; or (iii) proximity to a wide range of places or events (e.g. sports venues, hospitals, friends and relations, weddings, transport termini). Together, the three types of location accounted for about half of the main reasons cited. Facilities, service and a clean appearance were also very important but price was not a major reason for discriminating between motels due to some uniformity in the rates charged. Centrality, as might be expected, was the dominant factor for the central city motels; proximity to a particular venue or event was the key in the suburbs, while the location element was less clearly differentiated for the major route motels. The profiles of the three groups of motel guests also varied with the sharpest contrast occurring between the central city and suburban motels. Central city motels tended to attract a larger share of vacationers, many of whom were from overseas and on their first visit to Christchurch. Suburban guests, on the other hand, were visiting the city for a variety of reasons, they were more likely to be domestic and on a repeat visit.

Yang *et al.* (2018) analysed data from over 8000 online reviews of 220 hotels in Los Angeles using a mixed-effect ordered logit model to identify the determinants of guest satisfaction (not selection) with urban hotel locations. They classified all location-related factors into three categories: (i) accessibility to points of interest; (ii) transport convenience; and (iii) surrounding environment.

Their results suggest that significant determinants of hotel satisfaction in that city were accessibility to attractions, airports, universities and public transportation (but not to beaches and shopping attractions), as well as the surrounding environment variables of green spaces, bodies of water and local businesses. Moreover, the satisfaction effects varied by guest segment: accessibility to museums, theatres and amusement and theme parks, for example, was most valued by family travellers while accessibility to stadia and arenas was more important to those travelling with friends.

From a spatial structure perspective, the results of Pearce (1987b) and Yang *et al.* (2018) demonstrate the need to link accommodation selection and satisfaction and guest profiles directly to the distribution patterns of motels and hotels. This research is complemented by Shoval *et al.*'s (2011) study which links activity patterns to hotel location. Their study of tourists staying in four different locations in Hong Kong showed strong links between where the hotels were and where the guests spent their time. Hotel guests from all four hotels visited icon attractions such as the Peak and Stanley Market but much of their activity was concentrated in close proximity to their own hotel. Hotel location was also linked to the timing of visits and to whether guests primarily visited on foot or used public transport. Shoval *et al.* (2011, p. 1609) argued that 'hotel location plays a critical role' in this behaviour but the direction of causality is less clear; that is, the choice of hotel may have been influenced by where the guests wanted to spend their time rather than the hotel location determining what they did.

More generally, research on accommodation demand draws greater attention to the needs of different types of urban visitors, particularly business travellers, who rarely feature in the research on attractions and activities but may generate a significant share of commercial accommodation demand. In 2013, 51% of visitors staying in hotels in Barcelona were on holiday while 41% were visiting for professional reasons (business, fairs and congresses) (López Palomeque, 2015). Patterns of demand for other significant groups of urban visitors are often neglected as few studies go beyond the hotel sector. Table 7.1 illustrates some of the differences which may occur between destinations and visitor segments in terms of the relative use of different forms of accommodation by independent tourists. Wellington is New Zealand's capital, a major urban centre and gateway; Rotorua is a smaller, regional urban centre and destination renowned for its

Table 7.1. Patterns of accommodation use among independent visitors to Wellington and Rotorua. From Pearce and Schott (2005), Table 1.

	Rotorua			Wellington		
	International	Domestic	Total	International	Domestic	Total
	n = 651	*n* = 339	*n* = 900	*n* = 610	*n* = 278	*n* = 888
Type of accommodation	%	%	%	%	%	%
Hotel	10.9	7.1	9.6	22.8	22.7	22.7
Motel	19.5	31.6	23.6	8.5	19.1	11.8
Backpacker/hostel	30.7	3.8	21.5	26.4	7.6	20.5
B & B, homestay, farmstay	2.8	1.8	2.4	5.6	3.2	4.8
Friends or family	5.7	11.8	7.8	24.6	32.0	26.9
Luxury lodge	1.4	1.2	1.3	0.0	0.0	0.0
Campervan, camping ground	15.2	10.3	13.5	6.2	1.8	4.8
Other	0.8	5.0	2.2	2.8	1.8	2.5
No stay	0.8	5.0	2.2	2.8	1.8	2.5
	100.1	100.0	99.9	100.0	100.1	99.9

geothermal activity and Maori culture. Motels and backpackers/hostels are used more than hotels in Rotorua. Backpackers/hostels are also in demand in Wellington and in both cities are more popular among international visitors. Accommodation provided by friends and relatives is much more important in Wellington than Rotorua and in both destinations is used more by domestic than international visitors.

A survey of visitors to Warsaw revealed only a quarter chose hotels while a third stayed with friends and relatives, 16% stayed in hostels and 15% rented private apartments (Derek, 2018).

Morphological studies

Compared to research on attractions and activities, morphological studies of accommodation have a stronger and more explicit conceptual or theoretical basis. Yokeno's (1968) seminal theoretical study of urban hotel location drew on von Thunen's agricultural land-use model to propose a concentric zone model of urban land use based on land rent curves, with the cost of land decreasing with distance from the city centre and the willingness and ability to pay varying by type of land user. In Yokeno's model a hotel zone is located in the central city, between the innermost administrative centre of the CBD and the commercial zone. The underlying assumptions here are that the city is monocentric and that tourists are willing to pay more for a central location, thereby enabling hotels to support the higher price of land associated with accessibility to the centre. Moreover, the clustering of hotels in the city centre was said to enable the visitor to 'easily choose one according to his taste or budget' (Yokeno, 1968, p. 16). The external benefits to entrepreneurs of clustering, such as convenience of purchasing and provision of infrastructure, were also recognized. Only a single hotel zone is depicted in Yokeno's model although the existence of different types of hotels is acknowledged.

In contrast, Egan and Nield (2000) identified a spatial hierarchy of hotel types in major cities in the UK based on the proposition that potential revenue varies with distance from the city centre and that different bid-rent curves exist not only for different forms of land use but also for different types of hotels. Luxury hotels with higher revenues from affluent clients have steeper bid-rent curves than other types of hotel but may be outbid by retailing and business offices pushing them to the edge of the main central shopping areas. A second zone of hotels is classified as budget hotels of Type A, usually consisting of converted buildings located in secondary locations at the edge of the city centre where there is less competition from other uses. Business hotels of 3-star quality are located in the suburbs. Type B budget hotels are found at edge-of-city locations, usually at road junctions, where their less expensive room rates resulting from lower land costs attract the tourist trade and business travellers using the road network.

Developed in the context of historic cities, Ashworth and de Haan (1985) and Ashworth (1989) offered a more descriptive sixfold typology of urban hotel locations related to other urban land uses (Fig. 7.7). In particular, hotels in the historic centre and close to many heritage attractions (Type A) and Class D hotels (small or medium hotels in high-amenity or nice locations, e.g. overlooking a park) are complemented by others located in terms of some form of transport: railways (Type B); main access roads (Type C); and large modern hotels on motorway and airport interchanges on the urban periphery (Type F). The relative importance of the latter has varied over time as each mode of transport has increased (motor cars and planes) or decreased (railways). Type E hotels are found in a zone of overlap or transition between the historic city and the modern CBD which has developed as businesses have relocated from the historic centre and new ones have been established, a process depicted in a separate model of urban growth. According to Ashworth (1989, p. 98):

> In many medium sized historic towns it is this zone that is most likely to be the site for large purpose built hotels, usually part of larger hotel chains. Such sites offer pedestrian accessibility to both historic and modern commercial attractions and often

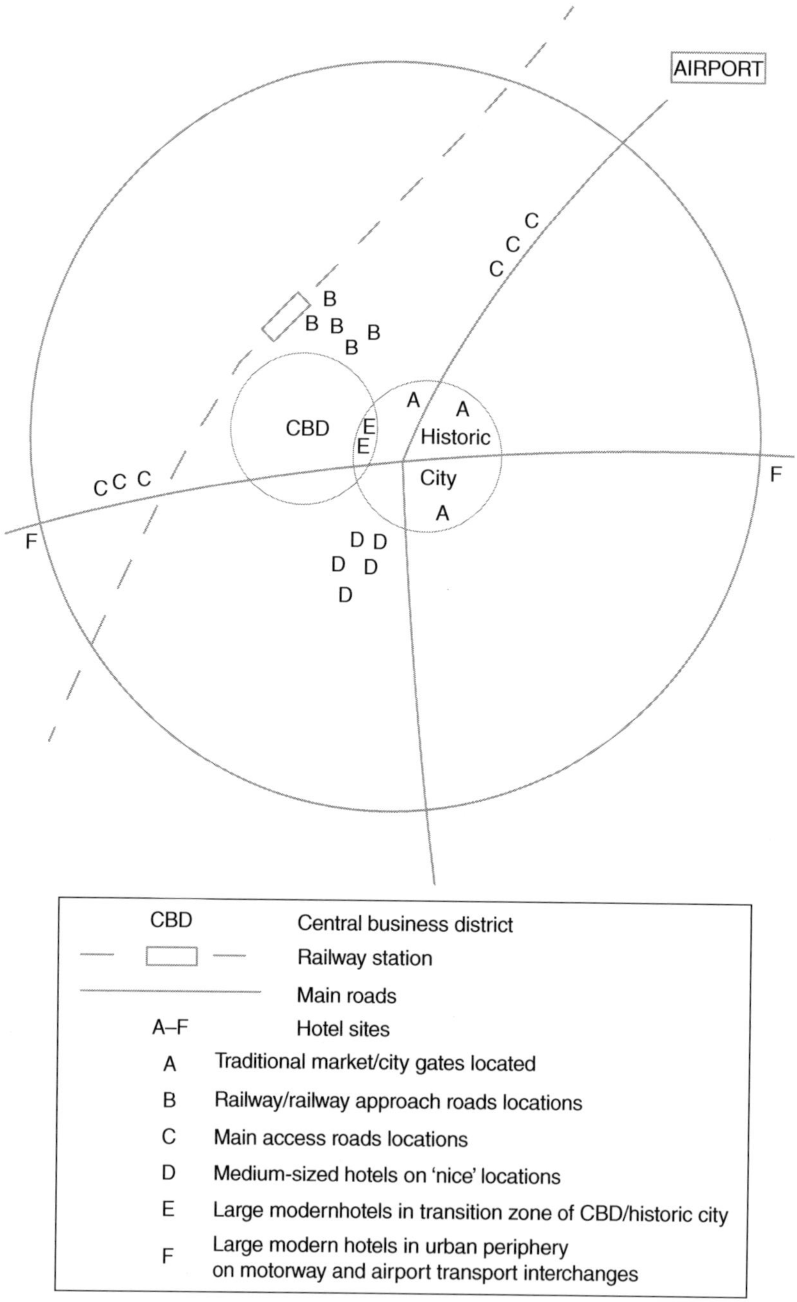

Fig. 7.7. Typology of urban hotel location. From Ashworth (1989).

> form part of prestigeous [*sic*], high-profile multifunctional development projects.

This notion of a zone of overlap, urban growth and the changing importance of the transport modes introduces important dynamic elements into the conceptualization of hotel location.

Hotels clustered around bus stations at the city's entrance gates are the dominant pattern in Aliagaoglu and Ugur's (2008) regional life model, based on an empirical study of Erzurum in Turkey. This model emphasizes the city's links with its surrounding region and travel to the city by regional residents to meet their commercial, administrative and health needs rather than for leisure travel.

Empirical studies lend varying support to these different models and also draw attention to other factors addressed less explicitly in them that influence the spatial structure of accommodation, notably the role of planning and land-use regulations (Burtenshaw *et al.*, 1981; Pearce, 1995a). Variations occur from case to case depending on the characteristics of the cities examined, the model applied and the approach adopted. Of particular relevance to the spatial structure of urban destinations are the relationships with other tourism sectors and other urban land-uses. In this regard, empirical studies show that while a central concentration of hotels is frequently found, central city hotels are not of such a density that they constitute a uniform single-use hotel zone similar to those that may occur in the seafront of coastal resorts or to the linear concentrations of financial institutions which may characterize other parts of the city. Rather, one or more clusters are usually found where hotels are interspersed with other commercial, administrative and sometimes residential functions (Pearce, 1995a).

In testing Yokeno's model, Shoval (2006) found that average room rates in Jerusalem were significantly higher for hotels located close to the city's main tourist district than those further away. This difference was shown to be a function of individual tourists being prepared to pay more for a central city hotel than tour operators handling organized tours who sought the cheaper rates of more peripheral hotels. Adam's (2013) analysis of the distribution of hotels in Kumasi, Ghana, revealed a spatial hierarchy of hotels along the lines suggested by Egan and Nield (2000); that is, the higher-rated hotels were concentrated in the core of the city while budget hotels and guest houses developed in more peripheral sites. However, few of the star-rated hotels were near the city's main attractions or close to the city's domestic airport. Room rates at airport hotels may be influenced not just by proximity to the airport but also by proximity to the CBD (Lee and Jang, 2011). In their study of selected US cities Lee and Jang showed that the proximity effect of CBDs to be greater than that of airports but the absolute size of the premium nevertheless was derived mainly from airports.

Research linking hotel development with patterns of urban growth provides support to the historic city model (Ashworth and de Haan, 1986; Ashworth, 1989). The specifics vary from city to city. Bégin (2000) found that a large proportion of smaller, cheaper hotels in Xiamen, China, were concentrated in the historic city but the centre of gravity of hotel distribution gradually shifted away from the old core as newer hotels were located in the transition zone between the old town and the new town, on the periphery of the urban area and around transport nodes: the train station, bus depot, a highway interchange and the airport. The major tourist sites did not attract much new hotel development. Bégin attributed this pattern to the availability of land for new hotels; congestion in the old town; building restrictions; and accessibility. In Madrid, the historic core contains a mix of economy, mid-price and luxury hotels but other modern hotels have developed to the north and northeast of the Puerta del Sol as new business and commercial activities developed there (Gutiérrez Ronco, 1980: Urtasun and Gutiérrez, 2006).

In a similar fashion, economic growth and globalization have resulted in the development and expansion of the modern CBD of Santiago ('Sanhattan'), in the districts of Providencia, Vitacura and Las Condes (Hidalgo *et al.*, 2016). Three main hotel clusters occur: (i) luxury hotels in the new commercial district of Providencia and Los Condes; (ii) two- and three-star hotels attracting individual family tourists located in the historic centre; and (iii) more peripheral two- and three-star hotels along the main highways and close to the airport (Fig. 7.8). Room rates peaked at 8 km from the city centre, not in the centre itself, in the hotels catering to business travellers in Providencia and Los Condes. Room size also tended to increase with distance from the historic centre.

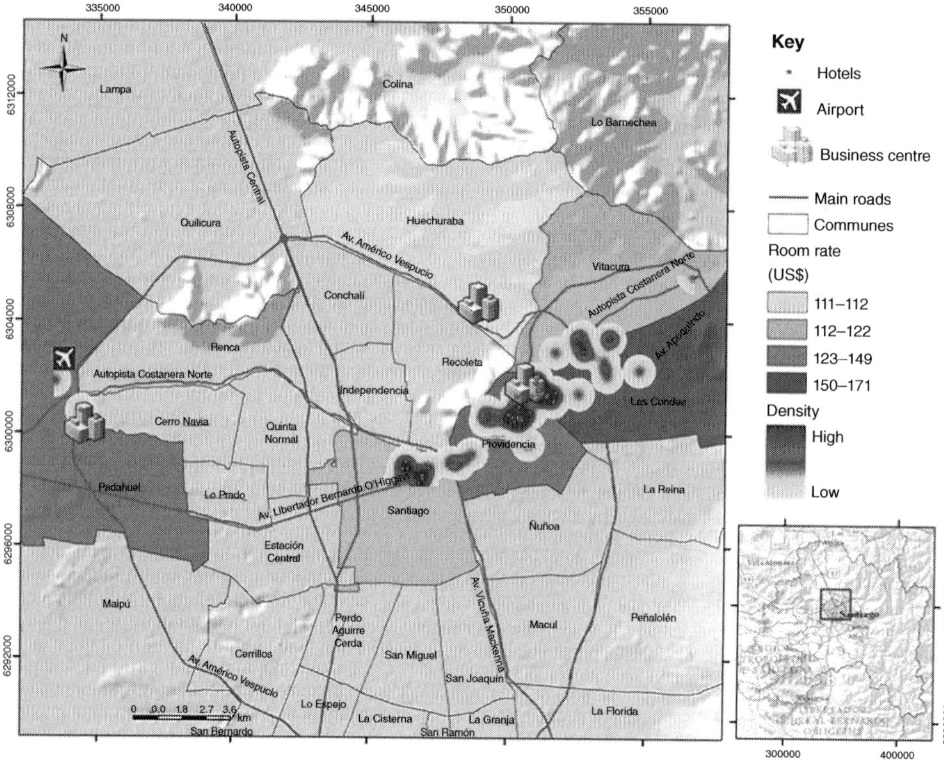

Fig. 7.8. Distribution of hotels and hotel rates in Santiago, Chile. From Hidalgo *et al.* (2016), Figure 1, with permission of the authors.

Celata (2012, p. 16) emphasized the recent growth in suburban hotels in Rome accompanying the urban sprawl that has occurred in that city:

> Suburban locations allow hotels to increase their internal economies of scale and to gain in profitability by growing in size and improving the range of services and comforts. Large multi-functional suburban hotels are particularly suited for segments of the tourism market – such as business travel, conference tourism, shopping tourism, event tourism – that are growing rapidly and seem very attractive to both private investors and planning authorities.

Such hotels are generally self-contained, not well integrated into their suburban surroundings, form part of global networks and contribute to what Celata calls the 'atomization of tourist space' in large cities.

Li *et al.* (2015) analysed explicitly the spatial association of upper-grade hotels, attractions, transport facilities and land-use characteristics in Hong Kong where there has been continuous development of hotels in the CBD and some expansion into surrounding districts. Commercial land-use type was shown to be the most significant and positive influence on the spatial distribution of upper-class hotels. The number of nearby attractions was negatively associated with the likelihood of establishing an upper-grade hotel. This negative association was attributed to the relative dearth of major historical and heritage landmarks, planning constraints and the nature of the shopping variable:

> Upper-grade hotels are more likely than lower-grade hotels to be closer to the commercial and business areas, rather than to attractions, suggesting that business facilities and shopping play an important role in determining the location of upper-grade hotels in Hong Kong. However, the negative coefficient of SAN, [the variable] which represents distance to shopping attractions,

Fig. 7.9. The Novotel hotel in Providencia, one of the many international chain hotels located in Santiago's new commercial districts. Photo: the author.

> suggests that super shopping centres impose exclusivity obligations and thus investors cannot change land use in surrounding areas to build upper-grade hotels.
> (Li *et al.*, 2015, p. 39)

Traffic land type was not significant, a function of Hong Kong's convenient public transport facilities. The determinants were also found to vary over time, with a suggestion that hotels may influence the built environment in which they are located; that is, a two-way relationship exists between hotels and surrounding land use.

Agglomeration

Another strand of research on the intra-urban distribution of hotels focuses on the impact of agglomeration in the hotel sector. The central argument here, following the seminal work of Marshall (1920) and industrial location studies, is that the clustering or co-location of hotels in itself results in positive externalities and brings benefits to both the consumer and the firm. For the consumer, the clustering of hotels is said to enhance the visibility of a district, thereby reducing search costs and providing assurance to the customer that this is an appropriate place to stay. Reduced search costs and lessened uncertainty in turn increase the likelihood of visitation and thus result in heightened demand for hotels located in the cluster or district (Baum and Haveman, 1997; Chung and Kalnins, 2001; Tsang and Yip, 2009). Heightened demand may also result from shared infrastructure (convention centres, shops and restaurants established close to high-star-ranking hotels) that can create value for customers (Baum and Haveman, 1997; Tsang and Yip, 2009). Individual properties can also benefit from overflow demand from neighbouring hotels. In addition, the clustering of hotels provides signals to prospective entrants about the feasibility of particular sites and allows firms to monitor their competitors more closely and to respond to specific competitor moves (Baum and Haveman, 1997; Canina *et al.*, 2005). These advantages may reduce or outweigh the localized competition that proximity to other hotels brings.

However, not all hotels contribute equally to generating these externalities or benefit from them to the same degree. Rather, researchers emphasize the importance of differentiating the hotel sector along a range of hotel characteristics (e.g. star rating, quality, size, price, chain affiliation, services) and the need for strategic decisions by owners to balance the extent to which they conform to or differ from other hotels in the same cluster and on what dimensions (Baum and Haveman, 1997; Chung and Kalnins, 2001; Canina *et al.*, 2005; Urtasun and Gutiérrez, 2006). Chung and Kalnins (2001), for example, argued that firms that reduce search costs for consumers would be those with greater room size and chain affiliations as these would be more likely to have greater advertising budgets.

The empirical evidence from these studies on the nature and extent of agglomeration externalities is mixed. Chung and Kalnins's (2001) research on the lodging industry in Texas indicated that agglomeration benefits were more likely to be present in rural rather than urban locations. They argued that the heterogeneity of urban demand gains to nearby hotels from a hotel's advertising would be less direct and that in any case the choice of urban hotels, compared to rural ones, was more likely to relate to proximity of an attraction or event rather than the number of hotels in the area. In contrast, Tsang and Yip (2009) showed that the agglomeration of high-star-ranking joint venture hotels in Beijing heightened demand and that all hotels benefited somewhat similarly from it. In another study in that city, Yang *et al.* (2012) demonstrated that the agglomeration effect varied by star rating and by ownership but not by scale. Luo and Yang (2016) found an inverted U-shaped relationship between location probability and incumbent hotel concentration for two- to five-star hotels in Beijing; that is, the relationship changed over time. A moderate concentration of hotels increased the probability of new hotel entrants up to a certain point after which the existence of more hotels lowered the probability of that location being selected. High-end hotel entrants had a lower turning-point, suggesting greater vulnerability to the presence of more hotels nearby.

Baum and Haveman's (1997) longitudinal study of Manhattan revealed that although the largest hotels there have always located in midtown Manhattan, where demand for hotel services from business activity and tourist attractions is densest, many other kinds of hotels are also located there. More homogenous hotel groups, typically of smaller and cheaper hotels, are found further from midtown. Their analysis of hotel size, price and location suggests trade-offs were being made between these attributes: 'Manhattan hotels locate sufficiently closely to neighbouring hotels on one product dimension (price) to benefit from agglomeration economies, but they differentiate on the other (size) to avoid direct competition and create potential complementary differences' (Baum and Haveman, 1997, p. 329).

In their longitudinal study perspective of changes in hotel location in Madrid, Urtasun and Gutiérrez (2006) considered competition between hotels in terms of geographic proximity (the ten nearest geographical neighbours) and price proximity (the ten nearest neighbours on price across all existing hotels). Importantly, their analysis takes account of existing conditions at the time the hotels were founded. They also controlled for exogenous variables by dividing the city into ten zones reflecting the way in which these and the hotels within them had developed. Modelling location at founding revealed a complex set of interactions between geographic location, price, size and services. In terms of location, Urtasun and Gutiérrez (2006, p. 397) concluded from their modelling that for both the geographic- and price-competition approaches that 'Madrid founders predicted greater benefits than costs by geographically agglomerating with competitors of similar size and services, but they predicted greater costs than benefits for geographic competition with similarly priced hotels.' Exogenous variables were more significant for the price-competition approach than for geographic competition, suggesting that agglomeration economies were associated more with the latter while the impact of area-specific characteristics other than localization economies are better reflected in the former.

In these studies the impacts of agglomeration have been estimated from different models based on hotel characteristics and locational data, not from directly investigating the motivations of the decision makers involved. Adam and Amuquandoh (2014) found that Kumasi hotel owners took account of four domains of hotel location factors: (i) economic; (ii) transportation; (iii) physical site characteristics; and (iv) socio-cultural characteristics. The most significant individual exogenous factors were: proximity to clients; proximity to CBD; good road network; good scenery; and a low crime rate in the neighbourhood. Adam and Mensah (2014) report separately on perceived agglomerative factors which influenced decision makers and found the most important to be: availability of related tourism businesses; benefiting from other hotels; availability of hotel clusters; and availability of similar hotels. The two sets of results appear to have been drawn from the same survey but as no explicit links are made between them the relative importance of the location and agglomerative factors cannot be assessed.

Other forms of accommodation

Differences occur not just between hotel types but also in the location of different forms of accommodation. Hotels in Christchurch exhibited the characteristic centralized clustering but many of the motels are distributed in a linear fashion along the major highway axes as well as in the suburbs (Pearce, 1995a). Camping grounds, whose densities and rent-paying ability are even lower than that of motels, were found in more peripheral locations while still being handy to the main access roads.

The increasing popularity of person-to-person accommodation is also transforming the spatial structure of urban tourist accommodation with the addition of new forms of rental accommodation such as Airbnb (Gutiérrez *et al.*, 2017). Generally located in existing housing stock, such accommodation has distinctive spatial patterns. In Barcelona, Airbnb listings are heavily concentrated in a concentric ring around the central hub of the Plaza de Cataluña, including very central residential districts not covered by hotels such as the Gothic Quarter and the area around the Sagrada Familia (Gutiérrez *et al.*, 2017). Hotels too have a central location but their pattern is more complex and structured by other urban features: many are concentrated along the Ramblas–Paseo de Gracia axis, in areas of business and finance such as along the Diagonal or in the coastal axis from the Barceloneta beach to the Forum. Comparison of the densities of hotels and Airbnb listings with the densities of tourist photos geo-located on Panaramio, taken as a measure of tourist interest, revealed a strong spatial association between both forms of accommodation and tourist attractions, especially in areas along the Ramblas and the Gothic Quarter and around the Paseo de Gracia, the Barcelona of Gaudi. However, the association with areas of tourist interest was greater for Airbnb, notably around the Sagrada Familia, while hotels recorded a greater number of outliers, a function of their orientation towards business travellers rather than the most visited attractions. Gutiérrez *et al.* also note that the density of both forms of accommodation exerts pressure on the city; hotels along the main hotel axis and Airbnb by expansion into residential areas, pushing up prices for rent and transforming businesses into those more heavily oriented to tourists than residents. In a more exploratory study in Paris, Heo *et al.* (2019) reported that hotels and Airbnb listings were not in direct competition and that their spatial patterns differed. Many hotel rooms were located in the northwest of the city, in the business district of the 8th, 9th and 17th *arrondissements*, whereas the leading *arrondissements* for Airbnb listings were the more heavily residential 11th and 18th.

Taking a social constructionist approach, Stors and Baltes (2018) argue that the advent of Airbnb in Berlin is not only providing alternative accommodation but also creating new tourist spaces through the messages and images of their neighbourhood that Airbnb hosts are posting on their listings. They compared the listings in two different neighbourhoods: one, an established 'tourist hotspot', City West; the other, an emergent hotspot in a predominantly residential neighbourhood, Kreuzkölln. In the former the hosts reproduced existing place images, such as those of the Kurfürstendamm and the Berlin Zoo, whereas in the latter everyday places were reinterpreted and endowed with new meanings as hosts encouraged their visitors to experience local life such as relaxing alongside the canal. The images portrayed in the City West listings largely corresponded to those promoted by Berlin's DMO while those of Kreuzkölln did not, presenting instead a more banal view of city life but one which was not without interest. The authors hypothesize that digital information-sharing platforms such as Airbnb are now more important in creating new urban tourism areas than the city's DMO.

Districts and Precincts

The city-wide research on attractions, activities and accommodation shows that these phenomena tend to be clustered in particular parts of cities, especially in or close to the city centre, and that some co-location occurs between these sectors and other destination elements, particularly transport. However, due to the scale and sectoral nature of this research these studies often lack detail on the structure of individual clusters and the co-location of other sectors such as restaurants and shopping. Except in some descriptive accounts of the historic city, little information in city-wide analyses is also generally provided on the local context, on the way in which tourism is inserted into the broader urban fabric which varies across the city. In contrast, research at a more localized level enables the structure and inter-sectoral relationships of particular areas within the city to be examined more closely. Various terms have been applied to such areas: tourist districts, tourist space, enclaves, bubbles and precincts. In some cases these terms are explicitly defined and/or identified by particular characteristics, in other cases the terms are used interchangeably.

Districts constitute an intermediate scale between the city as a whole and site-specific features such as individual attractions or buildings (Pearce, 2001). As Ashworth (1990) emphasizes, districts are constructs and thus may take different forms depending on whose perspective is taken. In the case of the Dutch city of Groningen these included that of the architect/historian; the legislative; the urban planner and manager; and the tourist historic city. The latter, according to Ashworth (1990, p. 151), is 'marked by selectivity and spatial concentration', a function of the visitor's time, mobility and concentration.

In one of the earliest reviews of tourism in cities, Burtenshaw *et al.* (1981, p. 172) drew attention to the spatial concentration of tourist activities and the concept of 'a "central tourist district" which occupies only a small proportion of the city's area but contains most of the tourist facilities.' Duhamel and Knafou (2007, p. 49) use the term Central Tourist District (CTD) to refer to the urban space of tourists and contend:

> Compared to the Central Business District, the CTD is not constituted by a single place and does not take a compact form: the central tourist space is generally much more spread out, even stretched out in a dominant direction. Moreover, in a metropolitan area such as Paris, the CTD is by definition discontinuous, being complemented by islands or satellites at a greater or lesser distance from the centre... the CTD is above all the area of recognised tourist practices made up of places to visit, stroll, buy, eat out, and in part a place to stay.

Building on Stansfield and Rickert's (1970) earlier coastal resort based concept of a Recreational Business District (see Chapter 8) and an empirical study of Niagara, Getz (1993, p. 597) developed a model of a Tourist Business District (TBD), the essence of which is:

> a synergistic relationship between CBD functions, tourist attractions, and essential services. Access into and within the TBD is critical, including safe and leisurely pedestrian access for tourists. The synergy must not only create a critical mass of attractions and services to encourage tourists to stay longer (preferably overnight), but must reinforce the image of a people-oriented place.

The emphasis here is on functions and the spatial structure of the TBD is not delineated.

In a similar vein, Judd (1993) emphasized the agglomerative nature of the components 'making up a tourist space', noting (p. 179) in his discussion of tourism in US cities that:

> Agglomeration economies apply to tourist districts not principally because concentration lowers costs or increases the efficiency of business transactions, but because a full panopoly of services and businesses is necessary to make the space maximally attractive to consumers of the tourist space.

In a subsequent paper Judd (2003, p. 27) elaborates on the enclavic nature of many of the urban redevelopment and revitalization projects which have incorporated a range of tourism elements:

> Especially in older cities, a tourist bubble took form around a cluster of facilities and amenities (new waterfronts, atrium hotels, festival malls, convention centres, sports stadiums, entertainment districts), a space or series of spaces segregated from the remainder of the city.

Such enclaves, he argues (p. 28), operate 'by enveloping visitors within an environment that floods their senses with the signs and symbols of consumption and play' but also by acting as 'total regulation' regimes directing their desire, consumption, time and movement. He acknowledges, however, that these enclaves formed but one part of the 'complex spatiality of urban tourism' which provides 'plenty of opportunity for visitors to escape the confines of enclosure' and (p. 25) and that 'when urban tourism is considered at the scale of the city, enclaves generally only capture some of the people, some of the time.'

Two other types of enclaves that occur in cities have been recognized in the literature: backpacker enclaves and religious enclaves (Cohen and Neal, 2012). The former are characterized by a clustering of small, low-budget guest houses or backpacker hostels catering to young, long-term, low-spending travellers and may be less bounded than other types of enclaves. Cohen and Neal (2012) provide a detailed example of religious tourist enclaves which they define (p. 572) as those that 'are based on the religious needs and preferences of observant tourists', in their case of Middle Eastern Muslim visitors to Bangkok.

Based on an analysis of the location and mix of firms belonging to the Orlando CVB (Convention and Visitors Bureau), Krakover and Wang (2008, p. 254) contend that the concentration of business-to-customer services in the theme park zone constitutes a very large bubble:

> Although certain sections of this area include nontourism businesses its focal points constitute places where tourists are most likely to meet other tourists not only at their hotels, restaurants, and attractions, but also at shopping malls and numerous outlet stores.

Business-to-business providers, such as those offering convention services, along with smaller-scale attractions and transportation businesses are located outside the theme park zone.

As a relatively recent activity in many cities, tourism is often overlain on urban districts distinguished by some more established identity, activities and form. Other types of tourist districts in addition to the historic city and the redevelopment districts described by Judd (2003) include the scared spaces found in pilgrimage cities with a long history of visitation like Santiago de Compostela, Lourdes and Jerusalem (Chadefaud, 1981; Schachar and Shoval, 1999; Castro Fernández *et al.*, 2016) and ethnic districts such as the Chinatowns and the Little Italys of American cities and the multi-ethnic districts promoted in Singapore (Yeoh and Huang, 1996; Chang, 1997). Initially these may have developed organically as tourist districts, subsequently being more actively promoted and managed as such by city authorities.

Pearce's (1998a) comparative case study of the spatial and functional structure of tourist districts in Paris focused on three districts whose underlying character, land-use and morphology differed:

1. the Île de la Cité, the historic core of the city, with well-defined physical limits;
2. Montmartre, a more peripheral, primarily residential quarter with a distinctive and colourful history and a place of religious significance; and
3. the Opéra quarter, a major upmarket shopping district.

Based on detailed land-use mapping, analysis of documents and key informant interviews, this research illustrated that in a large polycentric city tourism may develop in a range of contexts, that various types of tourist districts may emerge and that both similarities and differences might be observed in terms of spatial associations and functional linkages.

In each case a marked clustering of tourist-related services was evident and a certain level of synergy was apparent between the different sectors. This was especially the case with the emergence of tourist-oriented shops and services in close proximity to major attractions such as the Sacré-Coeur basilica in Montmartre and Notre Dame Cathedral on the Île de la Cité (Fig. 7.10) and with the linear concentration of similar businesses along the Rue de Rivoli. At the same time, the functional association between major and secondary attractions was not as strong as physical proximity alone might otherwise suggest. Particularly in the case of the Île de la Cité, and to a lesser extent in Montmartre, there was little spillover in demand as visitor numbers fall away sharply within a very short distance from the major nodes, respectively Notre Dame Cathedral and Sacré-Coeur (Fig. 7.1). This very pronounced distance decay factor might in large part be attributed to the nature of many of the city's historic and cultural attractions which were not purpose-built for tourism and whose location reflects some earlier or other extant function rather than the concentration of attractions appealing to the same or similar visitor segments. Entrance records suggest that the major visitor flows are more between the city's major attractions – the Notre Dame–Eiffel Tower–Louvre–Sacré-Coeur circuit – than between such nuclei and more immediately adjacent but less well known or frequented attractions such as the Sainte Chapelle and Conciergerie on the Île de la Cité. As a result, ease of accessibility between these major attractions is a key factor and transport nodes – coach parks for package tourists and metro stations for independent visitors – are key locations in each district. A varying degree of compatibility between tourism and other urban functions was found in the three areas, highlighting the significance of the underlying land use of each district and the role of historical factors. Tourism was seen to lie less easily in predominantly residential Montmartre, despite its Bohemian past, compared to the Opéra quarter where tourist-oriented shops and services appear to have been accommodated without conflict in the retail zone. In the Île de la Cité, well-entrenched administrative and other land-uses have limited ease of movement between tourist sites and restricted the spread of souvenir shops, cafés and restaurants.

Derek (2018) took the same approach in Warsaw and identified a similar pattern in the reconstructed Old Town. Attractions are scattered throughout the city but it is only in the Old Town that there is a co-location of tourism services and that a tourism precinct is identified. Over one-third of the services inventoried there were eating establishments, 31% were souvenir shops and 15% other tourist services such as museums, galleries, foreign exchange outlets and two hostels (but no hotels). These were clustered around the Old Town Market and adjoining streets. Marked differences in visitor numbers again suggested little functional association between attractions in the precinct and that flows of visitors occurred between the major attractions of the city.

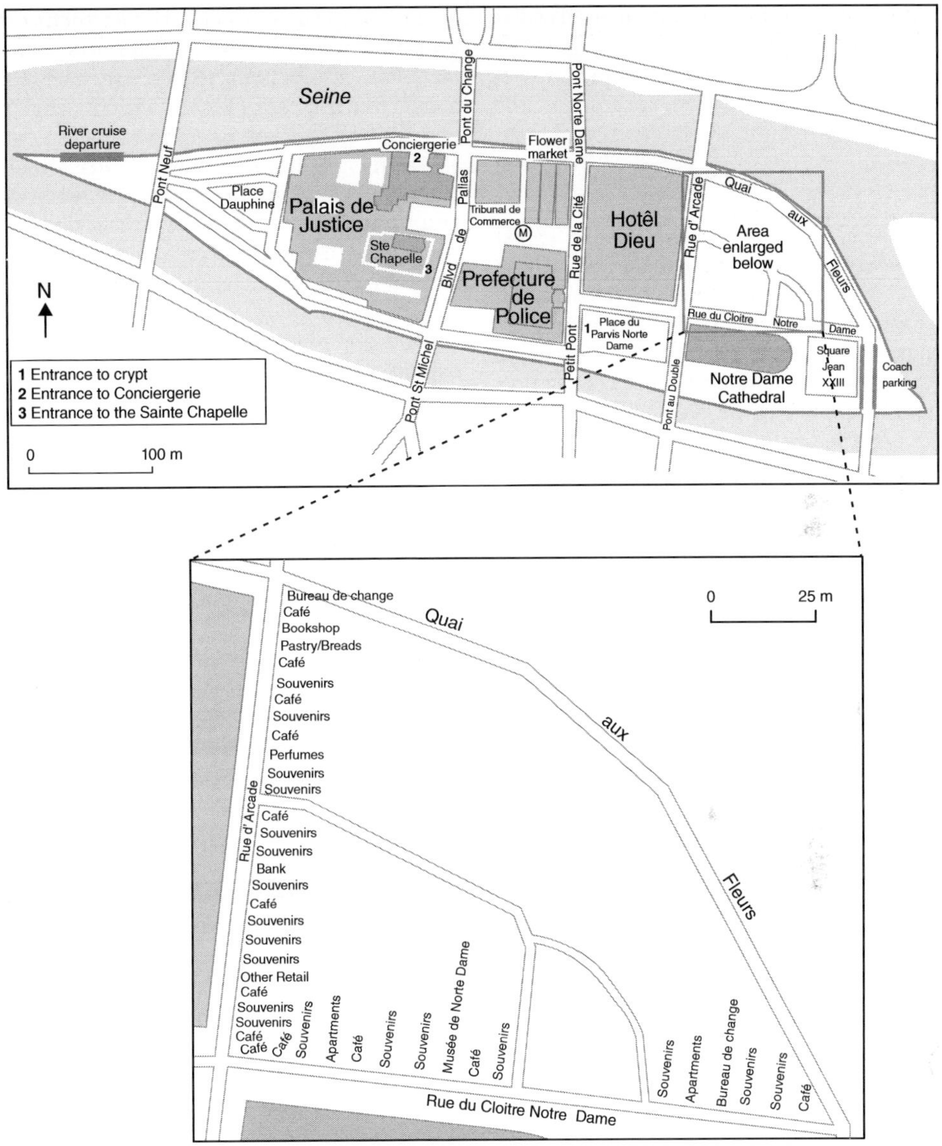

Fig. 7.10. Patterns of land use in the Île de la Cité, Paris. From Pearce (1998a), Figure 3.

The term 'tourism precinct' is analogous to tourist district, being defined by Hayllar and Griffin (2005, p. 517) as:

> A distinctive geographic area within a larger urban area, characterised by a concentration of tourist-related land uses, activities and visitation, with fairly definable boundaries. Such precincts generally possess a distinctive character by virtue of their mixture of activities and land uses, such as restaurants, attractions and nightlife, their physical or architectural fabric, especially the dominance of historic buildings, or their connection to a particular cultural or ethnic group within the city. Such characteristics also exist in combination.

Those using the term often take a more people-oriented approach compared to the supply-side emphasis of the tourist district studies (Hayllar *et al.*, 2008). Griffin *et al.* (2008) argue that precincts need to be considered in terms of the functions they perform for visitors and suggest these fall into three broad categories:

1. Facilitating functions: as easily negotiated spaces, precincts offer opportunities for visitors to meet up with others; to get their bearings and facilitate exploration; and to have short but satisfying experiences.
2. External or place-connecting functions: precincts are places which help tourists to relate to and understand 'the broader destination of the city, or even country, rather than just the precinct *per se*' (p. 50).
3. Internal or state of mind functions: precincts offer 'labile space' enabling urban tourists to satisfy such needs as freedom, novelty, escape from routine and social interaction.

Griffin *et al.* (2008) then offer a typology of tourism precincts based on the functions they might perform for visitors as meeting places, places of orientation, comfort zones, places of respite or refuge, play spaces, encounter zones, zones of intimacy, authenticity, distinctiveness and contrast. Moreover, most precincts would perform multiple functions. Their research on the experiences of precinct visitors identified three basic visitor types: explorers, browsers and samplers.

Organizational Configurations and Stakeholder Relationships

Agglomeration studies provide a spatial and economic perspective on the co-location of hotels but relatively little research has been undertaken which directly examines organizational configurations and relationships among tourism stakeholders in cities. The work that has been carried out has primarily taken a network perspective and focused on aspects of structure and behaviour.

In their study of sustainable urban tourism development, Timur and Getz (2008) used SNA to analyse the interconnectedness of stakeholders in Calgary, Victoria (British Columbia) and San Francisco. 'The goal of adapting the network perspective to tourism and hospitality industry', they state (p. 447), 'stems from the recognition that a destination is a system which consists of relations that are likely to influence destination stakeholders' opportunities, constraints, behaviour, or values.' The resultant graphs revealed differences in the network structures of the three cities. DMOs held a central position in each case, particularly in the Calgary network. Victoria had a more diverse group of central players which also included cultural attractions and hotels as well as recreational operators, motels, local institutions such as financial and consulting companies and the university. The network in San Francisco was sparser; there were fewer links between the different sectors and stakeholders and local government was perceived as a key player. As Timur and Getz (2008, p. 456) note, higher measures of centrality also reflect power in the network which they associated with 'access to or possession of critical resources.' This again varied from city to city. In Calgary the hotel sector was seen as an influential group; in Victoria stakeholders had strong ties to the cultural attractions sub-sector; and in San Francisco it was a broader group of attraction stakeholders.

While many urban DMOs are an extension of or are funded by local government, the public sector also has a wider role beyond supporting the marketing of cities. Given the importance of monuments, museums, art galleries and public open spaces, local government and different branches and agencies of national government may also play a major role in the management of urban attractions, in many cases not as their primary function (Pearce, 1998b). In addition, local government functions of land-use planning, regulation, provision of infrastructure and stimulation of economic growth mean city and town councils may play a significant role in shaping the spatial structure of urban destinations, whether by fostering the development of tourism or reacting to conflicts which its growth may bring (Pearce, 1998b, 2015b; Judd, 2003; López Palomeque, 2015). López Palomeque (2015) notes how the pressures and congestion which have resulted in recent years from the rapid expansion of tourism in Barcelona have brought new stakeholders into the picture, notably neighbourhood associations who have become increasingly active in making their voices heard by local authorities (Bruttomesso, 2018).

Pulido-Fernández and Merinero-Rodríguez's (2018) detailed comparative analysis of business relationships in 16 small-to-medium cities in inland Andalusia revealed differences in network structures between the cities and showed these were related to levels of tourism development as measured by a Tourism Index. The Tourism Index used is an official tax-based statistic reflecting the number of beds and annual occupancy rates in each city. This study is particularly valuable in that it systematically examines this relationship across 16 comparable cities in terms of three network indicators (density, centrality and intermediation) and four fields of tourism's 'productive process': planning and development; management; marketing/promotion; and distribution/vending. This is done with a series of graphs in which the three indicator values for each of the four parts of the process are plotted against the Tourism Index of the 16 cities.

Fig. 7.11. Evolution of the relational dynamics of stakeholders. From Pulido-Fernández and Merinero-Rodríguez (2018), Figure 6.

The results of this analysis showed a positive relationship between these network indicators and the level of tourism development across the 16 cities, leading Pulido-Fernández and Merinero-Rodríguez's (2018, p. 13) to conclude that 'when the operation of the stakeholder network improves, there is an increase in the level of tourism development at the destination.' This relationship was generally clearer in terms of planning and development and management. However, the importance of the four fields in the network structures varied among the cities depending on the level of tourism development (Fig. 7.11). Those with lower levels of development exhibited stronger ties associated with marketing/promotion whereas distribution/vending was more the focus of relationships in the more developed destinations. The authors interpreted this as a process of consolidation whereby stakeholders progressively shifted their relationship efforts as the destinations developed from marketing/promotion to planning and development, and then to management before concentrating on distribution/vending.

This study is a major advance in understanding network relationships within urban destinations and also raises some key issues which might be followed up in subsequent work. The authors identify a need to investigate the causes of different patterns of stakeholder behaviour and, secondly, to examine more broadly and in greater depth the links between network characteristics and levels of tourism development. In this latter regard, attention might also be directed at the direction of this relationship, that is, do stronger network ties result in more development or do stronger networks emerge in more established destinations?

The work of Stors and Baltes (2018) also raises issues of how multiple small hosts such as those listing rooms or properties on Airbnb and other such platforms fit into any destination stakeholder network. The types of networks associated with different functions also merits more research.

Strong networks do not necessarily exist in well-developed destinations. Gómez Diaz *et al.* (2017) set their case study of Oaxaca, Mexico, in the context of the evolution of clusters and then used SNA to examine relationships between a wide range of stakeholders in the city. These relationships were very weakly developed. The network density of those from the tourism industry (hotels, restaurants and travel agencies) was only 4.4%. Government agencies were the most central actors in the network. Gomez *et al.* concluded that in terms of development Oaxaca was at the consolidation stage of Butler's (1980) tourist area life cycle but as a tourism cluster the city was still in a gestation phase. This they attributed largely to the individualistic nature of the local tourism business owners.

In terms of distribution, Pearce *et al.* (2004, p. 208) concluded from their demand- and supply-side research in Wellington that 'we cannot speak ... of distribution channels for the city as such, but rather of distribution channels for the different segments and industry sectors that make up tourism at any urban destination.' The longest channels were found in the chain-owned hotels which had up to six layers of intermediaries between the individual property and the end user. Much of the hotel demand was from business travellers and corporate clients. The principal Australian tour operators tended to use only one or two hotels in the capital for their group tours but wholesaler catalogues offered a wider range of properties to cater for the more diverse individual traveller segment. Other forms of accommodation had simpler channel structures with many motels and motor inns depending more heavily on direct distribution to independent domestic travellers. Little 'bundling' of attractions took place as this sector largely adopted an 'at destination' approach to distribution whereby visitors, domestic or international, were targeted once they arrived in Wellington. Few attractions catered for the group tour market. Factors accounting for this included: the general absence of Wellington from many package tour itineraries; capacity issues which limit opportunities for coachloads of visitors; the specificity of the attractions, which does not lend itself to inclusion in a standardized series tour; the paucity of commissionable products; and short length of stay in the city.

Other research has sought to determine the factors that lead to the successful functioning of tourism networks. From their case study of a hotel marketing network in Östersund, Sweden, von Friedrichs Grängsjö and Gummesson (2006) identified a set of elements that built the social capital among the hoteliers that led to success for the group (Fig. 7.12). This includes basic principles of how to participate, the recognition of the need to balance individual and group interests and a seven-point code of conduct. A drive for action was found to be

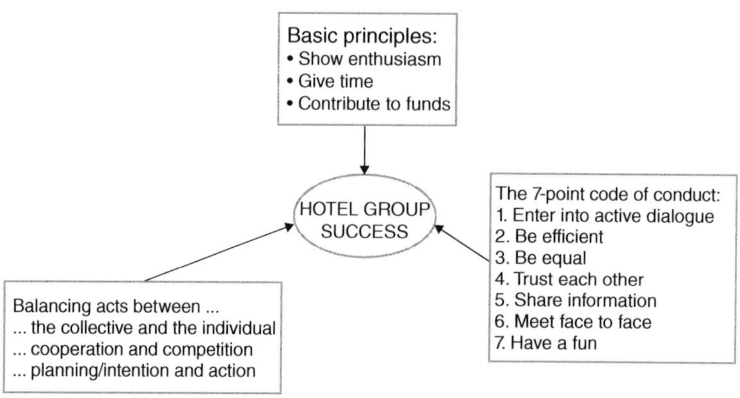

Fig. 7.12. Elements that build social capital with ensuing success for the hotel group in Östersund, Sweden. From von Friedrichs Grängsjö and Gummesson (2006), Figure 2.

more critical than plans and expressed intentions. In d'Angella and Go's (2009) study of Barcelona and Vienna, development of social capital through participation in the DMO network was also found to be a key factor in establishing strong relationships between the DMO and tourism firms. Social capital increased trust which in turn facilitated collaboration. Other facilitating factors were formal contracts, the firms' perception that mutual rewards through interactive engagement outweighed risks, especially financial risks, and the assessment that the DMO was effective in achieving strategic goals. Turisme de Barcelone strengthened relationships through an inclusive new product development strategy while Vienna's DMO fostered firm participation in decision-making processes. A lack of support from government agencies was identified as a key factor limiting cooperation among hotels in Brasilia (Miranda Júnior *et al.*, 2016). Government support was seen to be a way to reduce the costs of coordination and limit opportunistic behaviour. The modest levels of cooperation that did occur took the form of referrals and the informal and formal exchange of information. Coopetition, the ability to balance cooperation and competition, was the key factor which led to differences in the degree of integrated management in Foz do Iguaçu and Curitiba, Brazil (Chim-Miki *et al.*, 2019). Daskalopoulou and Petrou (2009) considered as successful performance of tourism firms in Petra, Greece, in terms of technical and scale efficiency and found there was much scope for improvement. The links between networking and successful performance were complex and varied. Positive effects resulted from cooperation among businesses and permanent residence in the area while participation in socio-cultural networks had a negative impact on business performance. Levels of technical efficiency improved where firms sought extra-local technical advice and support.

Discussion and Conclusions

A considerable amount of increasingly focused research using a variety of approaches and techniques has been undertaken on the structure of urban destinations in the past two decades. Increased specialization has led to a more detailed understanding of particular issues but there has been relatively little synthesis across topics and there is still a place for more descriptive but integrative accounts, such as that by López Palomeque (2015) on Barcelona. The research undertaken has been characterized by an imbalance of studies on spatial structure relative to organizational matters and an emphasis on tourists and their activities compared to other stakeholders and the relationships between these.

The spatial structure studies of attractions and accommodation show that tourism is concentrated in particular parts of the city, especially the centre; that distribution patterns vary from city to city and that changes are emerging with development of new forms of accommodation such as Airbnb; and the desire among some segments for more banal experiences than visiting iconic landmarks. The connections between attractions are being progressively revealed through the use of surveys and the application of new technologies to analyse the activity patterns of leisure visitors. These studies and those on tourist districts highlight the links between major attractions rather than between major attractions and nearby secondary ones. Much less is known, however, about how business travellers and other visitors spend their time and where they go. Surveys show location is a key factor for both leisure and business travellers in the selection of hotel accommodation but more could be done to establish which specific locational features are most important to them and especially to those making decisions about where hotels and other forms of accommodation are to be located. The distribution of hotels is largely explained indirectly and in aggregate, whether in morphological terms or in economic ones relating to bid rents and externalities associated with agglomeration. Much scope exists to complement such research with studies on how decisions about the location of individual hotels have been made.

Morphology is emerging as a critical factor in a number of areas, not just with regard to the distribution of attractions and accommodation but also in the way in which city form influences how visitors perceive and navigate the city. A key reason for this is that cities are multi-functional by nature and in general tourism is a comparatively recent function and an activity which is overlain on the existing urban fabric; that is, the spatial structure of urban destinations is in large part influenced by what has gone before. However, tourism has become a significant activity in particular districts or precincts whose morphology and functions are now being shaped by it. The co-location of tourism-related businesses is a feature of such areas but relatively little is known about the way in which businesses there or across the city as a whole interact with each other or with other stakeholders, especially public-sector actors. Network analysis is beginning to reveal some useful insights into these relationships but much remains to be done. The approach taken by Pulido-Fernández and Merinero-Rodríguez (2018) is particularly useful in this regard. With the exception of the network analyses and the externalities related to agglomeration, the work on urban destinations has not been cast in terms of the frameworks presented in Chapter 3. Nor has there been any attempt to explore empirically how the multiple flows and different nodal functions of destinations discussed in Chapter 2 impact on the spatial and organizational structure of urban destinations.

8 Structure of Coastal Resorts

The distinction between coastal cities and coastal resorts is not a sharp and distinct one but rather a matter of gradation. Due to their location, cities which lie on the coast may have a mix of urban functions, beaches and harbours which attract visitors in varying degrees. Survey data for holiday or leisure visitors in Sydney, for example, show they engaged in a diverse range of activities: 61.8% ate out or dined at a restaurant or café; 24.6% went shopping for pleasure; 22.7% went sightseeing or looked around; 16.2% visited friends and relatives; and 14.2% went to the beach (Destination NSW, 2017). As a result, the spatial structure of tourism in such cities may exhibit patterns influenced by factors associated with urban areas on the one hand and coastal resorts on the other. Hotels in the three South African coastal cities of Cape Town, Durban and Port Elizabeth, for example, are clustered in the CBD, at beachfront locales and other 'nice' suburban locations depending on the particular urban revitalization projects carried out and the importance of tourism in each city (Rogerson, 2012). Cao *et al.* (2016, p. 124) examined the distribution of floor area ratios in the Chinese city of Haikou and found that 'the newly developed mono-centric "city" model along the coast superposes on the previous one around the city centre as Haikou expands from a pure production city to a hybrid production and consumption city.' The focus in this chapter is on those destinations where tourism is a dominant function and the beach and coastal waters are a prime attraction. Their morphology is considered first and then their organizational structure is examined.

Morphology

Much research on destination morphology has focused on the form and function of coastal resorts, initially in Europe and North America and later in Asia and other parts of the world as the demand from domestic and international tourist has increased (Pearce, 1995a; Liu and Wall, 2009). As discussed in Chapters 6 and 7 and as other reviews of have shown (Pearce, 1995a; Bramwell, 2004), coastal tourism is not uniform in the way it has developed and the form it takes. Nevertheless, repeated commonalities are to be found in the spatial arrangement of basic structural features which demonstrate a close relationship between attractions, accommodation, transport and supporting services. Variations to general patterns occur from resort to resort as the result of the interplay of factors such as the physical setting, patterns of demand and the degree of planning.

Attractions, accommodation and supporting services

Proximity to the main attraction, the beach or seashore, is the dominant influence of the morphology of coastal resorts. Time-budget studies of holiday makers in coastal resorts show they spend between one-fifth and one-quarter of their non-sleeping time at the beach (Gaviría *et al.*, 1975; Hottola, 1992). A significant amount of time is also spent in and around their accommodation or in the resort or city. However, Bujosa *et al.* (2015, p. 79) stress that sun-and-beach tourists are not all alike in their spatial behaviour:

> the push effect [towards beaches] is more intense in locations characterized on average by a lower quality of accommodation establishments, which attract younger tourists with an education level under the mean, using means of transport different from the car, spending

> less money and fewer days in Mallorca, and lodging, in general, in establishments other than hotels. As a consequence, the beach-trip generation capacity is biased towards repeated, shorter trips to the closest beaches.

Much of the distinctiveness of coastal resorts, as Stansfield (1969) pointed out, is due to growth outwards from a coastal core being limited to approximately 180° as opposed to approximately a full 360° in most urban centres. The basic structure is linear in form and typically consists of a parallel association of the beach and maybe port, a promenade, road or railway, and a first line of buildings comprised of the densest and most expensive forms of accommodation and a core of tourist-related shops, bars and restaurants. A gradation of accommodation in terms of height, density and price occurs behind this seafront or *front de mer* as high-priced hotels and high-rise apartments give way to smaller and less expensive hotels and apartments, pensions, villas and camping grounds which merge with residential accommodation and other urban functions.

The gradation in land use and density away from the beach is in large part a response to economic forces, the land nearest the attraction commands the highest rent and thus generates a more intensive form of land use. After mapping major clusters of hotels on Majorca and Ibiza, Cirer-Costa (2013a) showed they were all linear in form, close to the beach and did not extend more than 200 to 300 m inland. The seafront hotels were on average larger and had a higher star rating than those located further back. 'Tourists', he observed (p. 17), 'want to be able to walk to the beach and they value the seafront location very highly. Well-located establishments can aspire to attract customers with higher purchasing power, because their product has higher added value.' This fact was well understood by early entrepreneurs and is reflected in the rectangular street layout of planned resorts in 19th- and early 20th-century resorts in Europe and North America, though in contrasting forms. Burnet (1963) discusses the 'rational' pattern of the speculative developments at Deauville and Le Touquet in France. In the latter, six arteries parallel to the shore resulted in a *front de mer* that stretched for more than 2 km with no villa more than 50 m from the beach. Stansfield (1978, p. 243) noted with regard to the street pattern of Atlantic City, New Jersey that: 'the simple right-angled grid pattern with its long, narrow blocks at right angles to the seafront ... [stretched] the desirable location boast, "one block to the beach".'

Various studies have measured different locational aspects of beach resort hotels, primarily through hedonic pricing. Cirer-Costa's (2013b) study showed rooms in Majorca with a sea view were 15% to 20% more expensive than others inside the same hotel. Alegre *et al.* (2013) calculated that having a room with a sea view increased the price for German packages on the island by 8.84% and for British ones by 6.85%. In a Mediterranean-wide study Fleischer (2012) showed prices for hotel rooms with a sea view were about 10% higher than those without. Latinopoulos (2018) found that a sea view corresponded on average to 4.85% of the room rate in Halkidiki, Greece but this effect showed considerable spatial variability, having no impact in some parts of the destination and double this rate in others.

Alegre *et al.* (2013) also reported a negative relationship between prices and the distance to the beach for German and British package holidays. For the British a positive relationship between prices and the distance to the centre of the tourist resort was observed but this relationship was negative in the case of the Germans. Distance from the airport was the only area-related attribute considered that proved relevant with a negative effect on prices being identified. In their analysis of a large sample of coastal hotels in Catalonia, Rigall-I-Torrent *et al.* (2011) found that a beach frontage increased hotel prices by 13% to 17%. Hotels close to a beach which had been awarded a Blue Flag had prices about 11.5% higher, everything else being held constant. Negative correlations were also observed between the prices of hotels and beach width, natural beaches and coarse sand beaches.

Proximity of hotels and the sea is taken to the extreme with the development of overwater bungalows, a form originating in French

Polynesia and now found in other tropical islands, such as the Maldives, where beachfront bungalows in the form of traditional *fares* have progressively been complemented by others on stilts built out into the lagoon (Fig. 8.1). According to Blondy (2016, p. 7) these bungalows are 'an expectation of tourists, a symbol of the direct, private, personal, even intimate, link with the lagoon' as well as a function of maritime concessions being easier to obtain than land due to the communal nature of land ownership there. Blondy shows how such structures have evolved on Bora Bora, extending first from the shores of the main island and then being developed on much smaller islands (*motus*) out in the lagoon. Later, canals were dug into the *motus* to open up new beachfront areas for accommodation. By 2012 almost half the hotel rooms in Bora Bora consisted of over-water bungalows.

Hotels, of course, are not the only form of accommodation in many coastal resorts, nor even the most dominant one. Second homes in various forms such as apartments or individual villas are common in many places and proximity to the beach remains a key determinant of resort structure. This is the case in many Brazilian resorts, such as Balneário Camboriú (Fig. 8.2), where real-estate speculation has resulted in the massive development of holiday homes in the form of dense, high-rise apartments fronting on to the central beach where prices per square metre are reported to be the highest in the country (Soares *et al.*, 2016). Such developments are not readily quantified due to the mix of permanent residences, apartments used as holiday homes by their owners and those that are rented out commercially. Around the turn of the present century Balneário Camboriú had a total of 20,000 beds in 110 mostly privately owned hotels and inns, with some 15,000 properties being rented out through 350 agencies (Costa *et al.*, 2006). However, the number of private holiday homes is not given. Using accommodation criteria, Hernández-Martín *et al.* (2016) identified nine micro-destinations in two municipalities in the south of Tenerife (Canary Islands). Significant differences occurred between them with regard to the number of apartments and hotel beds and the presence of permanent residents. Differences were also observed between the micro-destinations in terms of such demand factors as occupancy rates in apartments, expenditure, use of all-inclusive packages and country of origin. Along the coast of Catalonia, Campillo-Besses *et al.* (2004, p. 228) observed that:

> Unlike mass tourism development, which focused on a relatively small number of

Fig. 8.1. Over-water bungalows at the Intercontinental, Moorea: 'an expectation of tourists, a symbol of the direct, private, personal, even intimate, link with the lagoon' (Blondy, 2016, p. 7). Photo: the author.

Fig. 8.2. Balneário Camboriú, Brazil: 'real estate speculation has resulted in the massive development in the form of dense, high rise apartments fronting on to the central beach.' Photo: the author.

> destinations, second homes sprung up all along the coast, occupying cliff tops and steep slopes as well as beach locations and tagging onto the outskirts of existing towns.

Close examination of the beach zone reveals its more detailed structure and the different ways in which this may be interpreted. Figure 8.3 depicts a cross-section of the main beach zone in Pattaya, a major mature resort in Thailand, comprised of five parallel strips: (i) sea; (ii) beach; (iii) promenade; (iv) beach road; and (v) beachfront buildings and footpath (Longjit and Pearce, 2013). Each of these represents a range of different, often multiple and sometimes conflicting activities. The sea is variously used by bathers, holiday makers riding banana boats or taking trips to offshore islands. The beach is a place to sunbathe but it is also a strip where umbrellas and deck chairs are rented; where peddlers sell drinks and snacks. The promenade enables pedestrian access the length of the beach but peddlers also do business here and prostitutes may ply their trade in the evening.

Pearce and Kirk (1986) identified different forms of carrying capacity associated with specific resort zones (Fig. 8.4). In this model the dunes are represented as a zone of transition between the accommodation and service-sector zone and the beach but what is at issue is the environmental carrying capacity of the dune system, typified by their resistance to erosion through trampling, compared to the perceptual carrying capacity of the beach as measured by perceptions of crowding. The implications of this from a destination management perspective are that the assessment of the carrying capacity of any resort is more complex than might at first be supposed given the different environmental, perceptual and physical thresholds which might exist as one moves from the coastal waters, across the beach and dunes to the immediate hinterland; the corresponding range of assessment procedures required; and potentially divided administrative responsibilities for each of these zones.

Differences in physical site characteristics mean that not all of the features shown in Fig. 8.4 will be found in every resort. Some, as is the case with Pattaya, lack a dune system. Dewailly (1990) distinguished between the cross-sectional profiles of English, Belgian and Dutch resorts on the basis respectively of cliffs, developed dunes and protected dunes. Where one or more of the 3Ss (sun, sand and sea) is less attractive or reliable, promenades tend to be more significant, for example along the Belgian coast.

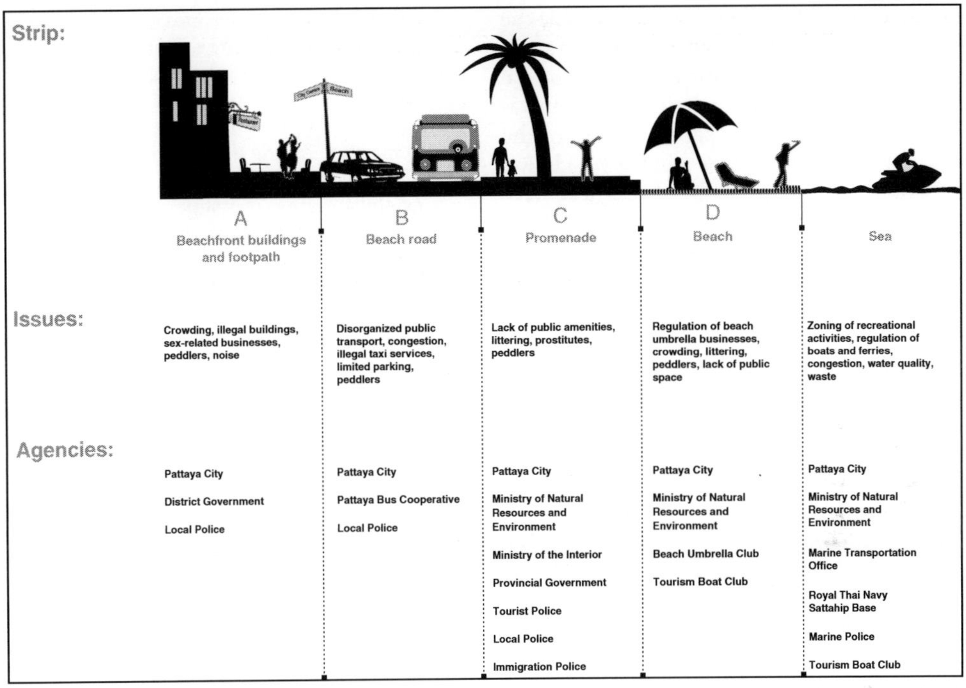

Fig. 8.3. Structure of the Pattaya beach zone: issues and agencies. From Longjit and Pearce (2013), Figure 3.

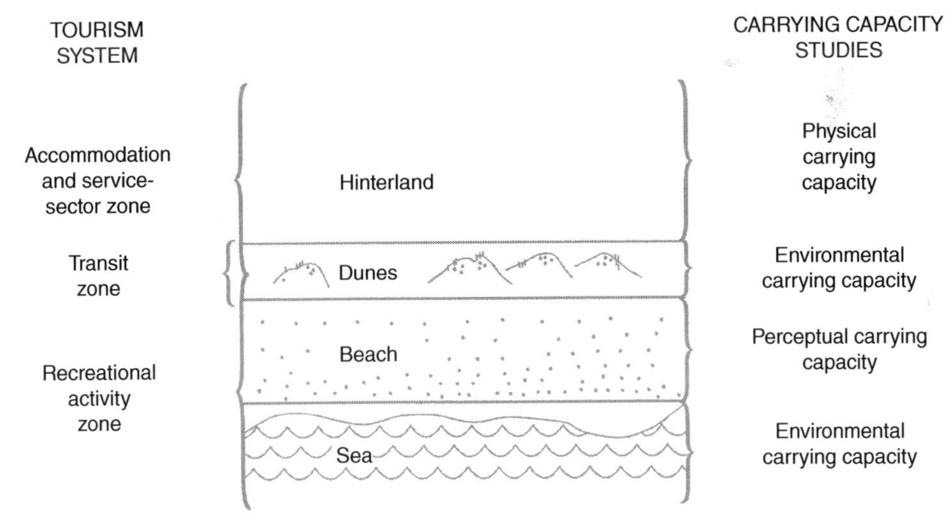

Fig. 8.4. Schematic representation of the spatial relationships between elements of the tourism system, the coastal environment and carrying capacity. From Pearce and Kirk (1986), Figure 1.

The renowned promenades of the French Riviera developed at a time when the winter season was the most important. Physical site characteristics may also influence the architecture of beachfront hotels (Pie i Ninot, 2013). Hotels on large sites tend to be built parallel to the beach; those on more limited sites often take a U form; and fish-bone structures at right-angles to the beach enable the most to be made of the most restricted sites.

Early studies of the morphology of coastal resorts identified a distinctive concentration of complementary services close to the point of arrival of visitors and the beach. Reflecting their beachfront location these were termed 'frontal amenities' by Barrett (1958) and the Recreational Business District (RBD) by Stansfield (1969) to distinguish them from the general commercial functions of the Central Business Districts (CBD). The more widely adopted term, RBD, was defined by Stansfield and Rickert (1970, p. 215) as 'the seasonally oriented linear aggregation of restaurants, various specialty food stands, candy stores and a varied array of novelty and souvenir shops which cater to visitors' leisurely shopping needs.' In their examples of two New Jersey seaside resorts the RBD is found along the boardwalk with the CBD and a major shopping thoroughfare located several blocks further back. Meyer-Arendt (1990) explains the location of RBDs in terms of tourist-oriented businesses capitalizing on the high volume of tourist throughflow. Stansfield and Rickert (1970, p. 219) also observed that 'the RBD is a social phenomenon as well an economic one.' In many Asian beach resorts the RBD often takes the form of shops and stalls, restaurants, rental-car agencies and other tourist-oriented businesses lining both sides of the street one block back from the beach behind the hotels (Pearce, 1995a). This may be complemented by hawker stalls selling souvenirs flanking one or more passage ways leading from this street on to the beach itself, as at Sanur and Kuta in Bali.

Cirer-Costa (2013a, p. 24) stresses the importance of these complementary services which are needed to complete a positive tourism experience in a coastal resort:

> Most tourists come in search of new experiences: they want to be able to walk around and shop, to eat and drink new things in new places, above all, they want to enjoy themselves inside a protected, relaxing environment, populated by other people with the same interests, even if they are of a different nationality or socioeconomic status. Only a dense web of small independent businesses can guarantee the existence of an offer diverse enough to satisfy all the clients, and only a large area with thousands of accommodation units can generate the demand to fill the theme parks, discotheques and other specialized leisure establishments. This superposition of business networks and clusters is the reason for the success of the tourist sector in the Balearic Islands.

Beaches are not the only attraction in many coastal resorts. The morphology of coastal resorts may also be shaped by the location of other features and recreational facilities and activities. Early coastal resorts in Europe often developed around small fishing ports with sites overlooking the port being valued as much for their charm as for their use for recreational boating. With the rise in recreational boating, marinas have been used to develop a second focus of activity to the beach, either to generate new demand for accommodation in existing centres or to spread prime real estate to sites in new centres. In Hawaii, the Costa del Sol and other places, golf courses have been used to develop resorts in greater depth, a pattern also found, for example, in some of the early French resorts (Burnet, 1963). Fairways are constructed behind the seafront hotels, creating an additional recreational attraction and a pleasant, open environment. The developer benefits from this strategy by enhancing the attractiveness of the hotels and broadening their clientele base, by reducing problems of seasonality, by direct returns from the golf course and, most significantly, by greatly enlarged real-estate opportunities, notably for the construction of villas and condominiums. Benidorm has seen substantial suburban development around the Terra Mitica theme park and associated developments (a golf course, upmarket hotel complex, Terra Natura, Agua Natura) but this has not lessened the influence of the traditional compact resort city as the driver of the destination's development (Soares *et al.*, 2016).

Nightlife activities may also play a major role in the appeal of coastal resorts and exhibit marked morphological characteristics. Such is the case in Pattaya which is renowned for its nightlife and has a reputation for sex tourism (Longjit and Pearce, 2013). Pattaya offers hundreds of beer bars, go-go bars, massage parlours, night clubs, cabarets and other forms of entertainment including mild-to-explicit sex-related services. Much of this activity is concentrated in Walking Street, a street which in the evening through to the early hours of the morning is reserved for pedestrians who flock to its many bars, restaurants, tailors, health and beauty shops, jewellery stores and souvenir shops (Fig. 8.5). It might be considered a less genteel version of an RBD. Clusters of other nightlife businesses, especially beer bars, are dispersed throughout the city, particularly along Beach Road (i.e. in

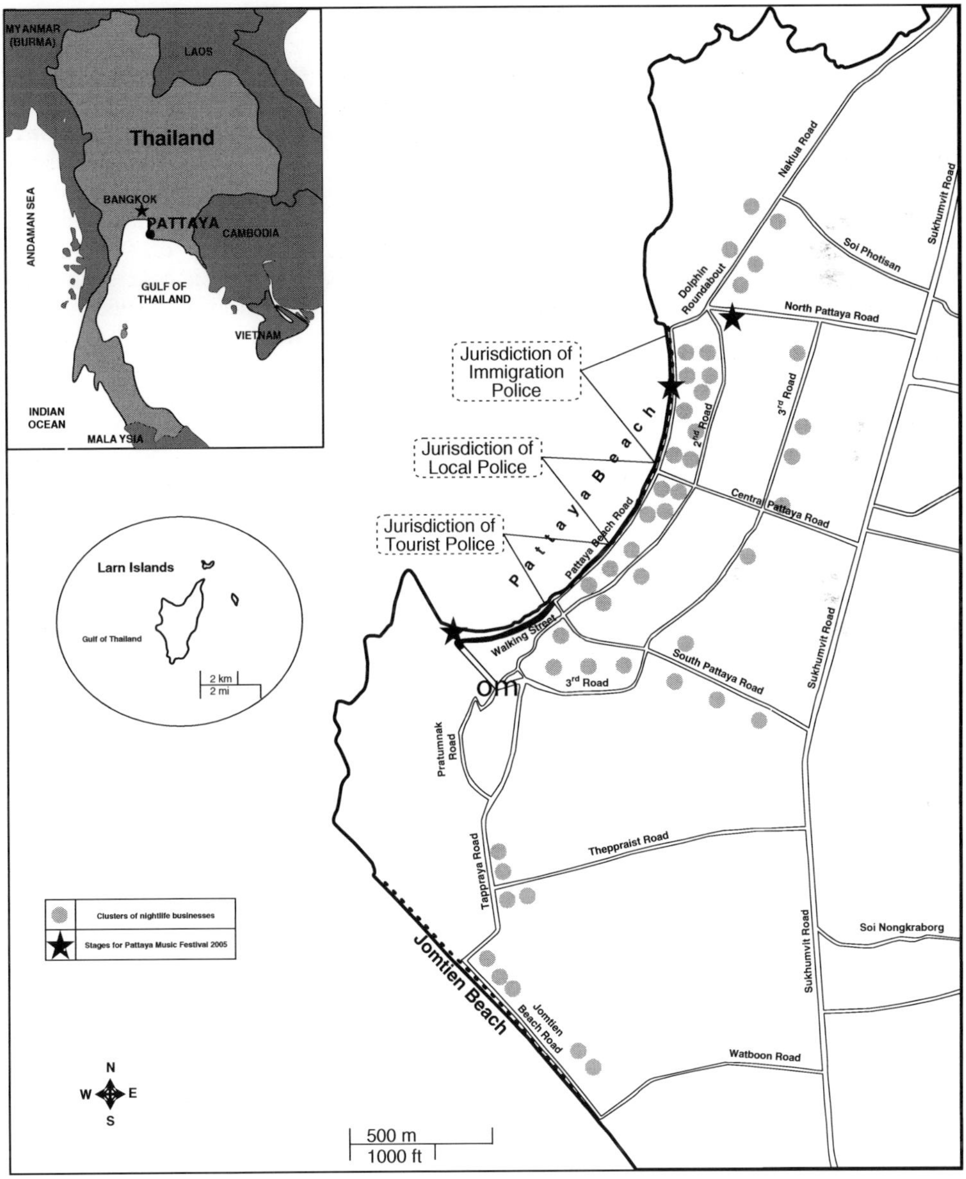

Fig. 8.5. Pattaya: location and distribution of nightlife business clusters and Pattaya Beach police jurisdictions. From Longjit and Pearce (2013), Figure 1.

strip A of the beach zone in Fig. 8.3) and a block back along 2nd Street as well as along other main roads in Central Pattaya, North Pattaya and in the Jomtien area. The clustering of such activities creates hustle and bustle which adds to the night-time atmosphere, provides choice and variety and leads to readily identifiable districts. As one of the city officials interviewed by Longjit and Pearce (2013, p. 171) commented: 'We do not need to promote sex related tourism; the tourists know by themselves which areas provide those kinds of services.'

Everitt *et al.* (2008) provide an interesting, broader-based approach to analysing the morphology of Puerto Vallarta, Mexico. First, they identified social areas in the city: the northern and southern coastal hotel zones and old centre were given over to 'mostly tourists' while other social groups clustered in distinct pockets throughout the city. They then mapped five tourism 'imprints' – hotels, the gay community, restaurants, art galleries and internet cafés – and found variations in where these were located. The distribution of hotels differed according to the type of hotel, ownership and clientele. Smaller traditional hotels were located away from the beach and attracted guests who could not afford to stay in expensive hotels or preferred not to whereas the large, modern chain hotels 'sell themselves in large part by being close to or adjacent to the high-rent beach areas, as well as often being "all inclusive" but at the same time *exclusive* of the local Mexican population' (Everitt *et al.*, 2008, pp. 94–95). The gay community was concentrated in the zona romantica, 'an "enclave' where the community has clustered together in order to maintain internal cohesion and provide mutual support for its members' (p. 96). Popular food restaurants catering to mass tourists were mostly close to the Malecón area of downtown while the higher-priced restaurants were further from the seashore.

Transport

The structural impact of transport on coastal resorts has long been recognized (Robinson, 1976; Stansfield, 1978). The advent of railways not only made coastal destinations accessible to a growing number of social classes but also resulted in early coastal resorts in England having a characteristic T-shape, 'the stem of the T being the main street leading from the railway station to the promenade' (Robinson (1976, p. 162). In Belgium, for similar reasons, Knokke and Blankenberg exhibit the same T-shape. The total reliance of Atlantic City, New Jersey, on railway access in its first six-to-seven decades resulted in a very compact pedestrian-oriented resort renowned for its seafront boardwalk and a dense and a well-ordered pattern of land use, which enabled visitors on foot unfamiliar with the place to readily find their way around (Stansfield, 1978).

The development of air charter tours in the Mediterranean from the 1960s and of air traffic to small islands in general has had a similar effect in delivering large numbers of visitors who subsequently become pedestrians once delivered by shuttle bus or taxi from the airport to their beach hotel. This factor has contributed significantly to the pattern described by Cirer-Costa (2013a) in the Balearic Islands and seen elsewhere where nearness on foot to the beach is paramount for tourists and hotel developers alike.

In between the rail and air eras the motor car had become dominant, especially along mainland coasts such as those in France or Spain, demand grew rapidly as the coast became much more accessible to a wider market. In established resorts the car often contributed to a lengthening of the *front de mer* or led to ribbon development along the coastal road in newer destinations. Coastal resorts along the Gulf of Mexico also developed a T-shape with the development of roads and causeways enabling access to beaches by car (Meyer-Arendt, 1990). Cars also facilitated development in depth of coastal destinations and opened up attractions in the hinterland to day visits and excursions. But within resorts cars also created traffic congestion and increased demand for parking. Railways and roads may act as barriers or hinder the free flow of foot traffic, especially if they lie between the accommodation and the beach, as in the case of regional or national roads and railways which followed the coastline and were already in place before the development of the destination.

Planned coastal developments

The explicit functionality of some of the early planned coastal resorts has been taken even further in some of those developed later in the 20th century. La Grande Motte was one of the master-planned resorts which formed part of the French government's ambitious programme in the 1960s to develop the coast of Languedoc-Roussillon for tourism (Pearce, 1978; Rieucau, 2000a, 2000b; Marez Lopez, 2013). The interplay of attractions, accommodation and transport is evident in the resultant morphology of the resort which is characterized by some innovative design principles but which also reflects the influence of demand factors and commercial realities experienced elsewhere.

Designed by Jean Balladur, La Grande Motte is oriented towards the beach with a major pleasure port acting as a central hub and zone of activity. Care has been taken with the external/internal road lay-out to facilitate access to the beach and port and to avoid some of the traffic congestion problems inherent in the parallel structure of the traditional *front de mer*. The major access road which links into the regional highway network runs parallel to the sea but at a distance of 200 m or so from the beach. Secondary roads terminating in car parks run perpendicular to this main road giving access to the beach. Such a 'comb' or 'glove' system removes traffic from the beach area while leaving it accessible from all parts of the resort. Landscaped paths encourage movement throughout La Grande Motte on foot or by cycle. The footpaths around the port flanked by the sidewalk cafés and restaurants assume the social functions of the traditional promenade which is absent from the beachfront itself.

Total bed capacity (initially 43,000 beds) was determined as a function of the capacity of the beach so as to limit overcrowding there. Distinct quartiers varying in type and density of accommodation were planned. To the south of the main access road, around the port and along the beachfront are the large pyramid-shaped apartment buildings which give La Grande Motte its distinctive image and identity (Fig. 8.6). The ground floors of these pyramids are often occupied by shops, restaurants and cafés. These apartment buildings become less dense and more conventional in form towards the interior of this quarter. Except for the Mercure, few of the hotels have been able to locate on the most prestigious sites. The centre of the resort is occupied by gardens, a recreational centre and administrative buildings. To the north of the access road villas separated the camping grounds in the northwest from a family holiday camp (subsequently closed) in the northeast.

The range of accommodation reflected the goal of making the resort accessible to diverse social groups but La Grande Motte's form and the distribution of the different quartiers were largely driven by economic factors similar to those seen elsewhere. After the bare land was acquired by the State, La Grande Motte was sub-divided and serviced by a group of local bodies who subsequently sold the developed lots to private enterprise for construction and sale. Although the costs of servicing were comparable throughout the resort, the price demanded for the sites varied according to the use to which it was to be put. In order to justify the higher prices for land for apartments, which were necessary to balance the development budget, the apartments were located on the prime beachfront and port sites, with the less dense forms of accommodation further inland. There was thus some spatial segregation of the different classes of holiday maker, although the patterns of circulation within the resort do reduce the actual effects of this. The high-rise pyramids also provided shelter for the relatively large green spaces created in their lee.

A second phase of construction in the 1980s saw the development of further apartments along the beachfront to the west of the port, primarily destined for holiday makers, while villas expanded in the northeast of La Grande Motte, many of which

Fig. 8.6. Pyramid-shaped apartment buildings characterize the architecture of the planned resort of La Grande Motte, France. The resort's layout facilitates access to the beach and protects the dune system. Photo: the author.

were occupied year-round by retirees or workers from nearby Montpellier. The attractions of the beach and port were complemented by the addition of a range of facilities: a casino (1969), convention centre (1983), a golf course (1986), spa centre (1988) and an aquatic centre (1993). These latter facilities supported expansion of housing in the northeast and some new hotel construction. Significant population increase has occurred. The resident population grew from 2,165 in 1975 to 8,546 in 2015. Visitation reaches a peak of around 120,000 in summer. These changes have progressively brought about a certain functional separation of La Grande Motte into a seasonal western zone primarily frequented by holiday residents in the summer and an eastern zone characterized by more year-round activity adjacent to the port, a range or shops and other services open 12 months of the year and a more significant permanent population to the northeast (Rieucau, 2000a, 2000b).

The 2017 land-use plan for La Grande Motte, the Projet d'Aménagement et de Développement Durables (PADD), makes provision for further changes but also stresses continuity with Balladur's key principles regarding the town's morphology (Ville de La Grande Motte, 2017). The PADD underlines the need to respect the distinctive architectural features of La Grande Motte and its landscaped open spaces; highlights the role of the port at the heart of the resort; and reinforces the importance of the beach zone and maintaining ready access to this. Provision is made for intensification of land use in the apartment zone, especially around the port which is to be expanded by 400 berths. Refurbishment and upgrading of the accommodation are also planned for to cater for growth of the permanent population and to respond to changes in demand from holiday makers.

Not all planned coastal resorts of course necessarily take the same form due to differences in the physical site and the ideas of the planners but their morphology nevertheless usually still reflects strong spatial relationships between the attractiveness of a beachfront, access and accommodation. This is clearly seen in those planned resorts developed on narrow sandy bars separating a lagoon from the sea as at Port Leucate-Bacarès (a sister project to La Grande Motte but with a different architect), La Manga del Mar Menor in Spain and Cancún in Mexico (Aldape Perez, 2013; Gómez de Salazar, 2013). In the first two development is very linear along two fronts, one facing the sea, the other overlooking the lagoon, with a central access road bisecting the resort. At Port Leucate-Bacarès the seaward side consists mainly of large high-rise apartments fronting directly on to the beach while marinas and small buildings on the water's edge characterize the zone overlooking the lagoon which is popular for water sports. Besides the central road, a canal running the length of the zone provides access by water and links ports at each end of the resort.

At Cancún, a *Centro Integralmente Planeado* or integrally planned tourism centre (Ambrosie, 2015) on the Caribbean coast of Mexico, there is a marked separation between the hotel zone on the coastal bar and the city where residential, commercial and other service functions are located. The access road, the Bulevar Kukuklán, follows the lagoon side of the bar with the accommodation, predominantly large five-star hotels, concentrated between it and the beach. Most of these hotels are U-shaped around a terrace of pools and bar, their height and U-shape maximizing the number of rooms with beach views. Aldape Perez (2013) criticized the absence of any form of promenade which might constitute a transversal link along the beachfront between the private built space of the hotels and the beach. Moreover, hotel construction and loss of the dunes have in places led to the degradation of the beach. While there are some commercial centres, museums, a convention centre and some restaurants along the hotel strip, the amount of complementary activities and facilities is relatively limited, something Aldape Perez (2013) largely attributed to the all-inclusive format of the hotels which essentially cater for all the guests' needs on-site. She also suggests more could be done to strengthen the urban character of the Bulevar Kukuklán to make it a zone of activity and intermixing.

Torres (2002, p. 100) argued that Cancún has evolved from an exclusive upscale resort into a tourist consumption space for the masses:

> It is the sheer numbers – the economies of scale and high concentration of so many large hotels – which contribute to the loss of exclusivity, rather than the class or quality of the hotels. Cancún's high-end hotels rival those of the most elite destinations, but, because of their large number and concentration, the price is driven down, allowing access to this typically exclusive 'class' by lower-end tourists.

Similarly, Murray (2007, p. 344) points to the increasing control of the market which has resulted in a move away from an exclusive resort aimed at the high end of the market to a 'mass-tourism model where large travel companies ... dominated travel by offering cheap charter/package deals' and the growth of the American spring-break market and that for winter holidays by middle- and even lower-income Americans.

Cancún has experienced difficulties in the residential zone due to the initial lack of provision for adequate residential housing and social services in a project which concentrated on the tourist zone and to the unforeseen scale of in-migration which the prospect of new jobs created, with the state government being unable to cope with the demand for basic services and infrastructure (Torres and Momsen, 2005a; Ambrosie, 2015). According to Murray (2007, p. 345): '[T]he more recent and more chaotic growth [including that along the Riviera Maya between Cancún and Tulum] can be seen as a unplanned extension of the original plan. It can also be seen in terms of a "transference" of control from FONATUR [the state tourism development agency] to market forces and a somewhat uneasy combination of regulation by three levels of Mexican government.'

Similar issues have arisen in other planned coastal resorts in Mexico. López-López *et al.* (2006) write of the social segregation of tourist space at Cabo San Lucas, characterized by a tourism section well-endowed with infrastructure and a residential section housing a local population of medium-to-low levels of income with few or no urban services. The main attractions of Cabo San Lucas are sport fishing and golf, the presence of which 'has led to a very specific use of the land that is manifested through spatial and economic segregation: first, these activities have a physical location that is "out of touch" with the local population and, second, the prohibitive costs of these activities ensure segregation' (López-López *et al.*, 2006, pp. 366–367).

Marinas are another form of coastal development whose functional form represents an attempt to create a particular holiday environment, in this case centred on recreational boating (Fig. 8.7). Marinas can take various forms; some just provide berths and associated boating facilities while others also incorporate various types of accommodation in a very functional arrangement (Jimenez Zurita, 2013). Port Camargue on the coast of Languedoc-Roussillon combines two forms, both of which integrate accommodation and attractions: a canal estate or *cité lacustre* and a basin overlooked by apartments (Pearce, 1978). The *cité lacustre* has been formed by creating a system of small islets on previously marshy ground. The extended length of quayside which this layout provides enables each of the dwelling units to have its own mooring berth. The dwellings separate pedestrian movement along the footpaths on the quayside from the vehicles which have access to the interior of each islet. In a similar fashion the apartment zone which surrounds the port complex is served by an interior access road. Such a plan, coupled with the design of the apartment blocks, allows the maximum of occupants not only an excellent view of the port or of the beach on the northern side, but also ready access on foot to them. In these ways the holiday maker has been brought into close contact with the attractions and an air of comparative calm has been created by the separation of vehicular, pedestrian and boat traffic. Similar principles have been applied in other destinations, such as on Denaru Island in Fiji (Xie *et al.*, 2013).

Organizational Structure of Coastal Resorts

Early research on structural aspects of mass tourism in coastal resorts in the Mediterranean

Fig. 8.7. Port Camargue, Languedoc-Roussillon, France. The functional form of marinas integrates accommodation and recreational boating. Copyright: iStock.

focused not so much on relationships within destinations but on external relationships through distribution channels. Researchers drew attention to issues of dependence on a handful of tour operators who dominated the package tour market and were able to control hotel prices and impose unfavourable conditions on hoteliers, especially in those areas with an over-supply of hotels (Buhalis, 2000b; Bramwell, 2004; Sharpley, 2004a). Conditions varied throughout the Mediterranean but this situation was particularly acute on small islands reliant on arrivals by air, such as Cyprus where one German tour operator controlled up to 30% of all arrivals (Sharpley, 2004a). In many respects these conditions were similar to those relating to the structure of tourism of island states in the Caribbean and the South Pacific where 'sun, sand, sea' tourism was developing (Hills and Lundgren, 1977; Britton, 1982). These studies, framed in dependency theory and core–periphery terms, pointed to the control exerted by metropolitan-based multi-national corporations over the international tourist industry.

The limited amount of research on coastal resorts that has been set in the industrial districts and clusters literature offers mixed results in terms of how they function and perform. These differences in part reflect the frameworks and approaches used as well as the characteristics of the resorts examined.

Marco-Lajara *et al.* (2014, 2016a, 2016b) framed their studies of the impact of agglomeration on various aspects of the performance of hotels in destinations along the Mediterranean coast of Spain in terms of industrial districts. Industrial districts were defined as Local Labour Systems (LLS) with LQs greater than 1 in terms of tourism employment in enterprises with less than 250 employees. In the first of these, 50 of the 53 LLSs for which relevant profitability data were available were deemed tourist districts by Marco-Lajara *et al.* (2014). This is not surprising given the concentration of tourism on the coast (Fig. 5.4) and that the denominator is employment at the national level. Hotels located in towns of greater agglomeration were found to be more profitable in terms of occupancy than those in towns of lower agglomeration, whereas the opposite resulted when hotel performance was measured in terms of ADR and RevPar. The authors attributed this effect to intra-destination competition whereby hotels lowered rates to increase occupancy, a practice leading to lower income. Using a similar procedure as in their previous study, Marco-Lajara *et al.* (2016a) identified 231 tourist municipalities belonging to 113 LLSs. Their results revealed negative and positive externalities related to the level of agglomeration. Increases in agglomeration were again associated with greater competition which induced price reductions and lowered incomes. In contrast, operating costs were shown to increase when levels of agglomeration were low but progressive cost reductions followed. greater levels of agglomeration. As a result, negative externalities are seen to exceed positive ones when levels of agglomeration are low but the opposite happens after a certain degree of agglomeration is reached, although the extent to which this occurs may be slight. Marco-Lajara *et al.* (2016a) concluded that their results supported the theory of tourist districts. In a related study, Marco-Lajara (2016b) examined externalities resulting from locations in which knowledge is obtained from other firms engaging in the same activity. Internal knowledge sources acquired over time and from labour mobility within tourist districts was shown to have the most impact on hotel profitability, along with knowledge acquired from universities and research centres. Although no rationale is given for their measure of employees' knowledge (the number of employees per room) they argued (p. 4860) that their findings 'confirm the fundamental premise about tourist districts.'

Studies drawing on surveys and qualitative data which have examined coastal resorts from the perspective of clusters have found little evidence to suggest that they function according to the cluster characteristics discussed in Chapter 3. In the case of Benidorm on Spain's Costa Blanca, Perles-Ribes *et al.* (2017, p. 1597) concluded that:

> the positive assessment of the concentration is not due so much to the

> marked predominance of competition between businesses (which stimulates continuous improvement and innovation) but to the conviction that complementarity between different companies of the value chain benefits the destination. Furthermore, neither cooperation between public or private agents ... nor shared goals exist, and both are necessary characteristics of a cluster. A more exhaustive analysis reveals the existence of a distinct individualism and a certain degree of secrecy when it comes to sharing sensitive or strategic information.

Innovation (p. 18) was largely in the form of 'incremental improvements of existing products/services rather than radically new ones.' The hotel and nightlife sectors were perceived to exert the most influence on the destination. Soares *et al.* (2016, p. 179) also refer to the incremental innovation which has occurred in the hotel sector in Benidorm and attribute this tendency to destination management characterized by 'a model of public–private coordination, more professional and less institutional', resulting from the creation of a specific entity, the Fundación Turismo Benidorm, supportive of the hotel sector.

Varisco (2004) found little joint strategic activity occurred among small firms in the Argentinian resort of Miramar, that with few exceptions no packages were developed there and marketed by wholesalers, and little innovation was evident. In Brazil the term *arranjos produtivos locais* (local productive arrangements) is analogous to that of clusters and has been used by Costa *et al.* (2012) to frame their investigation of the behaviour of businesses on Brazil's Costa dos Corais. This showed that although the majority of those businesses recognized the importance of cooperative relations the links between them were fragile, that there was no joint purchasing of goods and services, that few businesses sought help from others and that when they did so it was only to improve their own business rather than to benefit the destination as a whole. The only joint activity was commercializing their products with institutional support from the hotel association and SEBRAE, the national agency which supports medium and small businesses. Innovation was limited to increasing accommodation and not motivated by belonging to a cluster. Few traces of contractual relations, information exchange or collective action were evident in Balneário Camboriú (Costa *et al.*, 2006). The impact of the dominance of apartments in the resort was not explored.

Relationships in Cancún have been framed in different ways. Ambrosie (2015) focused on the establishment period of Cancún and set out to debunk what she as saw as the myth of the role of the state in the resort's creation by drawing on a wealth of archival material which she interpreted through the lenses of organizational theory institutionalization and fiscal sociology. She makes a strong case that 'Cancún was the brainchild of economics-trained central bankers inexperienced in tourism with a mandate to increase foreign revenue' (Ambrosie (2015, p. 65), that the initial years of the development project were chaotic and that the introduction of integrally planned resorts defied government economic development plans. Ambrosie argues that institutions are a combination of structures and agency and depicts how both these factors enabled the bankers to act in the way they did. Financing in the troubled period of the initial development, for example, involved both use of an infrequent common law public trust fund instrument and a situation whereby 'the tight Mexican banking alliance of técnicos implemented Cancún with relative autonomy ... now turning to their banking cronies to swap shares for Cancún's developed land' (Ambrosie, 2015, p. 68). She stresses (p. 77) that 'no major hotel chain invested in Cancún until the 1980s.' Moreover, Ambrosie (2015, p. 80) argues that some of the key features of Cancún have subsequently become institutionalized such that integrally planned resorts 'are now imprinted on Mexico's main tourism mechanisms' but 'rather than deterministic, linear or clean, the outcomes are a series of possibilities, some excluded and discarded while others are embraced and implemented.'

Torres and Momsen (2005b, p. 314) refer to the resort as Gringolandia, a name given to the place by locals, which they characterize as a 'dynamic "hybrid-space" in which elements of Mexican, American, and artificial Maya culture have been re-constituted for tourist consumption.' They examine Cancún's transnational space in terms of multi-scalar social relations which they depict (p. 321) as 'the continual flow of people, goods, capital, services, information, and ideas across international and national urban and rural borders that makes Cancún a truly transnational hybrid space' (Fig. 8.8). Figure 8.8 incorporates three interconnected transnational social fields: the international sphere, the urban Mexican sphere and the Quintana Roo rural periphery. The first of these refers to the flows of international tourists, second-home owners and retirees, together with foreign professionals, expatriate tourism-industry workers and Mexican elites with strong US ties. These flows are accompanied by other movements associated with foreign direct investment, franchise and management contracts, expatriation of profits and foreign exchange leakages associated with product imports. Half of the hotels in Cancún, representing almost three-quarters of all rooms, were under a franchise agreement, with 50% of those franchises being foreign. However, the majority of the hotels were Mexican-owned, primarily by firms or families in Mexico City, Merida and Monterrey. Such ownership and domestic franchises constitute a significant part of the urban Mexican sphere. Other components include immigration associated with various levels of tourism-industry-related employment; ownership of second homes and businesses by Mexican elites; imports of agricultural products and other goods and services via Mexico City; and the extraction of profits and taxes back to Mexican urban centres. The third sphere is comprised of less skilled seasonal workers from rural Quintana Roo; low volume, irregular seasonal provision of some basic agricultural products; and remittances of wages back to home villages.

The experience of master-planned coastal resorts in Languedoc-Roussillon and Cancún shows some critical organizational changes occur in the transition from development to destination management in such projects (Pearce, 2010). Initially, the development agency has a large if not total measure of control, it is able to enforce its plan and bring a unified sense of direction

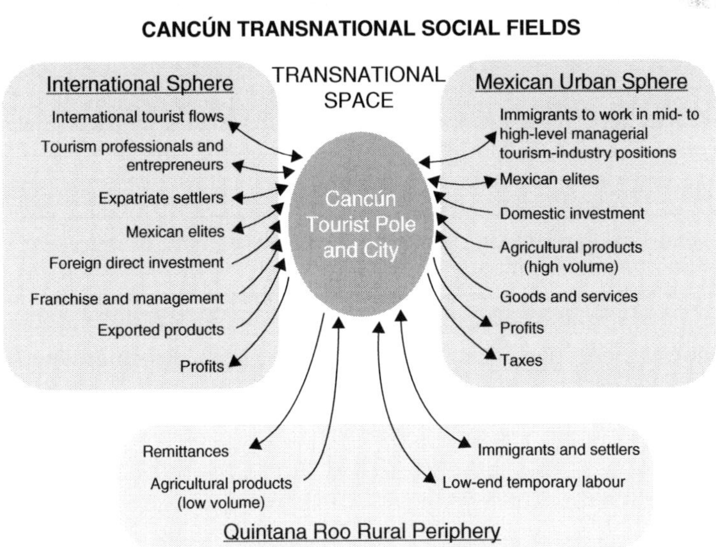

Fig. 8.8. Conceptualizing Gringolandia as transnational space: Cancún's social fields. Redrawn from Torres and Momsen (2005b), Figure 6.

through regulatory powers, the acquisition of land, as well as by funding infrastructure and possibly loans or incentives for construction. Subsequently the agency responsible for ensuring a comprehensive integrated approach to development is progressively replaced by an ever-broadening set of stakeholders: business owners and operators across a range of sectors; visitors, residents and workers; local, regional and national government agencies. Each of these groups of stakeholders has their own goals, interests and resources; some of these may be complementary, others may be in conflict. Property and resources are now redistributed to multiple owners who may be constrained by local authority rules and regulations but they will generally have a reasonable degree of freedom over how they operate. Accommodation providers, and other operators, for instance, are able to target the markets they feel fit and promote their businesses according to their own strategies. This in turn has implications for branding the destination. A second crucial change involves the funding regime. Over time capital development funding for infrastructure and other works which typically comes from central government or international aid agencies is replaced by the need for operational funding, particularly for the provision of utilities and public services. This will usually come from local authority rates or taxes and the budgets of government agencies. How well the local authority and other agencies are now funded, together with the powers that they have, will determine the level of services provided with regard to such essential areas as sewage disposal, water supply, public security and the management of protected areas. Inadequate ongoing resourcing and limited regulatory authority commonly create difficulties for effective management of a destination. In these regards, the transition from development to management has been more orderly in La Grande Motte than Cancún. The interdependence of the different sectors and actors which is seen to be an integral part of the planning and development phase does not stop there – it is a key to the successful management of a destination.

Drawing on responses from a range of stakeholders in Pattaya, Longjit and Pearce (2013, p. 169) defined destination management as 'the collaboration of relevant agencies responsible for providing multiple tourism products at the destination in a way to achieve common goals or destination goals.' When viewed in these terms they concluded that destination management was not being practised in Pattaya. Although a range of management activities were being undertaken, the agencies involved were generally carrying out their tasks independently in accordance with their broader institutional goals and responsibilities. Their activities may be complementary but the extent to which they are coordinated, integrated or collaborative is rather limited, most commonly to a series of formal or informal dyadic relationships. Tasks are differentiated but weakly integrated and spatial proximity alone does not necessarily ensure more integrated management. In strip A (Fig. 8.3), for example, there was no integration between Pattaya City which authorizes the construction of buildings for businesses and the district government which issues business licences. However, various forms of joint or collective management may occur. In most cases this happens through various dyadic relationships, some of which are more formal than others. The formal relationship between Pattaya City and the beach umbrella clubs (strip D) is one by which the city authorities regulate the clubs. The city and the clubs may also work together to deal with issues or problems which arise with individual umbrella businesses. The local police office is the main agency responsible for arresting and fining violators while the security staff of Pattaya City deals with peddlers (strip C). In the case of the night-time Safety and Prostitution Prevention Project for the Pattaya Beach promenade, different stretches of the promenade are policed under the respective jurisdictions of the immigration police, the local police and the tourist police (Fig. 8.5). Representatives from public agencies and local businesses form the Walking Street Committee which manages the daily activities of the street. While each has its own goals, management activities of the committee are directed towards a common goal, that of promoting the area as a major tourist attraction. In general, however, the different agencies managing aspects of tourism in Pattaya were not directing their efforts towards destination goals and there was no strategic approach linking them. Collaboration and cooperation appear to be more readily achieved where there is a more confined or specific management focus whether this is spatial (Walking Street), temporal (the Pattaya Music Festival) or issue based (beach security). Collective action in these areas contrasts markedly with the management (or lack thereof) of the more dispersed nightlife activity (Fig. 8.5).

Conclusions

A large number of studies have been undertaken on the morphology of coastal resorts, one of the earliest lines of research on the spatial structure of destinations. In part this is due to the worldwide popularity of coastal tourism but to a large extent interest in the topic might also be attributed to the distinctive morphology of such destinations which exhibits a close relationship between the resource base, the beach and sea, and the accommodation. It also reflects the influence of transport and the location of supporting services as well as local physical factors and other land uses. The attraction/accommodation relationship was recognized at an early stage, has been explored in a variety of contexts and continues to attract attention. More recent accounts go beyond description and employ hedonic pricing to measure the strength of this relationship. At the same time it is also a relationship which is taken for granted from the perspective of visitor behaviour as relatively little work other than some time-budget studies has been undertaken to analyse in detail how and where 'sun, sand and sea' tourists actually spend their time.

In comparison to the abundance of morphological studies, relatively little work has been carried out on the organizational structure of coastal resorts. There has been more interest in how coastal resorts have developed than in how they are subsequently organized and function. Research on their organizational configuration has primarily been framed in terms of agglomeration, industrial districts and clusters. The effects of agglomeration appear mixed and there is little evidence from the resorts studied that they are functioning as industrial districts or clusters in the manner discussed in Chapter 3. External relationships have been highlighted in some coastal destinations in the form of the control exerted by tour wholesalers over hotels and in the multi-scalar relations of Cancún's transnational space. The morphological studies have identified the importance of villas and private apartments but the inclusion of second-home owners and the role of the letting agencies which may rent out their properties have been neglected.

9 Structure of Ski Resorts and Rural and Natural Area Destinations

This chapter examines the structure of ski resorts and of rural and natural area destinations. In their general setting ski resorts share some characteristics with other destinations located in rural or natural areas. However, their resource base is more specific and their physical form and organizational structures tend to be quite distinctive. As a result, separate literature streams have developed dealing with ski resorts on the one hand and rural and natural area destinations on the other. Moreover, reflecting the diversity of rural tourism, the latter stream consists of a rather disparate set of studies. For these reasons ski resorts are dealt with first and the second part of the chapter provides examples of the spatial and organizational structure of rural and natural area destinations.

Ski Resorts

Research on the structure of ski resorts has examined both the morphological and organizational dimensions and to a lesser extent the relationships between the two. Key structuring factors for both dimensions have been the relationships between the ski lifts and trails – the basis of the attractions – and accommodation. Contextual factors have also played a major role, not only the physical site elements but also the economic and social contexts in which ski resorts have developed. Issues also arise relating to the spatial association of the resorts and the administrative areas in which they are located.

Morphology

French resorts

The morphology of ski resorts is to a large extent a function of the relationship between the location of the ski lifts and trails and the accommodation. This relationship is in turn influenced by the physical setting and existing patterns of settlement and land use (Préau, 1970; François *et al.*, 2012). The installation of the lifts is crucial for downhill skiing as the lifts provide access to the *pistes* or ski trails. Such access is further enhanced by proximity of the accommodation to the lifts. Lifts require substantial investment to develop and are costly to operate. The investment and operating costs are recouped in part by sales of lift passes. However, part of the investment in the ski lifts and their initial operation, in some cases a significant part, may be financed by the development and sale of accommodation (Vlès, 2010). Or, seen from another perspective, the opening up of a ski-field creates real-estate opportunities for developers (Fablet, 2013). These financial considerations strengthen the links between attractions and accommodation, not only in terms of the form and function of resorts but also with regard to issues of organization and governance. François *et al.* (2012) outline a series of steps for defining and mapping ski areas (*domaines skiables*) based on the spatial relationships of ski lifts and accommodation.

Préau (1968) considered a winter sports resort to be 'a complex organism' comprised of four sets of elements: (i) the site; (ii) the clientele; (iii) the plant (two distinct but complementary parts: accommodation and winter sports facilities, i.e. ski trails, lifts and other recreational facilities, notably skating rinks); and (iv) the local community. His fourfold typology of ski resorts in the French Alps was based on the site, the way in which winter sports were incorporated into the range of tourist activities and the demographic situation of the local population (e.g. declining or growing). 'The site', Préau (1968, p. 132) argued, 'constitutes a primary element in any typology: it alone triggers pressure from developers; its quality and capacity determine the characteristics of the resort; the local conditions dictate the relationships between the ski-field, the tourist accommodation zone and existing settlements'. His four categories, each with a distinct morphology, were:

1. Large resorts grafted on to sizeable communities which had experienced the growth of tourism in the summer months through sightseeing, climbing and *climatisme*; most notably Chamonix but also Megève, Morzine, St-Gervais and Villard-de-Lans. Here there was a clear geographical separation between the high-altitude ski-fields and the existing tourist accommodation in settlements on the valley floors. The ski-fields and settlements were linked together by high-capacity cable cars. Access from St-Gervais and Villard-de-Lans was more difficult.
2. Medium-sized resorts grafted on to large villages which had a similar history to those in the previous category but on a smaller scale. They had both a summer and a winter season and ski-fields at a lower altitude, generally between 900 m and 1200 m.
3. Purpose-built ski resorts created *ex-nihilo* higher up the mountains above existing settlements in locations determined by the physical characteristics of the site and snow conditions (e.g. Val d'Isère, Tignes, Serre-Chevalier, L'Alpe-d'Huez and Les Deux-Alpes). While variations occur these new ski resorts were characterized by having a high-altitude ski-field (a base at least at 1500 m) with accommodation at the very foot of the lifts and trails.
4. Secondary centres which had little tourism background; were at a lower altitude; were much more modestly equipped; and catered largely to beginners and weekend skiers. In some cases the small ski-fields were separate from the existing settlements.

Cumin (1970) distinguished between the large, purpose-built resorts in terms of their form and the extent to which they were planned. He identified three generations of resorts. A general lack of planning and coherence was the hallmark of the first generation (L'Alpe-d'Huez, Auron, La Clusaz). There was some overall concept in the second generation resorts (Courchevel, Chamrousse, Meribel) but these were nevertheless characterized by 'a group of small individual buildings stretched out in a slightly disorderly fashion along several kilometres of winding road' Cumin (1970, p. 50). The term '*station intégrée*' ('integrated resort') was applied to the then more recent third generation and referred to compact, coherent resorts undertaken by a single developer.

These large, purpose-built resorts were very much a product of their time and need to be seen in the context of the central government's Plan Neige which sought to stimulate a declining Alpine economy by developing skiing rapidly and on a large scale with State support, primarily with external capital and inputs (Préau, 1970; Wozniak, 2002; Achin and George-Marcelpoil, 2013). The model was one in which the State developed the access and a single large private developer built and operated the resort having obtained a concession to do so from local government.

Begun in the early 1960s, La Plagne in Savoy was the first of these 'integrated' resorts (Pearce, 1978). The heart of the resort is situated at 2000 m on a small shelf at the foot of an immense natural amphitheatre which serves as the main ski-field. By 1972 there were more than 5000 beds available, three-quarters of which were in studios or apartments and one-quarter in hotels or para-hotel establishments. The accommodation was compact in form, lying east/west. The resort is oriented north towards the *front de neige,* towards the zone of activity also known as the *grenouillère* (Fig. 9.1). In Cumin's (1970, p. 51) words, the *grenouillère* is 'the small, well-situated shelf receiving ample snowfall and sunshine, towards which the major pistes converge and from which the principal lifts radiate'. Consequently, skiers were able to leave their hotel or apartment building, already wearing their skis and immediately take a lift giving them access to much of the ski-field.

In terms of form, functionality was paramount. Accommodation separated the three means of mobility: on ski, on foot or by car. The access road terminated in car parks on the south side of the accommodation, skiers occupied the *grenouillère* to the north and

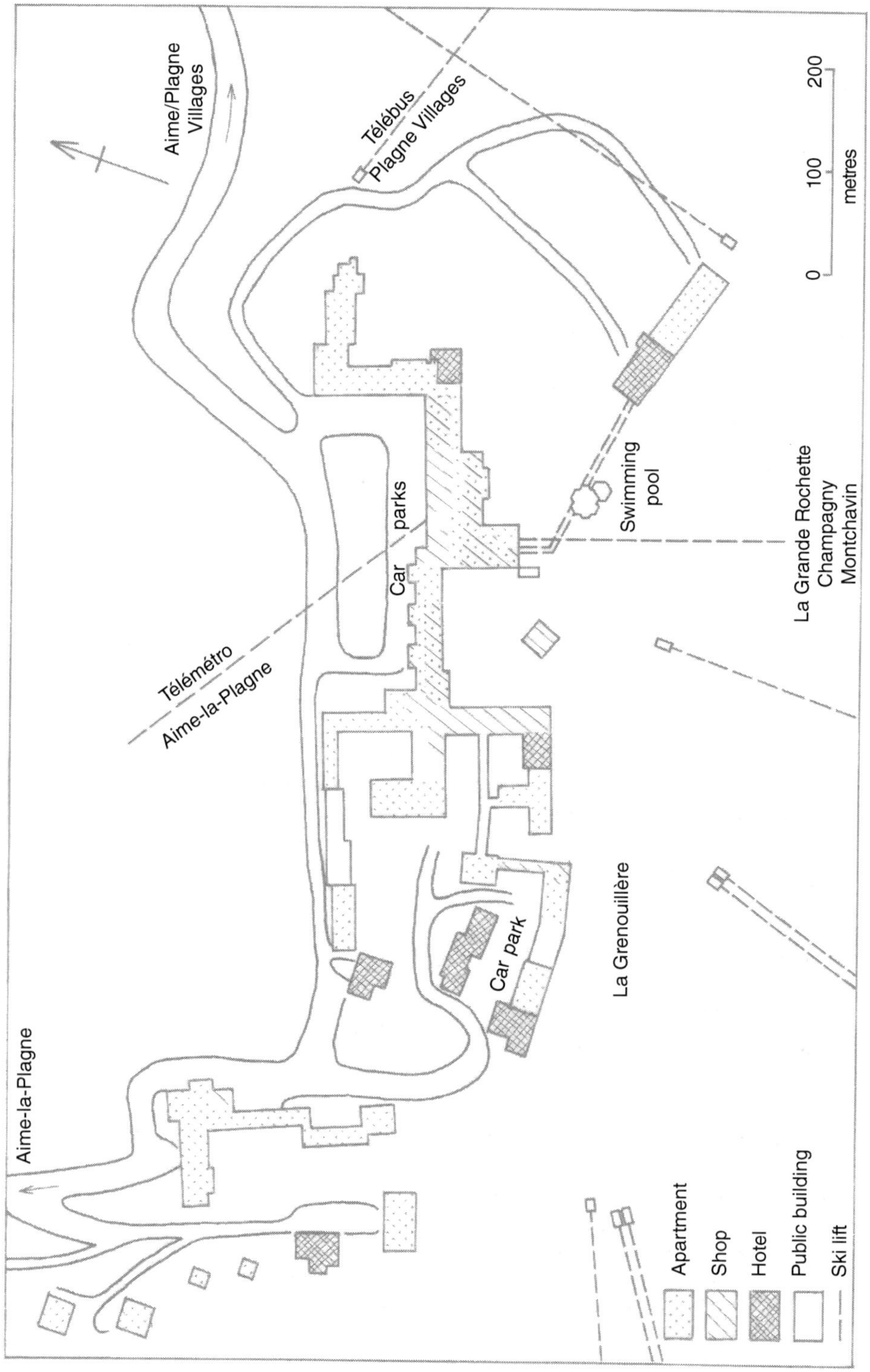

Fig. 9.1. La Plagne: morphology of an integrated ski resort. From Pearce (1978), Figure 5.

the compact form and a series of covered walk-ways and shopping galleries facilitated interior access by foot from one end of the resort to the other, an important consideration in view of the low temperatures experienced at 2000 m. Such linkages and lay-out brought about a veritable physical integration of the 30 or so buildings which constituted the resort. The concentration of buildings also led to more efficient central heating and easier management, limited the amount of land which had to be bought and lowered infrastructural costs. Furthermore, functionality by way of a balance between accommodation and lift capacity was built into the lease for La Plagne which specified that the hourly capacity of the lifts leaving the base of the resort had to be at least equal to two-fifths the number of beds available.

Subsequent expansion was by way of construction of smaller satellite resorts at a similar altitude (Aime 2000, Plagne Villages, Plagne Bellecôte, Belleplagne) and expansion of the lifts and pistes into adjoining basins. Extreme physical integration occurred at Aime 2000, a giant monolith capable of accommodating 2500 skiers in hotels and apartments and of housing under the same roof a full range of shops, restaurants, a cinema and ski school (Fig. 9.2). However, at Plagne Villages total integration was rejected in favour of a series of small units but a linear design was retained.

Functional integration was accompanied by financial integration as La Plagne and its high-altitude satellites were developed by the one company, the Paris-based Société d'Amenagement de La Plagne. However, a two-tier development process did take place as more modest and traditional projects involving small hotels, chalets and holiday camps and *gîtes* were undertaken around existing settlements lower down the mountain between 1000 and 1500 m at the initiative of the local communities (Longefoy, Montchavin, Champagny). These were connected to the upper ski basins by a series of linking lifts to form part of the overall complex of La Grande Plagne.

The functional form met with a strong response from a rapidly expanding market and was repeated in varying degrees at resorts

Fig. 9.2. Aime 2000: an integrated French ski resort. Copyright: Olivier Allamand.

such as Les Belleville and Les Arcs. However, the concept of integrated resorts was not without its critics, both at the time and later. The *front de neige* was criticized for its rigidity of form (Chappis, 1965). Vlès (2010, p. 41) is particularly damning of the architectural legacy of the 'functionalist paradigm' of the integrated resorts:

> There are not any real streets with a social life but rather routes for traffic, no meeting places but parking lots, 'products' and 'events' rather than cultural exchanges between summer holiday goers and residents. The integrated resorts are a line of buildings or sub-divisions which stretch out without a town centre.

Changing conditions led to a questioning of what a destination should be with regard to the smaller resorts at lower altitudes, those of the *moyenne montagne* (François and Billet, 2010; Achin and George-Marcelpoil, 2013). These resorts came under pressure as poor snow seasons, fragmentation of holiday taking and shorter stays, and local authority indebtedness created problems for a model based heavily on accommodation linked to ski-fields, both spatially and commercially. An initial response to upgrade lifts and increase artificial snow-making was followed by government policies and support to diversify products and clientele in order to spread demand more evenly throughout the year and reduce reliance on vulnerable ski-fields. Diversification included activities such as mountain biking, forest walks and nature appreciation, activities which could be done elsewhere. This in turn changes the scale and nature of the destinations, extending them beyond the villages immediately adjacent to the ski-fields to encompass a wider territory, for example, regional parks and other communes. The extent to which these changes have been implemented and visitor behaviour has evolved remains to be seen.

North American resorts

Variations on the village-based/purpose-built models of ski resort development occurred in North America (Dorward, 2006). Much of the initial growth of skiing there in the post-war period was based on existing mining towns such as Aspen, Breckenridge and Crested Butte or ranching towns like Steamboat Springs and Jackson, Wyoming where 'resort development has been laid over a rich and appealing frontier past, providing an ambience unavailable elsewhere in the world' (Dorward, 2006, p. 268). However, as pressures grew from the large numbers of visitors arriving by car these existing communities faced challenges in terms of insufficient resources, lack of infrastructure and resistance to change. This in turn gave rise in the 1960s to the development of purpose-built resorts such as Vail in Colorado and Whistler in British Columbia. Such resorts were often undertaken by a single private developer, with the Vancouver-based real-estate developer, Intrawest, being especially prominent.

Morphologically, these new master-planned resorts were developed around a strongly directed 'village' form. According to Dorward (2006, p. 271):

> This newest incarnation of the village form typically offers a carefully choreographed pedestrian network segregated from cars, a service and delivery system largely hidden from view, public spaces that are precisely oriented to sun and important landscape views, and a diversity of programmed activities to go along with shopping, euphemistically termed 'entertainment retail' by those in the industry. In this model, skiing is no longer the central attraction, but rather just one of many amenities in a smorgasbord of recreational and cultural activities, and the product is less a village than a small retail-oriented downtown.

Citing Clifford (2002), Dorward (2006, p. 273) argues that these resorts are:

> first and foremost, real estate commodities, deliberately segregated from the communities that service them and the messier functions that support them in order to maintain the high level of aesthetic quality and security that commands high prices. They are successful not simply because they are located in beautiful settings that happen to be relatively easy to access, but also because they are safe, benign and predictably comfortable.

Reviews of American ski maps (Field, 2010; Tait, 2010) show lifts fanning out from and trails converging on the valley-bottom resorts and day lodges on the slopes, with adjoining basins linked by a series of connecting lifts. Field (2010, p. 229) argued that: 'The key feature of the mountain environment for a skier or snowboarder is the interconnectivity of the lifts and trails combined with the level of difficulty.'

Mountain resorts in the American West occur in a variety of physical settings: at the end of a box canyon (Telluride); along a valley accessible from two sides (Vail Valley); or in a basin or wide valley accessible from multiple points (Steamboat Springs). Hartmann (2006, p. 294) summarized the general pattern thus:

> The resort town or its initial core area is usually situated on the south side of the stream [generally the lowest point of the valley floor], with direction toward the base of the mountain and/or the base of the main ski area. Extension areas of the core resort area follow usually parallel to the stream. They may be connected to another base of the (now extended) ski area. Alongside the base of the ski area(s) additional 'ski-in and ski-out' development may occur, mostly in the form of second homes, condominiums and/or hotels, and rental properties for incoming skiers/visitors. Closer to the main road or highway ... are either low/medium-cost residential areas or necessary infrastructure ... As the residential development goes up to 'higher grounds' home or condominium owners are able to overlook the valley, part of the core resort town, and the ski area. These are the more preferable locations.

Much of the growth in North American mountain resorts occurs 'down-valley' where many of the service elements are located and much of the resort workforce lives, giving rise to social and spatial segregation and inequalities. Hartman presented a model showing the interconnectedness and dynamic nature of the interrelationships between the core resort areas and the down-valleys. Dorward (2006, p. 276) noted:

> At some point early in the planning process, there are pivotal choices to be made about the desired relationship between a resort village and the resort community it spawns, how best to nurture a positive relationship, and how to effect the transition between the node of concentrated resort activity and the surrounding community.

Hartmann (2006) suggested that delineating the boundaries of mountain resorts is complicated by the longer distances covered by commuting resort employees and by the presence of amenity migrants who own a second home or time share but continue to have a place of residence elsewhere.

A Japanese example

In the Japanese tradition of detailed land-use mapping, Kureha (2014) plotted the developments that have occurred in the Niseko-Hirafu District following the increasing popularity of the resort with Australian skiers which lifted the resort out of the stagnation or decline being experienced elsewhere in Japan. Figure 9.3 shows a concentration of larger condominium buildings at the base of the ski-field with cottages, hotels and pensions infilling vacant lots elsewhere in the villages. The condominiums were built largely to cater to the demand for longer-stay accommodation from the Australians who typically spend a week or more in the resort compared to a night or two for the Japanese. The size and architectural styles of condominiums and cottages have introduced new features which do not conform to the landscapes of the traditional Japanese ski resort. The multiplicity of buildings in Niseko-Hirafu contrasts with the integration of La Plagne but the location of the newer, larger buildings closer to the ski-field in the former shows common functional linkages exist between lifts and accommodation.

Organizational configurations

Researchers analysing the development and morphology of ski resorts in the French Alps recognized that the different forms and processes which they were describing were the product of different organizational and

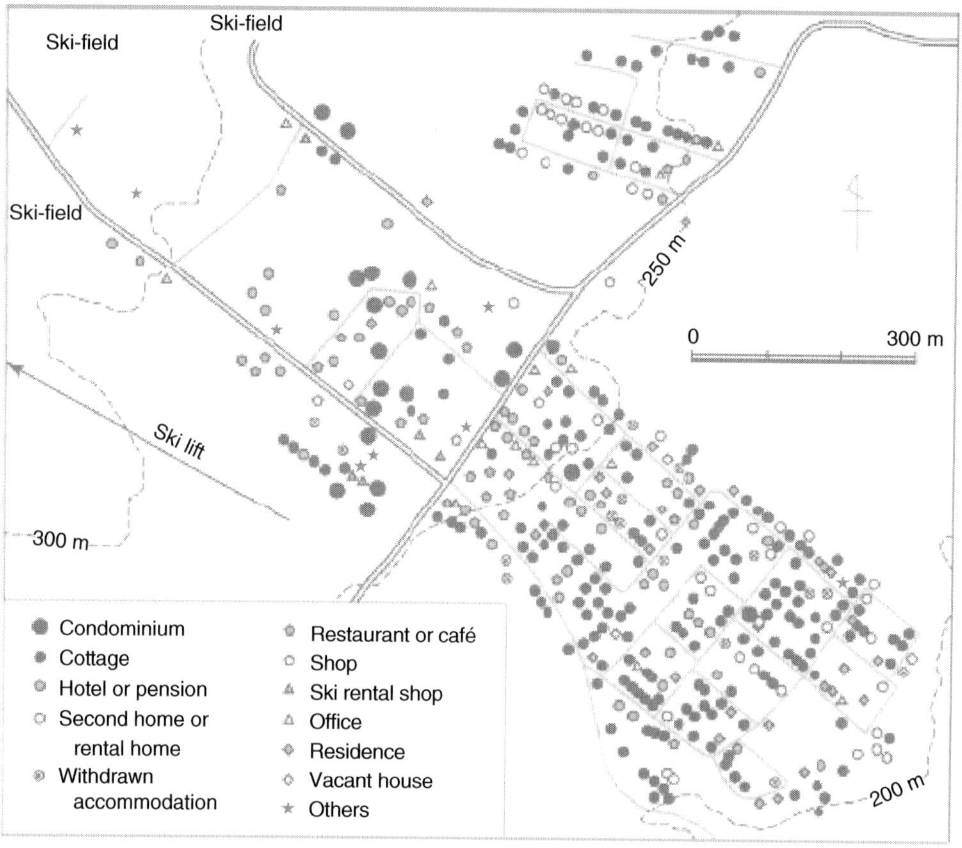

Fig. 9.3. Distribution of buildings and their use in Niseko-Hirafu District in 2012. From Kureha (2014), Figure 2, with permission of RCAPS Ritsumeikan Center for Asia Pacific Studies.

institutional structures. Préau (1970) presented two different scenarios for alpine tourist development which incorporated the features outlined earlier:

- The Chamonix type, based on a gradual process of development beginning with tourists' appreciation of the alpine environment and carried out in large part by the local community, with a few urban developers who played a complementary role operating larger hotels and providing capital for a mountain railway.
- The Les Belleville type, a large-scale operation which began with an image of the developed resort, was based on a technical evaluation of site and access characteristics and was undertaken with urban capital and knowhow. The only demands made on the local community were for its land and labour. This is analogous to the integrated resort.

Later French researchers have examined the organizational configurations of ski resorts in terms of different forms of proximity. Bocquet (2008) states that external relations are often underestimated and argues that a ski resort might also be considered as a stage in the value chain of a mountain holiday in which ski-lift operators, accommodation providers and other local actors are linked downstream to suppliers and upstream to tourists by tour operators, travel agencies and local distributors. From this perspective the resort is 'a territory which is not postulated at the outset insofar as it is the outcome of productive and institutional relationships between local and non-local actors' (Bocquet, 2008, p. 164). Bocquet then uses Torre and Rallet's (2005) proximity matrix (Table 1.2) to analyse the competitiveness of winter sports resorts in France based on the degree of inter-firm coordination associated with both geographical and organizational proximity (Table 9.1). She carried out semi-structured interviews with representatives in different parts of the value chain and local stakeholders in eight resorts in Savoy. The degree of physical proximity is that perceived by the respondents. Organized proximity is assessed in terms of a logic of belonging and a logic of similarity based on shared language, representations or beliefs. In Quadrant 1 the respondents associated the resort with the village whereas in Quadrant 2 the physical boundaries of the resort were seen to extend beyond the village limits due to an extended closely connected network of local and sectoral actors. The example in Quadrant 3 is that of a small 'integrated' resort (integrated in the sense that a single private operator was responsible for the lifts, a reservation office, some accommodation, retail outlets and restaurants) with

Table 9.1. Geographic and organized proximity of ski resorts in the French Alps. From Bocquet (2008), Table 4.

Combination of spatial scales	Geographic proximity	Organized proximity
Geographic proximity	1) Destination resort Aussois and Les Aillons Margériaz *Physical proximity* restricted to limits of the village *Organized proximity*: - logic of belonging: lack of relationships with external actors weak relationships between actors located within the resort - logic of similarity: lack of shared representations 2 of 8 cases	2) Territorial resort Val d'Isère, La Norma, Courchevel, Tignes *Physical proximity* exceeds the geographic limits of the resort *Organized proximity*: - logic of belonging: strong relationships with external actors (cooperating relationships) medium intensity of relationships between actors located within the resort - logic of similarity: existence of shared language and codes between different actors. Emergence of a common vision for the resort 4 of 8 cases
Organized proximity	3) 'Low-cost' product resort Val Fréjus *Geographic proximity* restricted to limits of the resort *Organized proximity*: - logic of belonging: weak relationships between actors located within the resort (competition) strong interaction between actors located in the resort and external actors (dependency relationship) - logic of similarity: lack of shared representations of activities, divergent short and long term visions for the resort 1 of 8 cases	4) Branded resort Arc 1950 *Geographic proximity* delimited by the boundaries of the firm in charge of the resort *Organized proximity*: - logic of belonging: strong interaction between the actors located within the resort (quasi-integration) - logic of similarity: shared representations that conform to the organizational culture of the group in charge of the resort 1 of 8 cases

an emphasis on profit maximization through linkages with external tour operators and correspondingly little involvement of local actors. Commercial activities between the resort operator and the tour operators were based on trust, requiring little face-to-face interaction. The example in Quadrant 4 is also an integrated resort, an upmarket one owned and managed by a global group. The resort is managed and marketed as a brand rather than as a locality; it is '"aterritorialized", run under the rules and routines shared by the stakeholders, defined by the group at a global scale and reproduced in different sites' (Bocquet, 2008, p. 172). While the resorts in Quadrant 2, which reflect characteristics associated with clusters and similar 'ideal' models, accounted for half of Bocquet's small sample, use of Torre and Rallet's matrix draws attention to a variety of other situations which may exist when a more nuanced approach to proximity is taken.

Related ideas are put forward by Marcelpoil and François (2008) who stress the role of local government in the governance of French ski resorts given their responsibility for the management of ski lifts. In 80% of cases the operation of the lifts has been delegated. As a result, multiple actors from the public and private sectors along with civil society may be involved. Marcelpoil and François (2008, p. 181) consider ski resorts as local tourism systems or districts and suggest that:

> certain resorts can be read as constructs which display a real articulation between the diverse dimensions and result in the emergence of a territorial ambience and a local tourism culture while in other tourism areas there has been no such chemistry and in its extreme form the resort can be seen as the simple juxtaposition of public and private actors.

This diversity is explained in terms of differences in organizational proximity and cognitive proximity. The former refers to that which binds the actors together through formal relations such as contracts. The latter relates to a less formal but common understanding which provides a logic for the action to be taken. It is comparable to the logic of similarity in Table 8.1. Organizational proximity is said to facilitate the exchange of information and knowledge and thereby develop norms and common values. Conversely, close cognitive proximity will engender co-operation and develop organizational proximity. Marcelpoil and François argue that differences in the degree of cognitive proximity have contributed to the way in which certain French resorts were developed as a common project between the various parties while in others the lack of a common understanding hindered development.

Other writers have also focused on issues of management and governance and on the ways the different public- and private-sector stakeholders come together and interact in a variety of organizational and institutional arrangements. This turn in direction was signalled in Flagestad and Hope's (2001, p. 449) study in which they defined a winter sports destination as 'a geographical, economic and social unit consisting of all those firms, organisations, activities, areas and installations which are intended to serve the specific needs of winter sports tourists'. Drawing analogies with multi-division firms, they developed their ideas on the organizational structures of destinations with a two fold community/corporate classification (Fig. 9.4) and a conceptual model of a winter sports destination for strategic management analysis (Fig. 9.5). The community model, said to represent a typical European resort, consists of multiple 'specialised individual independent business units', none of which is dominant. Local government institutions or a local tourist board/DMO supported by them play a significant role in the management of the destination which is concerned with such issues as environmental sustainability, planning, product development and marketing. In contrast, the typical American resort is founded on a corporate model, characterized by a dominant multi-divisional corporation with business units dealing with particular ski operations (lifts, ski school, retail, food and beverage on the mountain) and some of the accommodation capacity. This corporation has a major influence on both the operation of the resort and on community-related development of the

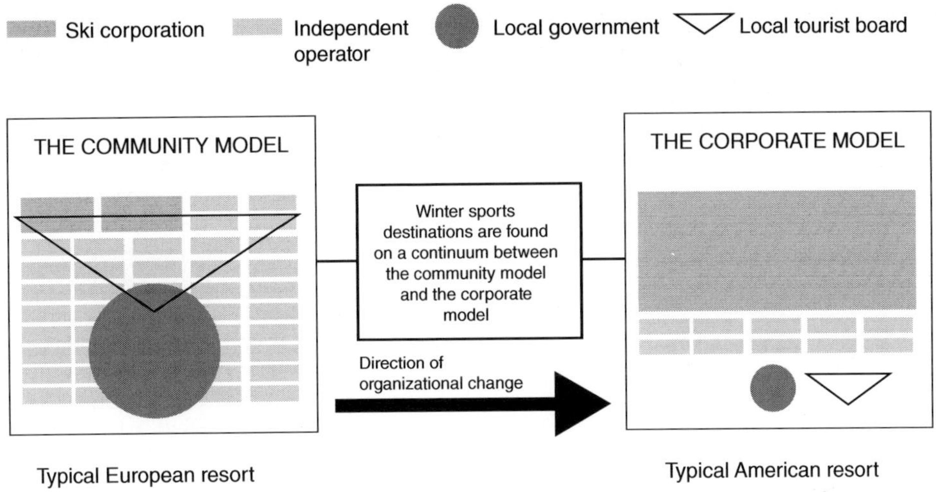

Fig. 9.4. Destination organizational structures: the community model and the corporate model. From Flagestad and Hope (2001), Figure 4.

destination. Local government and the local tourist board have much lesser roles than in the community model. These are much generalized models which sit at the opposite ends of a continuum. As the preceding section has shown, considerable variety exists within Europe. Although framed differently, Fig. 9.4 is not dissimilar in many respects to Préau's (1970) Chamonix/Les Belleville classification.

Figure 9.5 portrays the different products of the value creating process for winter sports destinations. The authors suggest the analogy of the multi-divisional firm can be applied to different organizational structures for the destination. Management activities are split into two levels: operational and strategic. Operational level activities deal with delivery of the destination product, supporting services and other value creating services related to the natural, cultural and social environment. According to Flagestad and Hope (2001, p. 456) the two main strategic tasks are 'management of the *configuration* of the destination tourism product (the mix of "business units")' and 'the management of *sustainable development* (the sustainability elements)'. Strategic success of the destination is said to increase when configuration management is directed *inter alia* at creating positive externalities and preventing negative ones and at achieving optimum market fit. Sustainability management creates externality value for all primary activities and other aspects of a destination's strategic performance. Flagestad and Hope hypothesized that the corporate model is better placed to create customer satisfaction through the control it exerts over service provision than the less coordinated community one but that the latter may perform better on the sustainablity dimensions through stronger stakeholder involvement. They also suggest that where two dominant corporations are involved destination performance will be better due to the competion this situation brings.

Kylänen and Rusko (2011) also distinguish between strategic and operational levels, not just in terms of what is being done at each level but also which actors are involved and whether their behaviour is competitive, cooperative or involves coopetition. Strategic level destination activities typically involve 'planning, zoning and remarkable investment decisions of the actors of provinces, municipalities, DMOs or the largest firms ... and on the operational level the actors are mainly individual firms and workers, or groups of them' (Kylänen and Rusko, 2011, p. 200). Moreover, drawing on their ethnographic research in the Finnish

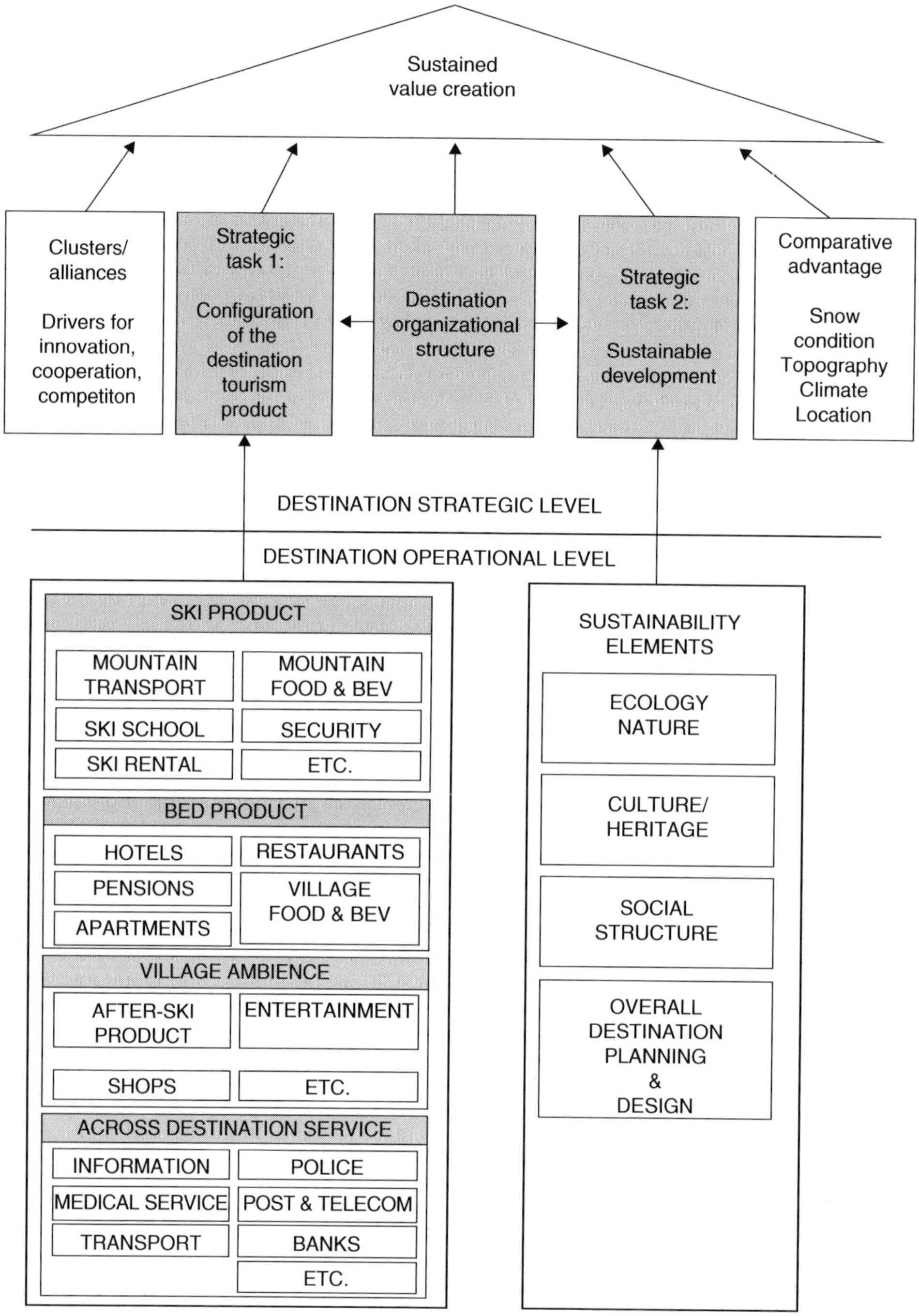

Fig. 9.5. Conceptual model of a winter sports destination for strategic management. From Flagestad and Hope (2001), Figure 6.

destination of Pyhä-Luosto (three municipalities, two resorts, one destination), they suggest that behaviour at one level may generate a different form of behaviour at the other. Strategic marketing of the destination, for example, may be accompanied by competitive behaviour on an operational level. Much behaviour may be intentional but this is not necessarily the case: '[I]n addition to co-location of the actors, the proximity of public, semi-public and private organisations creates a strategic-operational environment where coopetition may be coincidental and serendipitous' (Kylänen and Rusko, 2011, p. 201).

In North America, Gill and Williams (2006) observed a growing corporate consolidation of ski resorts and reported that at that time three large publicly traded companies had acquired interests in 23 ski resorts there. Mergers and consolidation were aimed at 'integrating investments in on-mountain capital improvements with the development of controlled mountainside real estate, facilities, and four-season resort villages' (Gill and Williams, 2006, pp. 28–29). Citing Clifford (2002), the authors continue (p. 29): 'skiing is no longer the key to corporate profit, but rather serves as an amenity to sell real estate'. This real-estate drive also explains the 'village' type developments discussed in the previous section. Gill and Williams also stress that place matters and warn against overgeneralization from case studies as their analysis showed relations between the corporate developer and the community in Whistler, British Columbia were more cordial than in other research which concluded that negative impacts result when corporations control mountain activities (Rothman, 1998; Clifford, 2002).

The cordial relations in Whistler were attributed to two factors: strong regulatory control over the comprehensively planned resort community at Whistler and the building of trust between the developer and the community. This resulted in part from senior management being Whistler residents but also from the gradual shift from confrontational to collaborative relationships between a strong local NGO representing community interests and the corporation (Ponsford *et al.*, 2006, p. 563):

> The current order of engagement, characterized by participation and partnerships, appears to posit W-B [the corporation] and the community stakeholders as collaborators in the development and operation of Whistler. The community still requires W-B to act as the primary economic driver in the region and W-B still requires a social licence to operate. Each group's power – be it derived from allocative or authoritative resources – appears to have achieved balance, instead of posing a threat to the other.

However in Mt Tremblant, Quebec, such a balance did not develop as the same corporation, Intrawest, did not meet with a similarly strong group of stakeholder and difficulties emerged in obtaining a social licence to operate there. These two cases affirmed for Ponsford *et al.* (2006, p. 566) the assertions of both stakeholder and resource-dependency theory that:

> Corporations will only engage community stakeholders who present themselves as possessors of resources the firm requires to gain competitive advantage in the global marketplace. Stakeholders who are neither powerful enough to revoke the social licence to operate nor able to present effective solutions-based alternatives for the firm are excluded from the decision-making process.

Ongoing consolidation in the ownership and operation of North American ski-fields has occurred. In 2018 the Alterra Mountain Company, a privately owned hospitality conglomerate headquartered in Denver, owned and operated 14 ski-fields, including some previously owned by Intrawest. Ten American ski-fields, including Vail, along with Whistler Blackcomb in Canada and Perisher in Australia, were owned and operated by the Vail Resorts Management Company which has its headquarters in Broomfield, Colorado (Table 9.2). The company was structured in three separate but integrated units: one was responsible for the ski-field operations (lift tickets, ski and snowboard schools, dining, retail and rental); a second was concerned with hotels and condominiums in proximity to the ski resorts, as well

Table 9.2. Vail Resorts' American ski resorts, 2018. From www.vailresorts.com (accessed 21 May 2018).

Resort	Uphill capacity (/h)	Vertical drop (m)	Trails	Lifts
Park City	83,027	976	185	41
Vail	59,069	1052	193	31
Heavenly	52,000	1075	97	29
Breckenridge	46,800	1036	187	34
Beaver Creek	43,221	1018	150	25
Northstar	37,891	694	100	20
Keystone	33,564	954	130	20
Afton Alps	20,000	107	50	22
Kirkwood	19,905	610	87	15
Mt Brighton	17,200	405	26	16

as three destination resorts in Grand Teton National Park and golf courses; while the third developed real estate in and around the resort communities.

Gajdošík *et al.* (2017) used SNA to test Flagestad and Hope's (2001) community/corporative model empirically in two leading ski resorts in Slovakia: the High Tatras and Liptov–Jasna. They found that over the period 1995–2015 the organizational structure of both resorts evolved from a community one centred on local tourism associations and municipalities to a more corporate one as a large, powerful, multi-divisional firm, TMR, emerged to dominate activities and competition (Fig. 9.6). This private stakeholder not only controls the core product through ownership of all the cable cars and lifts, ski rentals, ski schools and on-slope après-ski bars but also owns restaurants and accommodation and has taken a lead role in product innovation. Similar situations are reported with the Weise Arena Gruppe in the Swiss resort of Flims Laax and Skistar which operates resorts in Norway and Sweden. In the case of the Swedish resort of Äre, Nordin and Svensson (2007, p. 58) noted a shift in public to private power, observing that:

> It seems that on the one hand Skistar may be a driver of development and innovation, but on the other hand its strong position runs the risk of indirectly limiting other actors. The uneven balance of power among the private businesses has, moreover, improved with the establishment of Holiday Club, which is another strong private corporation.

They also identified a spirit of coopetition and the role of local government.

Rural and Natural Area Destinations

As Chapters 5 and 6 have shown, tourist destinations have developed in a range of geographical contexts which might be broadly defined as rural destinations. In some respects their spatial and organizational structure is best dealt with at a regional or sub-regional scale as rural and natural area tourism in many instances is rather dispersed and destinations of this type generally do not exhibit the same concentrated forms as urban destinations and coastal and ski resorts. As a result, boundaries between the local and the regional may become blurred. In addition, as the focus moves to the local scale analysis of the spatial and organizational structure of rural and natural area destinations becomes very challenging due to the diversity of forms such destinations take and the prevalence of many very small actors, including individual farmers. As Lane (2009), Sharpley (2004b) and others point out, there is no commonly accepted definition of what rural tourism is. Rural may be seen as a counterpoint to urban on a variety of social, economic, physical and other dimensions though that too may be an over-simplification in the absence of any clear dichotomy. Some conceptualizations require an agricultural base or connection; what may be referred to as agri-tourism and is characterized by some form of farm-stay or production-based activity such as visits to

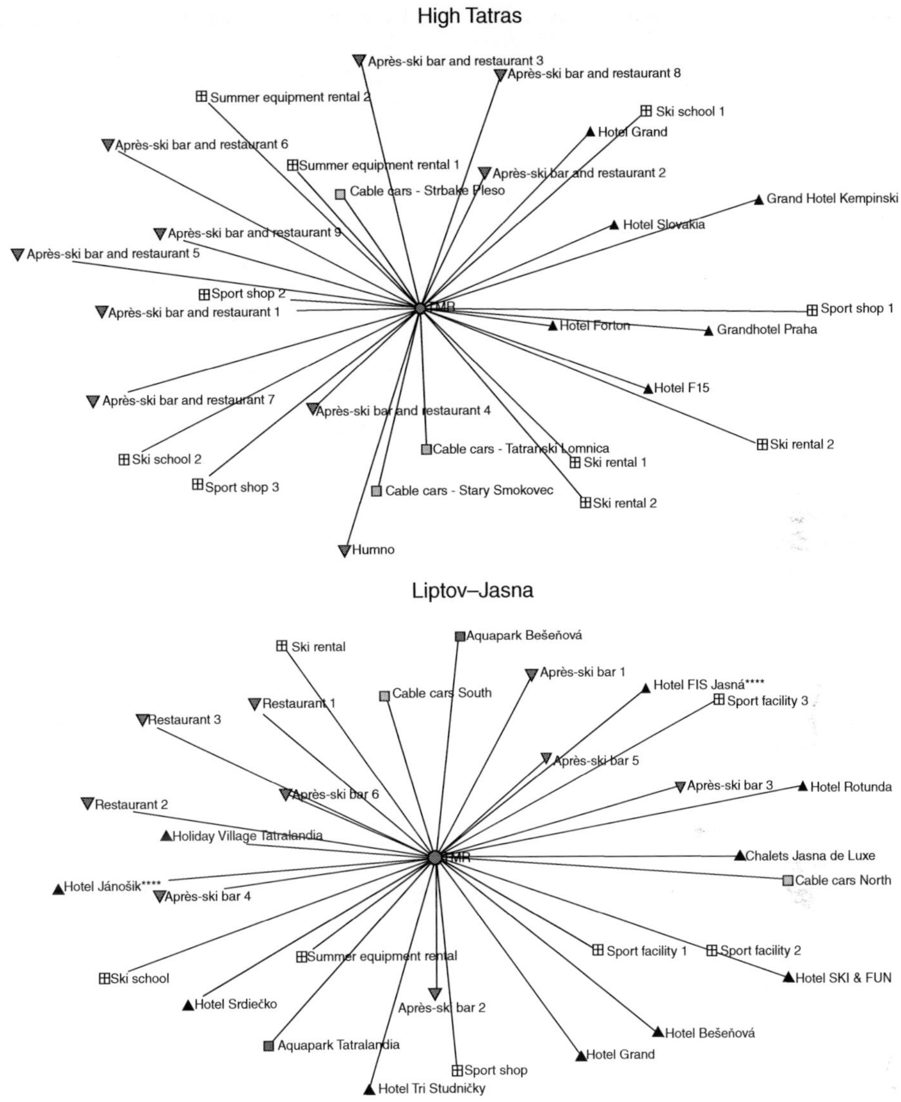

Fig. 9.6. Ownership of resources by the dominant private stakeholder in the High Tatras and Liptov–Jasna. From Gajdošík *et al.* (2017), Figure 4.

wineries and wine tasting. In other instances a distinction is made between rural and agri-tourism (Przezbórska, 2003). Rural tourism can also be considered as involving a range of activities not associated with farming that might take place in the countryside such as tramping, biking and fishing. In other instances tourism may develop around a major natural attraction in a rural area. Pavlovich's (2003) oft-cited network study, for example, centred on changing relationships related to the management of a major glow worm cave and nearby adventure tourism activities in a predominantly farming area. Much natural area tourism might be included at the remote, less-developed end of the rural tourism spectrum but in many cases it is treated in separate and often non-convergent literatures, especially when the focus is on national parks and other protected areas (Butler and Boyd, 2000; Newsome *et al.*, 2013). Given the diversity of rural and natural area destinations and range of research undertaken no attempt is made within the scope of this book to derive generalized patterns of their structure as local destinations. Rather, selected examples are given to illustrate some of the structural features that occur.

Spatial structure

Lane and Kastenholz's (2015) keyword analysis of rural tourism indicates relatively little work on spatial structure or morphology in that literature. In research on rural tourism data on farm accommodation or second homes may be aggregated up to a provincial or regional level (Przezbórska, 2003) but they are rarely available at a finer scale to show the localized distribution of such phenomena. Villages and small towns may also constitute important features of rural and natural area tourism, whether as attractions in their own right or bases from which to explore the surrounding countryside. One of the most explicit studies of the spatial structure of rural settlements experiencing the growth of tourism is that by Xi *et al.* (2015) who carried out a detailed analysis of land-use change and functions in three villages in the Yesampo tourism area of China. In particular, this study shows the influence of location on the internal structure of the three villages which are at varying lengths from the entrance of Bailixia canyon, the main tourist attraction in the region: Gougezhuang (2 km away), Liujiahe (4 km) and Shangzhuang (6 km). Figure 9.7 shows that Gougezhuang has undergone the most intensive development, particularly on land overlooking the river, and has been totally transformed by tourism. A wide range of tourist services have been developed there, the floor area ratio has increased the most and the village has experienced a significant loss of landscape authenticity. In contrast, Shangzhuang has been much less affected on all these measures while Liujiahe constitutes an intermediate case.

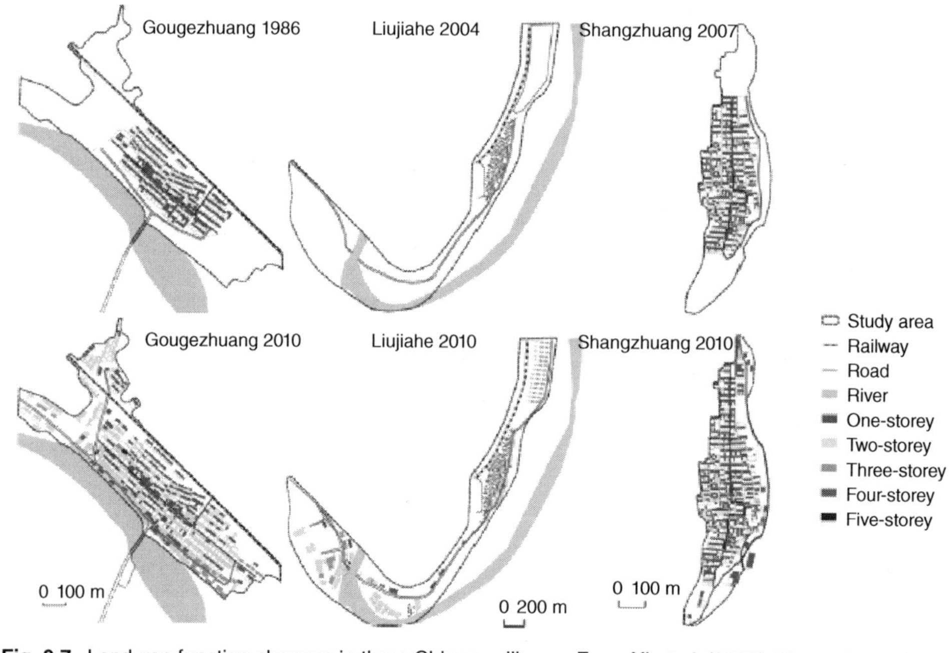

Fig. 9.7. Land-use function changes in three Chinese villages. From Xi *et al.* (2015), Figure 3, reprinted with permission from Springer, copyright, 2015.

Local patterns of visitation in rural areas are often linear in nature. This is especially the case with themed trails or routes, such as food and wine trails or routes linking historical sites. These may develop organically but they are also strategically promoted to increase demand for a particular product or locality by creating a stronger image and more substantial attraction through spatial and themed association. According to Macleod (2017, p. 424):

> As linear visitor attractions, trails organise the environments that they inhabit and create new tourist spaces using selected themes and stories. The trail's task is therefore to create a sequence of sites to be visited and in doing so it creates a sense of order and a structured, chronological narrative. This means of producing space creates a unifying frame through which the diversity of place can be viewed.

Xu *et al.* (2016, p. 171) provide a detailed analysis of the geospatial character of wine trails in North Carolina in terms of spatial pattern ('dictated by trail total length, number of comprising nodal tourism attractions, and the relative distance between two nearby attractions'), connectivity and accessibility and three measures of tourism attributes 'deemed critical to evaluate wine trails' and wineries' services': comprehensiveness, dominance and complementariness. In the context of destinations, the complementary nature of the services different wineries along a trail offer is particularly relevant 'so that the trail as a unity can provide a greater variety of services to tourists' (Xu *et al.*, 2016, p. 172). Xu *et al.* identified three types of spatial patterns based on variations in the relative distance and dispersion of wineries along the nine wine trails they examined: (i) centre clustered; (ii) trailhead clustered (i.e. concentrated at either end of the trail); and (iii) evenly spread. Most of the trails were characterized by clustered trailheads, high levels of accessibility, moderate comprehensiveness, low dominance and moderate complementariness.

A classic hub-and-spoke visitation pattern is evident in San Pedro de Atacama. The Atacama Desert is one of Chile's major destinations for international tourists outside the Santiago/Valparaiso area (Table 5.1). San Pedro de Atacama is formally classified as a mature, rural and nature destination (Table 2.2). In 2017 it attracted an estimated 173,800 visitors. The town serves as the base from which day or half-day trips are made, radiating out to a range of dramatic landscape features including the Tatio Geyser (98 km away), the Laguna Miscanti (117 km), the Valle de La Luna (16 km, Fig. 9.8) and to

Fig. 9.8. The Valle de la Luna (Craters of the Moon): one of the dramatic landscapes visited on half-day excursions from San Pedro de Atacama, Chile. Photo: the author.

other sites such as the Puritama hot springs (33 km), wildlife viewing on the Salar de Atacama (62 km) and local cultural attractions like the church at Socaire (90 km). Some visitors make their way to these sites independently but most use the services of one of the many tourist operators based in San Pedro, either making a booking through international distribution channels or selecting operators and accommodation once they have arrived in town (Fig. 9.9). San Pedro also provides accommodation, restaurants, souvenir shops and currency exchanges offices. Most of the natural features are located in some form of protected area where no accommodation is permitted and facilities are generally limited to ticket and information offices and toilets.

Many national parks have been established not only to protect natural landscapes, fauna and flora but also to provide for the public's access to and enjoyment of them (Butler and Boyd, 2000). Location of accommodation, transport and supporting facilities relative to protected areas is a key consideration in their management so as to avoid or mitigate land-use conflict. Miller *et al.* (2018, p. 11) propose the concept of a *protected destination system* (PDS) which is composed of a *protected area* and a *gateway region* 'that includes human communities adjacent (or with quick access) to, the protected area'. Nature predominates in the protected areas; gateway regions or localities are peopled places characterized by commerce and infrastructure.

In many places, however, significant development has occurred within national parks, such as at Banff in Banff National Park, Canada, which receives more than 3.5 million

Fig. 9.9. San Pedro de Atacama is the hub from which tour operators offer 'spoke' excursions radiating out to surrounding natural and cultural features. Photo: the author.

visitors a year (Dearden, 2000). Due to a variety of physical and historical factors, different situations exist in New Zealand in terms of the location of national parks relative to access routes, gateway communities and facilities for visitors. Some national parks are traversed by state highways (Westland/Te Tai Pouhni, Arthurs Pass), some are at road end (Aoraki/Mount Cook, Fiordland) and others have several local access roads (Tongariro). The parks have well-developed hut networks but commercial accommodation within these is generally limited. Hotels are located within Aoraki/Mount Cook and Tongariro National Parks but as visitor demand has grown these have been complemented by facilities located outside the park boundaries or in nearby towns. Westland/Tai Poutini National Park is serviced by the small communities of Franz Josef and Fox Glacier which lie on the State highway outside of but immediately adjacent to the park boundaries. Zoning within national parks is commonly used to protect the natural environment and provide for different types of visitor use (Newsome *et al.*, 2013).

Organizational structures

Rural development agencies and other institutions have seen a need for collaboration among small farms and rural enterprises and have promoted the concept of clusters as a means of developing economies of scale to make tourism a viable option for combatting rural decline (Czajkowski and Cunha, 2010; Partalidou and Koutsou, 2012; Pejanović *et al.*, 2017). This raises the issue of whether clusters can be developed from the top-down or from the bottom-up and whether formal ties are more effective than informal ones. The evidence in support of institutional clusters is variable. Members joined tourism clusters formed in the Danube region of Serbia to increase their profit but few significant results were achieved. Nor had mutual trust and a greater degree of cohesion among members been attained. Partalidou and Koutsou (2012) found that many of those who belonged to rural tourism clusters in northern Greece did so primarily to be eligible for other benefits from the LEADER+ programme. This study again identified trust as a key factor and showed this was stronger in terms of relationships with the official bodies than with fellow members-entrepreneurs. Czajkowski and Cunha (2010) added social networks to Porter's Diamond model as a framework to investigate a rural tourism network in the municipality of Colombo, in the state of Paraná, Brazil. Their results indicated that the organization and coordination of a cooperation network resulted from the work of the Municipal Secretary for Tourism, the Municipal Tourism Board, and the Circuito Italiano de Turismo Rural Association of Businesses.

Montero and Parra (2001) refer to San Pedro de Atacama as an ecotourism cluster. Although they do not go on to analyse tourism in the area using an explicit cluster framework their discussion suggests that the destination was still in a formative phase and not yet functioning as a cluster in the ways outlined in Chapter 3. Little cooperation was evident among the providers, there was not much coordination among the multiple government agencies which characterize Chilean administration and local leadership was lacking. Economic growth was occurring but negative externalities were emerging in the form of an inadequate infrastructure, a lack of regulation, an absence of strategy, pressure on natural sites and growing social divisions between the local Atacameños and *afuerinos* (outsiders). Progress in some of these matters is reported by de la Maza (2018, p. 102) who noted that in 2013 the municipal government 'created an institutional framework to help organize, plan and execute tourism activities in the town. The main objective of the program was to strengthen cooperation between public and business sectors and indigenous communities, and improve the town's capacity to receive tourists.' State agencies responsible for protected areas have shifted the management of a number of tourist attractions to Atacameño communities who charge an entry fee for doing so, for example at the Valle de la Luna. This has led to greater regulation of the tour operators and more attention to conservation.

Two other Latin American nature-based clusters depicted by Barbosa and Zamboni (2001) and Acuña Ortega and Brugnoli (2007) have nature and/or attractions at the heart of their framework. Barbosa and Zamboni's (2001) framework for Bonito, Brazil, takes the form of a set of concentric circles around the attractions core: successively trade (hotels, transport, guides, etc.); social actors (business associations and local government); supra-local supporting organizations (public-sector federal organizations, ecotourism and environmental NGOs and educational institutions); and an outer ring of the urban and rural context.

In the more detailed framework of the ecotourism cluster of Monteverde, Costa Rica, developed by Acuña Ortega and Brugnoli (2007), the historic nucleus is the Monteverde Reserve, a protected area and the original centre of attraction. This has progressively been complemented by the development of other nature-based attractions (e.g. butterfly house, serpentario, orchid gardens), adventure activities (e.g. canopy walk, off-roading) and agro-ecotourism, which have led to the spatial expansion of the cluster's nucleus. This growth, facilitated by the leadership and organizational capacity of local institutions, has been accompanied by the development of productive chains, both backwards to accommodation, tour operators and hospitality and forwards to cultural and educational tourism as well as other businesses such as internet cafés, bookshops and furniture manufacturing. Complementary upstream and downstream productive chains have also accompanied and enabled growth of the cluster. Downstream activities include shops, banking and finance, transport, utilities, health and security services. Upstream, activities related to traditional agro-industrial activities have been a source of land, capital and business skills which have enabled the transition to ecotourism activities.

Acuña Ortega and Brugnoli (2007, p. 123) describe the structure of the Monteverde cluster as 'clearly a network in which the businesses, implicitly and together, contribute to the commercial success of the cluster itself'. It is (p. 127) 'relatively developed and mature with a clear orientation towards long term specialization, product and process innovation and environmental management as a business strategy'. Success is also attributed to product/service variety, external links and the ability to adopt foreign ideas. In contrast, Barbosa and Zamboni (2001) concluded that Bonito was more of a potential cluster, one limited at that stage by such factors as the lack of any tradition of working together, the incipient level of business knowhow and the short time span over which tourism had developed.

Viken and Aarsaether's (2013) analysis of the situation at the North Cape, located on the northernmost point on mainland Norway, portrays a dichotomous market and organizational structure at the destination. The North Cape cliff attraction, together with the main hotels in the municipal centre, is owned by Rica, the largest Norwegian hotel chain. Cruise-ship passengers and long-distance coach tourists make up 80% of the quarter of a million visitors who go the cliff attraction each year. These segments are serviced by a mass, standardized Fordist system, one dominated by Rica and supported by local bus companies and guide services, several cruise-ship operators, the coastal steamer company and a multitude of national and international coach tourism operators. Tourists travelling independently and making their own arrangements constitute the other 20%. This segment is in the main catered for by small-scale businesses (e.g. activities, camp sites, small hotels), although some independent tourists choose to stay in more upmarket accommodation. The growth in small-scale ventures catering to independent travellers and niche markets less suited to the larger companies, such as angling, has led to a more diversified destination in which the two systems co-exist and influence each other. Viken and Aarsaether (2013, p. 51) sum up the structure that has evolved in these terms:

> only very few travel to this destination without visiting the main attraction, North Cape cliff, and in that respect all tourists purchase services from the monopolist on the site. In general, however, more tourists are opting for additional attractions,

> providing business opportunities for the other parts of the tourism industry. Thus, the two parts of the tourism industry tend to become more integrated ... we traced very few signs of outright competitive attitudes within the tourism field. From a theoretical point of view, we claim to have witnessed a move from traditional Fordist production principles to a more flexible neo-Fordist inspired business attitude on the part of the dominant actor, while many of the small operators have moved from a pre-Fordist, 'lodgings-only' mode to a post-Fordist mode, typically providing unique and exotic experiences in which the tourists are expected to participate. The North Cape tourism challenges seem to have paved the ground for extensive exchanges and reciprocity between autonomous actors. As far as we have seen it, this integration is more or less based on informal community-type mutual adaptation than on contractual collaboration or on the participation of formal networks. However, the role of the public facilitators should not be neglected.

Protected natural areas are commonly administered by some state agency, typically a department of conservation, a forest service or some other land manager. Where tourism has developed in such areas, issues arise relating to who is responsible for managing them and any commercial tourism operations that take place there. Organizational structures vary throughout the world but a common situation is one in which the responsible agency undertakes overall visitor management while commercial operations such as commercial guiding and adventure activities are carried out by private-sector operations which have been granted a concession or licence to do so for a fee and under specified conditions. Camping grounds or huts are often managed by the responsible agency but other accommodation such as lodges and hotels may be run by private enterprises under a fixed-term lease. After a detailed study of visitor management in two New Zealand national parks Sowman and Pearce (2000, p. 240) concluded: 'Visitor management in national parks is a complex and challenging activity due to the exceptional physical and or cultural settings, the varied demands which arise in providing for a satisfying visitor experience while protecting the resource base, and the range of parties involved.' They offer a conceptual framework for an integrated approach to visitor management in national parks which might also be used to explore organizational structures but stress the need to take account of local factors and conditions. In one of the early organizational studies in this field, Selin and Beason (1991) discovered a general lack of cooperative relations between the US Forest Service, chambers of commerce and tourism associations adjacent to an Arkansas National Forest, but found (p. 648) that 'dyads that were geographically close exhibited more cooperative relations than dyads that were geographically farther away'.

Conclusions

Research on the structure of ski resorts is characterized by the attention given to both their physical form and organizational structure and by the recognition that the two are closely connected. The relationships between accommodation and the ski lifts and trails – the basis of the attractions – are the dominant influence on both the spatial and organizational structure of ski resorts. Variations occur but in general terms the location and morphology of the ski resorts discussed have been dominated by the drive to link accommodation with ski slopes. This is seen both in terms of the location of resorts – moving ever higher up the mountain in the case of the French resorts – and by physical forms which facilitate ready access to the lifts and trails and which progressively extend the networks of these. These general morphological characteristics are modified by the particular physical setting and the nature of the existing settlement pattern, factors which themselves are closely interrelated. This is especially apparent in the progressive location of French ski resorts at higher altitude and the creation there of integrated resorts which contrast with those of the village-based resorts of the *moyenne montagne*.

The nature of ski-field operations is such that invariably there is a single operator, that is, a monopoly exists in this sector. Two types of organizational structure commonly occur. Given the substantial investment and operational costs involved with ski-fields, their development is often offset by or undertaken in order to generate large-scale real-estate opportunities. As a result, the ski-field developer and operator becomes the dominant player in the resort in terms of both the attractions and the accommodation, a structure apparent in the large French resorts, the American corporate model and the Slovak examples. In other instances, particularly in France where they have a legally mandated authority to do so, local government may take a lead role in developing and operating lifts, or do so by default, and a broader-based form of community development occurs. The corporate/community distinction (Fig. 9.4), however, represents two points on a spectrum which contains a wide range of intermediate cases.

The relationships between the morphology of ski resorts and their organizational configuration have been stressed with regard to the stylized 'village form' of many North America resorts and differences in the patterns of location, form and organizational structure of resorts in the French Alps. Greater diversity was illustrated in Table 9.1 which shows the different types of geographic and organized proximity found in a selection of French ski resorts and how variations in their territorial extent are associated with different patterns of stakeholder relationships, including those with external parties. Some instances are found in the table of cluster-like behaviour but a wider variety of situations is also evident here and in other research. In general, most of the ski resort studies do not reference the frameworks discussed in Chapter 3 and suggest other approaches are needed to understand their structure.

For the reasons outlined earlier it is more difficult at this stage to arrive at generalizations regarding the structure of rural and natural area destinations. The work of Xi *et al.* (2015) and Xu *et al.* (2016) shows how research on the spatial structure of different forms of rural tourism might be approached. The small scale of many rural businesses has led to attempts to link them spatially and organizationally, as in the case of themed trails or the establishment of rural tourism clusters. Their establishment, however, has been challenging, especially in terms of developing trust between members. The development of tourism in national parks and other protected areas has created issues associated with the location of tourism facilities relative to environmental features and relationships between land management agencies responsible for these areas and the operators of these facilities and providers of commercial visitor services. Moreover, as the North Cape example showed, not all natural area operations are managed by public-sector agencies or are small scale. Viken and Aarsaether's (2013) study demonstrates the benefits of a detailed, theorized case study which highlights the influence of local conditions and experiences. At the same time, the specifics of the case illustrate the challenges of synthesizing the characteristics of organizational structures in natural areas.

10 Spatial and Organizational Structure of Destinations: a Synthesis

As the preceding chapters have shown, a wide range of findings on the spatial and organizational structure of destinations have resulted from the studies undertaken for a variety of purposes using many diverse approaches in different parts of the globe. The literatures dealing with the central topic are very fragmented as research has become increasingly specialized. As a result, it has become progressively more difficult to see the big picture. This book has addressed this issue by synthesizing research in this field.

Synthesis is about identifying key points and characteristics and making the connections between them so that overall patterns emerge more clearly and our understanding of specific points is increased by seeing how they fit into the larger scheme of things. This is achieved by identifying commonalities and differences and establishing the relationships between patterns and processes. Due to the current state of knowledge this process does not always lend itself to making definitive statements about some aspects of the structure of destinations but by drawing attention to where gaps exist it also identifies avenues for targeted future research. Synthesis requires a large measure of abstraction and generalization. The process inevitably means detail is lost, that there are always exceptions to the generalizations advanced and that despite the wide net used here to identify relevant related research other studies that have not been included will present additional information and insights. A first level of synthesis has been undertaken in earlier chapters: of perspectives, frameworks and theories, methodological issues and empirical results relating to national and regional destinations and different types of local destinations. This chapter takes the synthesis a stage further by drawing together the key themes from these chapters to arrive at a more integrated set of findings and conclusions concerning first the spatial and then the organizational structure of destinations. Relationships between the two forms of structure are then discussed. For common or general points reference is made back to the relevant chapters; for more specific points or exceptions particular studies are cited.

Spatial Structure

As introduced in Chapter 1 and elaborated on in subsequent chapters, the spatial structure of destinations concerns the arrangement of different tourism phenomena in space; how and why those arrangements have come about; and how knowledge of that arrangement helps us to understand what destinations are and how they function. Three fundamental dimensions of spatial structure are location, spatial interaction and distribution. Although commonly dealt with separately, these three dimensions are closely interrelated. The spatial structure of a given destination will reflect the interrelationships between them as well as the underlying geography of the country, region or local place in question.

Location

Co-location within and between sectors is repeated in varying degrees at each scale. These patterns suggest common factors and processes are at work. These are summarized in Fig. 10.1. A fundamental feature of destinations is that tourism involves the consumption *in situ* of a set of interrelated services: attractions and activities, accommodation, transport and supporting services (Chapters 2 and 3). Due to the interdependent nature of these services their utility is higher when they are consumed in one place (Terhorst and Erkuş-Öztürk, 2011). This results in the co-location of multiple services which are needed for the successful functioning of destinations. The basic issue in terms of co-location is which one is the anchor service/ sector; that is, the one towards which or around which the others gravitate and are located. The sector that most commonly anchors co-location is attractions and activities. From the visitor's perspective, attractions and activities are fundamental services; they provide the 'energizing power' of tourism systems (Gunn, 1979, p. 48). Woodside and King's (2001) purchase-consumption framework of leisure travel behaviour shows choices about destinations, activities and attractions precede and trigger those relating to accommodation, transport mode and route, retail purchases, eating out and moving around the destination. Visitors value proximity to the attractions they wish to see and prefer to be close to the activities that they want to do. Location is a key factor in guests' selection of urban hotels and in coastal resorts visitors are prepared to pay higher rates for hotels close to the beach or with a sea view (Chapters 7 and 8). Providers in turn seek to satisfy this demand by locating their services and facilities close to the attractions and to the places where activities are carried out.

There are, however, numerous types of attractions and visitors engage in a variety of activities which may be categorized in different ways. From the perspective of co-location the following interrelated classes are especially relevant.

Fixed or footloose: Fixed attractions are those based on essentially immovable physical or cultural resources, such as beaches, skifields, natural and cultural landscapes, historical buildings and monuments. Footloose attractions are those which are purposefully created in locations which may be subject to certain physical conditions, planning objectives and constraints as well as to demand considerations but there is some inherent flexibility in where they are sited, such as golf courses, museums, casinos, sports arenas, theme parks, shopping complexes.

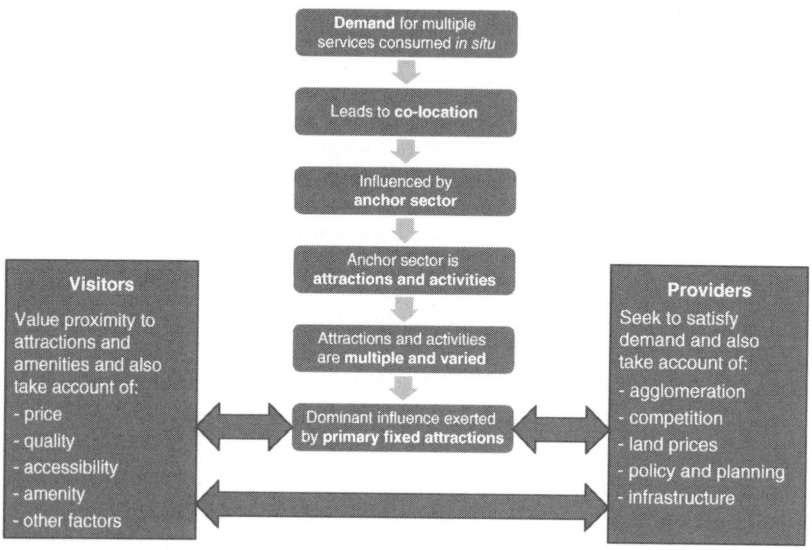

Fig. 10.1. Factors influencing co-location in destinations.

Primary or secondary: Primary attractions or activities will appeal to the majority of visitors and/or be the ones they spend the most time at or on. They are the big 'must see', 'must do' attractions and activities. These will normally be accompanied by secondary attractions and activities which have lesser appeal and are visited or engaged in as a complement to the main programme or by smaller, more specific market segments.

Initial or subsequent: As destinations develop and demand grows and broadens the initial attractions are subsequently complemented by, perhaps even superseded by, others. The initial attractions are usually the fixed, primary ones which in time are followed by a succession of footloose, secondary ones. Nightlife entertainment and activity expands in coastal resorts, golf courses and theme parks are developed on their inland margins. Variations to the general pattern are to be found; for instance, in the 'village' form of American ski resorts where Dorward (2006) argues that 'entertainment retail' might be considered the dominant attraction

Destinations at various scales and stages of development will offer a variety of attractions and activities classified in these ways. Patterns of co-location will be significantly influenced by and organized as a function of the location of the initial, fixed, primary attractions and activities. In contrast to these, the buildings and facilities of the other sectors – accommodation, transport, supporting services – are purpose-built and although they may be subject to site constraints they are essentially footloose and can be located as a function of the fixed, primary attractions. As the research on urban tourist districts shows, supporting services are very spatially dependent, not just on co-location with regard to attractions but also in respect to accommodation and transport. Within countries, regions and at a local level the nature and distribution of tourism resources that serve as primary attractions or provide the base for activities thus play a dominant role in shaping the spatial structure of destinations. Figures 8.1, 8.2, 8.6, 8.7, 9.1 and 9.2 portray different manifestations of an extremely close relationship between attractions and accommodation.

This is not to take an overly deterministic view as these resources must first be exploited and marketed. The beaches of Languedoc-Roussillon had little more than local usage until the creation of the regional strategy which led to the construction of La Grande Motte and other large new resorts and pleasure ports along that coast. It was not until the Plan Neige was implemented that the higher-altitude slopes of the French Alps were developed as the ski-fields of the new integrated resorts. But when these operations were undertaken, both driven by a very functionalist approach, it was proximity to the beaches and slopes respectively that drove the location of the other services.

Changes may occur over time as primary attractions evolve or new ones are created. This is most evident in cities where some displacement of the CBD has occurred or urban redevelopment projects have been undertaken, as here the underlying attractions are contemporary and created and much activity is related to business travel. Where the primary attractions are physical resources such as beaches and ski slopes large scale re-orientation away from these is rare. Moreover, technical complementarity in cities is generally less than in coastal and ski resorts as the greater diversity of products and services in the former means more substitutes for each other and thus more alternative locations exist there (Terhorst and Erkuş-Öztürk, 2011).

Proximity to attractions and activities, however, is not the only factor which comes into play for both visitors and providers. In selecting accommodation or places to eat out visitors also take into account other interrelated factors such as price, service quality, accessibility and amenity. Some will seek the hustle and bustle which dense concentrations of attractions and activities bring; others will long for the peace and quiet of more isolated places. Trade-offs may therefore be made, for example, selecting a more distant but cheaper hotel or one which is further away, more expensive but in a nicer setting. In addition to ease of access to attractions and activities within a destination, external accessibility is also important, as seen in the clusters of accommodation close to gateways associated with various forms of transport: airport and railway terminus hotels, strip motels along major entrance routes. Moreover, the attractions and activities that appeal or are the most significant will vary by type of visitor with the result that proximity to a variety of features will be viable locations for other services depending on which segments are targeted. Distinctions between the motivations, activities and needs of different segments are essential, especially between leisure and non-leisure visitors. At a national or regional level much of the demand for hotel accommodation in cities comes from business travellers while at the local level such demand is a major factor in the location of hotels close to the current CBD where meetings are held and business is transacted. The relationship of proximity to activity still holds here; it is just that the activity of business travellers is different to that of holiday makers and may occur in other places.

Providers of accommodation, supporting services and footloose secondary attractions seek to satisfy demand by taking account of customer preferences in these respects but will also take into account other interrelated factors when making locational decisions.

Agglomeration externalities

Most attention has been given to the benefits of agglomeration at both a local and regional scale, particularly for hotels. The emphasis here shifts from questions of resource endowment and proximity to attractions and activities to the positive externalities which arise from locating close to other hotels and other activities. Local level studies suggest heightened demand results from reduced search costs and assurance for customers and overflow from adjacent properties. In addition, agglomeration allows for closer monitoring of competitors and provides signals about the feasibility of sites. The benefits of agglomeration may also change over time and vary between different types of accommodation. Destinations having a large share of non-hotel accommodation will generate demand for more supporting services than resorts characterized by all-inclusive hotels. Agglomeration is also evident in the clustering of souvenir and gift shops, cafés and restaurants and other tourist-oriented services such as foreign exchange outlets close to major attractions, a process which contributes to the emergence of tourist districts within cities. Two location forces are again at play: being close to the attractions to draw on passing customers and attracting customers through the greater choice which multiple services located in one place offer (Love for Variety Theorem; see Andergassen *et al*., 2013). Research at a regional level suggests growth and performance are variously affected by localization-driven agglomeration (i.e. inter-industry specialization, in this case tourism) and urbanization-driven agglomeration; the former may be more apparent in coastal and mountainous regions and the latter in major urban regions but more work is required to establish the generality of this pattern. Negative externalities associated with agglomeration include congestion, loss of amenity and pressure on infrastructure but little attention has been given to these from a location perspective although they are recognized in impact studies.

The advantages of agglomeration may reduce or outweigh competition for customers and staff from other nearby providers in the same sector. Firms may seek agglomeration benefits by co-locating with those which are similar in some respects but different on other attributes, thereby obtaining a competitive advantage (Baum and Haveman, 1997).

Land prices and availability

The ability to support land prices varies across different types of services so that competition also arises with respect to the price of land, both within and between tourism sectors as well as between tourism and other forms of land use. From a tourism perspective land prices generally decline with distance from the major attractions; in terms of competition from other sectors, the latter may be able to support higher prices which factors such as the benefits of centrality also bring to them. The nature and extent of land-use competition will vary according to the type of destination: in urban areas competition will occur with other tertiary sectors as well as within the tourism sector; in coastal areas it may primarily be within and between tourism sectors or between residential housing and second homes.

Policy and planning

Incentives to develop tourism as part of regional development schemes or redevelopment projects may affect the location of destinations or their internal structure. To be realistic, however, these will need to be underpinned by some form of attraction or have the potential to develop them. In other instances planning constraints may limit forms of development, such as restricting nearby construction to protect and conserve natural or cultural attractions or containing entertainment districts so as to reduce conflicts with other forms of land use.

Shared infrastructure

Particularly in terms of transport facilities and public utilities, shared infrastructure may also reduce costs for individual businesses.

In addition to these general features patterns of co-location in particular destinations will be conditioned by their individual physical setting and by the existing and evolving built environment in the cases where tourism has developed within an existing settlement.

Spatial interaction

Other research, particularly research which takes a 'travel to' approach to destinations, contends that destinations cannot be understood in isolation. Rather, attention must also be given to examining patterns and processes of spatial interaction and analysing the relationships between different places, including other destinations and origins. These studies have been rather unequal in terms of their focus. Applications of GIS, GPS and the analysis of big data sets have seen a marked increase in the analysis of tourist movements within local level urban destinations and the links these form between attractions. Network analysis has proved a useful means of establishing regional structures through examining linkages between local destinations, identifying the impact of position and associated nodal functions. The limited amount of research on circuit travel at a national scale has focused on international visitors; research on domestic travellers has primarily involved origin-linkage analyses. Inter-scale interactions remain largely unexplored. Regional network studies, for example, do not consider activity patterns within different types of nodes in the network.

In terms of spatial interaction, two basic types of destination nodes might be distinguished based on the travel patterns and motivations of the visitors they receive. Mono-destinations are those places which constitute the sole focus of an overnight trip; they are only connected to the origin (though possibly accessed through a gateway). Circuit destinations are visited in association with one or more other places, as part of a short or extended itinerary. These may be sub-divided on the basis of the various functions they might perform on the circuit: gateways, hubs, primary or secondary destinations. Destinations at a national and regional scale may consist of a set of local level mono- or circuit destinations or, as is more common, a combination of the two. Moreover, some local destinations will have multiple functions; that is, they will be visited in varying degrees by a mix or mono-destination and circuit travellers. Similar concepts might be envisioned within local destinations where the nodes are attractions or clusters of attractions.

International tourists are by definition visiting another country but most are only visiting a small part of it. Circuit tourism may be popular with some segments and more common in some countries than others but intranational circuit travel is generally rather limited (Chapter 5). Many international tourists visit only one or a few regions or local destinations; commonly, though not necessarily, the same ones. The degree of dispersal may depend in part upon the number and distribution of international gateways with travel onwards from these being subject to the friction of distance. Long-haul tourists tend to travel more extensively within the countries they are visiting than those from neighbouring countries, often attempting to see as much as possible in a one-off trip. Where circuit travel has been analysed at a national level distinct patterns appear which link gateways, primary and secondary destinations. Overlap occurs in these functions. Major urban areas frequently serve as both gateway and primary destination for circuit travellers and this, coupled with the amount of mono-destination traffic they receive, will contribute to a hierarchy of nodes.

National origin–destination studies show domestic tourist travel exhibits a marked distance decay effect with the result that much domestic tourism occurs in the home region or involves travel to neighbouring regions. Such studies also show regions act both as origins and destinations, the balance between these two functions depending in large part on their population size and tourism resources.

As Chapter 6 has shown, network studies have begun to reveal regional structures, identifying the links, sometimes of great complexity, between places and their different nodal functions within the network: gateways; hubs; primary, secondary, peripheral and touring destinations. Some functions are in part due to the relative position of places in the network. Hub-and-spoke travel may also occur. Where domestic travel is concerned some nodes, perhaps most, will act both as origins and destinations. Network studies tend to omit mono-destination visitors and thus understate the role of primary destinations, often major urban centres, within regions. In general, much more research is required to establish clearly the extent to which various regional nodes are interconnected and together function as a nodal regional destination. The sparse evidence available suggests that the proportion of regional mono- and circuit-destination visits or single- and multi-island visits varies by such profile characteristics as nationality, purpose of trip and first/repeat visit, but that the extent of intraregional travel is rather limited.

Local level studies are providing an increasing amount of detail on the ways in which tourist movements within urban destinations link attractions in discernible patterns (Chapter 7). The general pattern is one of intensive activity within the central city where tourists move between iconic landmarks, occasionally visiting some further out as well as taking in various secondary attractions. The monumental, however, is also complemented to some extent by the banal. Inclusion of shopping among the activities undertaken adds other clusters which may broaden this pattern. Patterns of spatial behaviour vary by visitor segment. Research to date has focused on first-time and repeat visitors and differences between nationalities. Repeaters tend to be more selective in terms of the places they visit and take in attractions and places which offer new experiences while also revisiting favourite places. Urban destinations are made up of different sorts of places and spaces which are shared in varying degrees with local residents. In some cases this has led to the identification of different functions in particular parts of the city.

Compared to spatial interaction research at a regional or national scale, more attention has been given to explaining patterns of behaviour within cities. Urban morphologies, time/space budgets and mental maps emerge as key factors among those travelling for pleasure. Urban destinations, however, attract a wide range of segments and little research has been done on other types of visitors. There has been little attempt to compare explicitly the spatial behaviour of those who are making a city or town the sole object of their trip and those visiting it as part of a larger circuit. Their intra-urban patterns, however, may not be that different as similar time constraints may apply in each case given that much mono-destination travel to cities is in the form of 'short break' trips and like the stays of circuit travellers these are of short duration. In contrast, stays at coastal and ski resorts generally involve mono-destination trips and longer stays but detailed research on activity patterns within such destinations is lacking. The progressive extension of ski-lift networks suggests certain groups of skiers have active and extensive patterns of activity. Tourists at coastal resorts may take a day trip into the hinterland but tend to remain within the resort. Some rural tourist travel may be linear in nature and hub-and-spoke patterns also occur but more research is needed to cover the diversity of rural and natural area destinations.

Distribution

The empirical evidence presented and reviewed in Chapters 5 to 8 clearly shows that tourism is not distributed uniformly or randomly at a national, regional or local level. Rather, distinct spatial concentrations occur within countries, within regions and within local destinations. Chapters 5 and 6 show that the extent of supply-side concentration within countries is in part a function of the measures used: the narrower the range of measures applied the more spatially concentrated the distribution of supply generally appears. Important distinctions might also be made between absolute and relative measures of distribution.

Country-wide supply-side distribution reflects a combination of international and domestic demand. International demand within countries is geographically more concentrated than domestic demand but may become more dispersed over time as markets mature. The concentration of supply is intensified in those areas where domestic and international demand coincides but is more dispersed in areas where domestic demand is dominant. The hierarchy of concentrations which are evident at a national scale tend to reflect a mix of urban destinations, with varying levels of concentrations in one or more of the other physical settings depending on the underlying geography of the country concerned.

Distribution patterns within formal administrative regional units suggest international demand is again more concentrated than domestic and that business travel is more concentrated than leisure travel. Consideration of the distribution of tourism at both a national and regional level (Chapters 5 and 6) does, however, raise questions about what constitutes a tourist 'region' or regional destination. The examples in Chapter 5 that use local level data (municipalities or LLSs) show that tourism is concentrated at this scale or in clusters of these local units rather than at a larger regional scale. The majority of the studies examined in Chapter 6 concern the analysis of distribution within the largest sub-national administrative units (states in the USA and Australia), which are broken down to tourist regions. Tourism within these larger territorial units is also concentrated geographically, a hierarchy of local destinations exists and spatial variations occur with regard to the types of local destinations, their structure, functions, levels of dynamism and types of visitors they attract (Figs 6.1–6.4; Tables 6.2 and 6.3). Major urban centres commonly form the apex of the hierarchy, supported by other primary and secondary destinations. The issue here is whether these local level concentrations, or even clusters of them, are simply independent sub-systems within a larger territorial administrative unit or whether they constitute nodes that are functionally linked and form part of a larger, regional system or destination whose boundaries are defined by such links. Where coastal tourism is dominant the distribution of supply and demand is predominantly linear, with a highly developed but differentiated coastal zone complemented by smaller, scattered localizations in its hinterland. In alpine areas variations occur with altitude and access.

At the local level, visitor activity, attractions and accommodation are concentrated in particular parts of urban destinations, especially the central city, and different types of activity spaces, districts and precincts can be identified. Variations occur from city to city: some cities have more or fewer clusters than others; some will depend primarily on heritage attractions, others will offer a wider range of districts (e.g. ethnic districts, redeveloped waterfronts, 'bubbles' of entertainment or sporting facilities). The linear development of coastal tourism at a regional level is repeated within coastal resorts to give them a distinctive morphology. This characteristically consists of a series of parallel strips of accommodation which exhibit a gradation in terms of height, density and price away from the beachfront and a core of tourist-related shops, bars and restaurants. Local variations to this general morphology result from different physical settings and existing settlement patterns. Differences in mountainous environments, patterns of alpine settlement and development processes have influenced the extent to which the physical association between ski-fields, accommodation and supporting facilities that characterizes the morphology of ski resorts varies with altitude and from country to country.

Differences also occur between these destinations in terms of the physical elements which are distributed, not just the urban, coastal and alpine attractions whose different attributes are difficult to standardize, quantify and compare. Second homes and rental apartments are significant components of accommodation in many coastal and ski resorts but this is not the case with urban destinations where commercial accommodation, particularly hotels, and stays with friends and relatives dominate. Camping grounds are especially common in coastal and rural areas; they are often located on the outskirts of cities but are not usually a feature of ski resorts. Airbnb accommodation is found in different types of destinations but has been given particular attention in cities through its spread into residential areas. Different types of facilities are used by different visitor segments. Much urban hotel demand comes from business travellers; most visits to urban attractions come from leisure tourists or VFR visitors rather than those travelling for business whose activity patterns are not explored. These differences tend to be overlooked or taken for granted in studies related to the distribution of accommodation, attractions and activities and need to be given greater consideration in explaining patterns.

These three dimensions are closely interrelated and together create the distinctive features of the spatial structure of destinations at different scales or in different settings. Geographic concentrations of tourism phenomena, the development of a hierarchy of local and regional destinations and the tendency for international demand to be more concentrated than domestic tourism result from this interplay. Concentration is an outcome commonly produced by the co-location of multiple tourism services around initial, primary fixed attractions. Such attractions are limited and found in varying quality in particular places. Some places whose attractions alone would not justify a separate trip are visited largely by tourists en route to other destinations. Attractions and activities that are sought by international visitors, who are generally investing more time, money and effort to reach the destination than domestic travellers, are likely to be more distinctive or of higher quality and thus fewer in number. International travel will also be channelled through a limited number of gateways from which the onward flow of visitors will be constrained by a distance decay effect. At the same time, the friction of distance limits much domestic travel to regions around or near to the major urban areas, the source of much domestic demand.

Other factors in addition to an uneven distribution of tourism resources play a role in the development of a hierarchy of destinations, notably economies of scale, position and access, local development initiatives, agglomeration processes and the urban hierarchy. In particular, greater attention might be given to the nature of the urban hierarchy within countries and regions and the multiple roles that large cities play, not only as destinations but also as gateways, stopovers and origins; roles which become clearer when issues of co-location are complemented by a focus on spatial interaction. Cities' functions as important transportation nodes are in large part a consequence of broader processes of urban agglomeration, processes which also generate a significant amount of non-leisure travel. Greater clarity is needed in this regard in terms of distinguishing the wider range of visitor segments which urban destinations attract and how that affects the distribution of supply-side phenomena within them, particularly of hotels.

In general, greater consideration might be given to the notion that places may have multiple functions, to the concepts of mono-destinations and circuit destinations and to how these may vary according to different segments, at different places and at different times. In this latter regard it is useful to recall Rodrigue *et al.*'s (2006, p. 272) definition of spatial structure relating to the '*cumulative* locations of infrastructure, economic activities and their relations' (emphasis added). While many of the examples discussed in the earlier chapters deal with patterns at a particular point in time, others show how patterns evolve due to changes in such factors as modes of transport, tourist preferences, local and central government policies, company strategies and processes of agglomeration.

In summary, from a spatial structure perspective destinations might be viewed as geographic concentrations of tourism phenomena characterized by the co-location of multiple, interdependent visitor services close to primary, fixed attractions and activities. National and regional destinations consist of hierarchical sets of local level mono-destinations, circuit destinations or some mix of the two depending on the roles they play for different segments and the travel patterns of those visitors. Circuit destinations of different types are connected in varying degrees by flows of circuit travellers.

Organizational Structure

The organizational structure of destinations was introduced in Chapter 1 in terms of the various configurations and ways in which and the extent to which multiple actors, collectively and individually, behave and interact to fulfil the needs and produce the range of services and experiences sought by tourists in a given place. Individual organizational studies of destinations set within a particular conceptual or theoretical framework have almost invariably been framed in terms of a single conceptualization and give little or no consideration to other approaches (Chapter 3). These studies give varying weight to particular structural features but common points did emerge from the review which could be categorized along the three dimensions set out in Table 3.1. In summary, Table 3.1 suggests that destinations have distinctive place and space characteristics and can be viewed as a mode of production, as some form of meta-agent characterized by the co-location of multiple, mutually dependent providers who supply visitors with all the products and services they need. While continuing to compete individually, firms also cooperate due to the complementary nature of these services which creates a situation of interdependence and results from the spatial proximity which co-location brings. The co-location and synergies of multiple interrelated firms are also said to create positive externalities. Cooperation is based largely on trust, shared values, a common resource base and efficiencies achieved through economies of scale. Levels of cooperation and coordination may vary from destination to destination, be formal or informal, supported by different degrees of institutionalization and focus on both strategic and operational matters. The overall emphasis is on the nature and configuration of the relationships between multiple, small and medium-sized firms. Destinations are dynamic and evolve due to a variety of driving factors.

Some disagreement occurs; notably that co-location does not necessarily result in cooperation and that it is the tourists, not the providers, who co-create their experiences and who are the ones who select and bring together the services they require (Saraniemi and Kylänen, 2011; Clergeau and Violier, 2013; Briassoulis, 2017). Baerenholdt *et al.* (2004) argued that viewing destinations as networks is rather tenuous because different types of networks can be identified, only some of which are territorially based, and because businesses can belong to several types simultaneously. Some of those networks include external partners, a point made by several other writers with respect to distribution channels and value chains (Bocquet, 2008; Fabry, 2009).

A number of the studies discussed in Chapters 1 and 3 are theoretical in nature. Others go on to apply or test their concepts empirically. Such studies and other empirical research discussed in Chapters 5 to 9 and which is framed in terms of a particular conceptualization provide some evidence of the extent to which destinations are structured and function in the ways the frameworks suggest. While some of the empirical work is comparative in nature there has been little systematic attempt to examine whether organizational structures vary from one type of destination to another or in terms of scale in order to see what commonalities or differences exist, why they occur and to draw these findings together.

The number of examples and the range of approaches limit generalization but studies which have examined the inner workings of destinations using surveys, qualitative research and SNA show varying levels of empirical support for the conceptual frameworks in question. The coastal resorts analysed did not appear to be functioning as clusters. Inter-firm relationships as measured by such factors as joint strategic activity, packaging of products and exchange of information were generally not well developed (Varisco, 2004; Costa *et al.*, 2006, 2012; Perles-Ribes *et al.*, 2017). Some of the nature-based and urban destinations were seen to exhibit characteristics of clusters (Huybers and Bennett, 2003; Jackson and Murphy, 2006; Acuña Ortega and Brugnoli, 2007) but others showed few signs of co-operation and lacked well developed levels of interaction (Barbosa and Zamboni, 2001; Miranda Júnior *et al.*, 2016; Gómez Diaz *et al.*, 2017). Attempts to develop rural clusters also met with challenges in these regards (Czajkowski and Cunha, 2010; Partalidou and Koutsou, 2012; Pejanović *et al.*, 2017). In general, these empirical examples lend support to Fabry's (2015) contention that destinations are not clusters but destinations may host clusters.

Mottiar and Ryan (2006) make a strong case for Killarney functioning as an industrial district as does Cirer-Costa (2013, 2014a, 2014b) for Majorca and the Balearics. These studies stress the relationships between firms and the ways in which tourism is embedded in the local or island environment. Local embeddedness, however, does not always occur as Marcelpoil and François (2008) noted with respect to French ski resorts.

SNA has proved a useful means of examining the strength and configuration of relationships among destination stakeholders but has generally been used with reference to a single or a few dimensions rather than as a means of providing a complete picture of interaction within a destination. SNA is especially useful as a tool for comparative research as it more readily allows a standard approach to be applied across several destinations. Comparative network studies have drawn attention to differences between destinations, network structures and activities undertaken. Timur and Getz (2008) identified variations in influential stakeholders in three North American cities and the degree of centrality in their networks. Pulido-Fernández and Merinero-Rodríguez (2018) showed levels of tourism development in small to medium cities in inland Andalusia were related to differences in network structure and the type of activity engaged in by network members. Differences exist in the institutional networks for rural tourism in Ireland and France and in the domains in which they are active (Cawley *et al.*, 2007). The network characteristics of high- and medium-altitude ski resorts in France differ, with the networks formed by the resorts varying according to the type of innovation developed and the network characteristic (Favre-Bonté *et al.*, 2016, 2019). Favre-Bonté *et al.* (2016) also suggested that ski resorts could no longer be treated as localized productions systems but needed to form part of much more geographically dispersed networks in order to innovate.

Other research to determine structural and behavioural factors that lead to the successful functioning of tourism networks has identified trust, the development of social capital and a willingness to balance individual and collective interests as key elements. The concept of coopetition was seen to have relevance to both firms and DMOs (von Friedrichs Grängsjö and Gummesson,

2006; d'Angella and Go, 2009; Daskalopoulou and Petrou, 2009; Nguyen and Pearce, 2015; Chim-Miki *et al.*, 2019). Van der Zee and Vanneste (2015) concluded from their review of network collaboration in tourism destinations that the empirical evidence of the benefits of networking was limited and far from conclusive in terms of the four major and largely non-convergent strands of research they identified: business networks, policy networks, 'coopeting' networks and network configuration.

An underlying assumption in many of these studies, especially those framed in terms of industrial districts, is that the providers at a destination primarily consist of a group of similar SMEs. Differences between these and the presence and role of large actors or focal firms have often been ignored. There has been little attempt in the empirical studies to explicitly examine the market structure of destinations with regard to buyer and seller concentration, product differentiation, ease of entry and associated questions of perfect competition, monopolies or oligopolies (Akehurst, 1984; Bull, 1995). Pan (2005) found market concentration in rooms could significantly improve international tourist hotels' profitability in Taiwanese hotels. Scale economies and market size were identified as key factors driving the evolution of the market structure of water towns in China as assessed in terms of their core heritage zones operated by a government-owned tourism management company (Ma *et al.*, 2015). Other research focused on market structure has tended to come from commercial reports rather than academic studies (Horwath HTL, 2018; MarketLine, 2018). These studies and reports tend to focus on a particular sector or product rather than across the destination as a whole. When included in studies of urban destinations and coastal resorts product differentiation has been shown to be a key feature which enables different types of co-located accommodation to co-exist (Baum and Haveman, 1997; Urtasun and Gutiérrez, 2006; Cirer-Costa, 2013a). Erkuş-Öztürk (2009) demonstrated the role of firm size as the key factor in international and local connections. Chapter 9 showed monopolies exist with ski-field operations and may also occur also with natural area operations but this is not generally the case with urban and rural tourism. Even when the primary, fixed attraction is a monopoly, opportunities may arise for smaller ventures to cater for niche markets or develop complementary products nearby (Viken and Aarsaether, 2013). However, many of the organizational studies do not take account of patterns of demand, that is, who is providing for which segments and how might this influence the patterns of interaction? In particular, one might anticipate different degrees of interaction between those firms catering primarily to leisure and non-leisure segments. At a national level, conditions of market entry and product differentiation are seen most clearly in the research on the evolution of air services and rise of low-cost carriers (Rubin and Joy, 2005; Reynolds-Feighan, 2010, 2018). Greater incorporation of these issues in destination research would add to our knowledge of how destinations are structured and how firms within them behave. More explicit consideration of product differentiation, for example, might shed further light on co-location and the nature and extent of coopetition while greater attention to supplier concentration would add to our understanding of the structure and workings of networks.

Commercialization and bundling

Not surprisingly, studies set in the organizational literature tend to emphasize commercial operations and the firms involved, with some also acknowledging the role of the public authorities and/or the community. For certain forms of tourism and types of destinations, however, this neglects significant non-commercial sectors: in some situations visitors essentially look after themselves in varying degrees without requiring the services of a commercial provider. Visitors can, for instance, use public open spaces or relax with a good book; travel by private car; stay in holiday homes or with friends and relatives; and self-cater using supplies they have brought from home. Such visitors might be termed self-providers. Moreover, visitors using commercial services do not necessarily require everything they need to be bundled together and delivered in a single package; they may prefer to make their own arrangements, selecting and purchasing what they need from providers, where and when they need it (Pearce, 2008). A broader and more comprehensive approach to the organizational structure of destinations could thus begin by asking what needs do visitors have, what services do they require when and where, and who provides them? In this way demand would be linked explicitly to supply (to the providers) via the services needed, with allowance being made for self-provision.

Two key structural dimensions in such an approach are the degree of commercialization in any sector, sub-sector or destination and the extent to which services in them are bundled together in some way or delivered separately (Fig. 10.2). Demand and supply in a destination can be arrayed along a commercialization axis from completely self-sufficient (no use is made at all of commercial services) to wholly commercial (all the visitor's needs are met by commercial providers). On the bundling dimension, demand ranges from the fully independent tourist whose services are all purchased and provided by separate suppliers to the completely dependent tourist for whom all the services are provided either in an all-inclusive package or a complete set of customized arrangements. Completely self-sufficient visitors might also be considered to be at the latter end of the bundling dimension in the sense that they themselves are taking care of everything they need for their holiday.

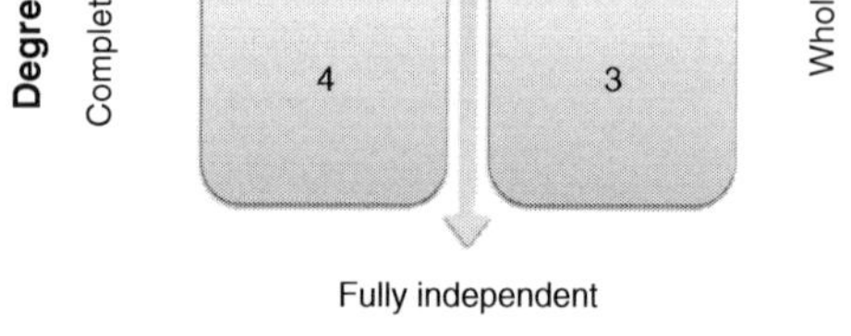

Fig. 10.2. Commercialization/bundling matrix for destinations.

The demand and supply situation at these extremes might then be translated into different types of destinations. The extreme case in Quadrant 1 might be that of a domestic coastal resort consisting entirely of holiday homes, with visitors arriving by car, spending their time at a public beach and catering for themselves. In Quadrant 2 the extreme case might be an exclusive coastal resort consisting of high-class hotels and private beaches where everything, including all transport, is provided in an all-inclusive package. The Quadrant 3 extreme is similar to that in Quadrant 2 except no packages would be on offer and tourists would purchase everything separately. No extreme would exist in Quadrant 4 as a holiday away from home could not be taken under the no self-bundling at all condition.

While examples of the extreme types of destinations might be found, demand in most destinations would more likely consist of a mix of segments having a variety of different needs provided by a range of suppliers. This in turn raises a series of interrelated questions with implications for the organizational structure of destinations.

- If services in a destination are only partially commercialized, in which sectors are services predominantly commercial or provided privately?
- Who is providing the commercial services?
- What are the relationships between these sectors and the suppliers concerned?
- If services in a destination are partially bundled, which ones are predominantly bundled and which are mainly independent?
- Who does the bundling?
- And, importantly in terms of synthesizing, are there common responses to these

questions or do variations occur across destinations?
- How and why do these variations occur?

Differences between coastal and ski resorts, and in particular the attractions and accommodation sectors in these, illustrate some of these points. As with their morphology, access to the attractions is again a determining factor. In contrast to beach-goers who generally, but not always, have free access to the beach, skiers have to pay to access and use the slopes. As discussed in Chapter 9, ski-field operations are monopolistic in nature and the close operational and financial interconnection between attractions and accommodation has stimulated a significant research thread on the organizational structure of ski resorts from a range of organizational and managerial perspectives which has revealed a range of destination types (Fig. 9.4 and Table 9.1).

Real-estate promotion and development, not only of hotels but also of holiday homes (Figs 8.2 and 8.6), may drive the development of coastal resorts as well but generally there is no dominant attractions operator *per se* leading this process. Rather, such activity is fostered by improvements in infrastructure and external access, whether undertaken by the public or private sectors. Rampant real-estate development, either by individuals or commercial developers, may also result from a lack of effective planning controls. Closer links between sectors may occur with privatized beaches and with marinas when an enterprise or agency develops both the accommodation and mooring facilities, either in a pleasure port or as individual moorings on a canal estate. Similar relationships are found with the development of holiday housing estates around golf courses on the inland side of coastal resorts. Where second homes predominate there will thus be a large number of visitors who are staying in non-commercial accommodation, a sector ignored by models derived from the organizational literature. In these cases a major change in the structure of the destination may occur with the transition from the development phase dominated by the interplay between a limited number of developers and government agencies and a subsequent phase characterized by many individual owners and local government. Resort apartments and villas may also be rented out through letting agencies, actors who are largely absent in the literature but who play a key role in linking private providers with holiday makers. When hotels constitute a significant proportion of the rooms available, major hoteliers may be the dominant actors not only in the accommodation sector but in the destination as a whole.

Urban destinations differ significantly from coastal and ski resorts in these regards. First, there is generally little second-home development in towns and cities but due to their resident population size a relatively large degree of non-commercial accommodation may be provided privately by those hosting visiting friends and relations (Table 7.1). Privately owned accommodation is increasingly being commercialized through platforms such as Airbnb, though this is not restricted to urban destinations. Second, as Chapter 7 showed, urban destinations offer a wide range of substitutable attractions and activities, many of which are commonly in the public domain and managed by public authorities. Much urban tourist activity, such as wandering the streets and squares and looking at the sights, may involve little direct commercialism with expenditure being incurred to a greater extent on supporting services in the hospitality and retail sectors. As a result, the market structure of urban destinations may be very fragmented; there may be no single dominant sector, sub-sector or enterprise across the destination; the DMO may assume a leadership role and local government becomes a key provider and destination manager.

The market structure of many rural destinations may also be very fragmented due to the small size of the enterprises involved and these may have to work together to create attractions such as wine trails with sufficient scale to attract visitors. Protected areas such as national parks are often characterized by the roles which state or regional land management agencies have in managing the resource base which attracts visitors to the area and in regulating the development and operation of commercial ventures. But, as noted with the North Cape example, private-sector monopolies on natural area attractions do exist.

The bundling or packaging dimension has attracted more attention. The issue here is who combines the different destination products and services required by visitors: the suppliers, intermediaries or the visitors themselves. Although intermediation is generally seen from a marketing perspective it has also been included in some conceptualizations of destinations as a key function which bundles products and services, links providers and visitors and connects the destination to markets (Ejarque, 2005; Bocquet, 2008; Andergassen *et al.*, 2013). Particular emphasis has been given to tour operators' control of destination offerings in small islands and some coastal regions in the Mediterranean (Chapter 8). Destination firms may also be portrayed as collectively creating 'a bundle of separate products and services that the consumer effectively purchases as a single item' (Michael, 2007, p. 39); a view that is challenged by those for whom the key actors are not the firms but the tourists who co-create their own experiences as they select and purchase the services they need (Saraniemi and Kylänen, 2011; Clergeau and Violier, 2013) or activate multiple tourism assemblages (Briassoulis, 2017). The advent of online sales and the development of platforms such as Airbnb have facilitated interaction between visitors and suppliers. As noted earlier with reference to Fig. 10.2, the two extremes – all-inclusive package/fully independent - are found in terms of both destinations and visitor segments. However, bundling or independence is not necessarily an 'either/or' situation. Often it is a question of degree and a more nuanced approach is needed, one that considers the nature and extent of bundling and examines which services are offered and purchased jointly and which are sold and bought separately. For instance, visitors might buy a fly/drive package which combines air transport to a destination with the use of a rental car once they arrive but which leaves them free to make their own accommodation arrangements and decide on the activities which interest them. Or the package might combine air transport and hotel accommodation or involve a ski package which offers accommodation, a lift pass and ski hire after the skiers make their own way to the resort.

Empirical research shows differences in the type of travel arrangements occur by origin and clearly show the role of wholly or partially independent travellers should not be overlooked. Zhang *et al.* (2004) distinguished between a sample of outbound leisure travellers from Hong Kong according to whether they took an all-inclusive package (65.1%), a basic package tour (hotel plus air ticket) (8.2%) or engaged in independent travel (non-package tour) (26.7%). Pearce *et al.* (2009) segmented New Zealand outbound travellers into three groups depending on the way their travel arrangements were made: independent (69%); package (11%); and an intermediate category, 'package plus' (20%), composed of those for whom part of their travel was packaged and part was arranged independently. Comparable data from a destination perspective are less readily available. However, many destinations might be expected to contain a mix of visitors segmented by travel arrangements with variations occurring according to their relative dependence on all-inclusive package tourists, independent travellers or some intermediate group.

Ejarque (2005) distinguished between those destinations characterized by package tourism and dependency on tour operators from 'diffuse' destinations lacking cooperation and coordination among suppliers and a clear image in the market and integrated destinations where the different stakeholders have been forced to come together to remain competitive. He provided some examples of each but many destinations might attract a mix of visitor types based on the extent of bundling that occurs in different segments. Figure 7.11 suggests there is an evolutionary dimension to stakeholder relationships as the distribution and vending function is foremost in the most highly developed destinations of Pulido-Fernández and Merinero-Rodríguez's (2018) study. More comparative studies are now needed to identify how and where variations in the extent of bundling and intermediation occur. The links between visitors and providers also need to be explored in more detail. In the case of diffuse destinations visitors are essentially obliged to select and assemble the services they need while in dependent destinations it is more a question of choosing a particular package.

More research also needs to be directed at which sectors are bundled together to a greater or lesser degree. Research in New Zealand shows that although attractions are combined and packaged with other elements in group tours it is often the sector where little bundling occurs and where many attractions and activities are bought by visitors once at a destination and where many providers favour 'at destination' distribution (Pearce *et al.*, 2004; Pearce and Schott, 2005; Pearce and Tan, 2006).

This focus on the degree of commercialization and extent of bundling not only provides a further means of differentiating between destinations but in doing so also draws attention to two groups of stakeholders whose roles are often understated or completely absent in some of the destination frameworks presented in Chapter 3: tourists and intermediaries. In varying degrees tourists may not only be consumers but also self-providers, catering for many if not all of their needs themselves. When they require additional services they may select and buy what they need from the suppliers at the destination. In other instances this bundling is carried out by intermediaries (outbound or inbound tour operators, travel agencies or perhaps local tourist offices). Both forms of bundling tend to put into question the notion underlying many of the Chapter 3 frameworks which suggest that the destination is acting as some form of aggregate entity or meta-agent offering a complete visitor experience (Huybers and Bennett, 2003; Michael, 2007). Certainly a range of interrelated products and services are being offered in most destinations by multiple providers and a degree of interaction may occur between them in the delivery of the total visitor experience, for example, accommodation providers referring visitors to attraction and activity operators. However, the assembling of these products and services, a critical end-stage function, might more commonly be done by one of the other two stakeholders, either the tourist or an intermediary, one that is often located outside the destination.

Bundling is only possible though if those products and services are in place and their existence is known. Other destination functions will facilitate this, such as planning, product innovation and joint marketing. These functions may in turn also be affected by a destination's organizational structure and more closely exhibit patterns and behaviours suggested by the frameworks in Chapter 3. This is especially the case in joint marketing efforts to achieve economies of scale and develop a stronger and more cohesive image in marketing the destination, collaborative efforts generally coordinated by a DMO. Effective plans, whether for tourism development, management or marketing, also benefit from input from a wide range of destination stakeholders working together. Networking, including not only internal but also external partners, may stimulate innovation in product development (Tables 6.6 and 6.7). Local values may govern resource use. Supportive public policies may provide the infrastructure, funding and regulatory frameworks needed for tourism businesses to operate successfully and in a sustainable manner. The interaction and cooperation suggested by the various frameworks are integral to all these functions but as many of the empirical examples show such behaviour is not a given and does not necessarily occur.

A way forward

The literature reviewed has shown few cases where the empirical examples clearly and fully demonstrate the characteristics of the ideal forms associated with particular conceptualizations of destinations. Several factors might explain this. First, it may be that reaching the conditions ascribed to a cluster or industrial district is an end state that takes time to achieve and that few of the cases discussed represented mature destinations but rather were still in their formative stages when studied and still evolving. Networks are somewhat different in this regard as there is less of an emphasis on an ideal type of network and most network studies focus on a particular type of network function (e.g. innovation, policy-making or governance) and do not claim to provide a comprehensive picture of a destination. Second, it may be that the wrong framework has been applied to a particular destination or type of destination with the result that it does not reflect the features expected. However, the analysis of different types of destinations – urban, coastal, alpine, rural – in different countries shows no real tendency for any particular type to display the characteristics of any specific form more or less than another, be it a cluster, industrial district or network. The diversity of cases may suggest features of an assemblage may apply but there are as yet insufficient empirical examples of this concept to draw any conclusions in this regard. Third, there is the more fundamental issue of whether the application of some framework from outside the domain of tourism is appropriate and whether destinations might be expected to share their characteristics.

Even when little evidence is found to show a destination is not structured and functioning as the ideal form of a given concept would suggest, no consideration is generally given as to whether it might be better conceived of or framed in some other way. Framing a destination in terms of a single concept provides some theoretical and analytical structure but also constrains the scope of the research and does not allow for the possibility that different types of organizational structures occur in different settings. The ski resort research illustrates this issue well: some ski resorts exhibit some characteristics of networks, industrial districts and clusters but when other approaches are adopted different types of destinations are also revealed. Little provision is made in the Chapter 3 frameworks for the corporate American model presented by Flagestad and Hope (2001) and illustrated by the examples in Chapter 9, nor for the range of situations identified in the French Alps by Bocquet (2008) when she applied Torre and Rallett's (2005) spatial proximity/organizational proximity matrix (Table 9.1). It may be that organizational structures are more complex than spatial structure, that destinations cannot be characterized by just one set of relationships and represented by a single organizational configuration.

One way forward in this situation is to take a more open approach to understanding the organizational structure of destinations by addressing a set of key questions which draw on ideas put forward in these frameworks without being constrained by them. These questions are also informed by issues raised in Chapters 1 and 2, by the empirical studies discussed in Chapters 5 to 9 and points raised in the earlier sections of this chapter.

Demand

A first set of questions relates to the patterns of demand at destinations, a topic which is the focus of destination marketing research but one which merits more explicit attention in organizational studies. Not all visitors have the same motivations, are seeking the same experiences, share the same travel patterns and need and/or can afford the same services. As a result, differences might be anticipated in who is catering for them and how, factors that will influence organizational structure. Who then are our destination's visitors and what do they want? These questions can be addressed in many ways and the specifics will vary from destination to destination but the earlier chapters suggest a number of basic features need to be borne in mind:

- What is the balance between leisure and non-leisure visitors?
- What is the mix of domestic and international demand?
- What is the proportion of mono-destination to circuit-destination visitors and what are the travel patterns which link origins to destinations and one destination to another?
- What are the motivations of each of these segments and what experiences are they seeking?

- What products and services do they need to meet these motivations and obtain these experiences?
- How and where do they want these products and services provided? In particular, what is the mix of bundled and commercialized products and services?
- What factors determine these patterns?

Variations in the patterns of demand and the need for multiple services will increase the likelihood that a range of services will be offered by multiple providers offering different products and services. This in turn raises three further sets of questions.

Products and services

- What is the range of products and services offered?
- How differentiated are they in terms of: quality, price, location, orientation to groups or individuals, level of standardization/customization, degree of technical complementarity?
- Is there an anchor product or dominant sector? If so, which one?

The range of products and services provided will not be a function of demand alone but will also depend on the destination's resource base, the ability of local and external entrepreneurs to develop and exploit those resource and other factors such as conditions of access and government policies.

Actors

The likely presence of multiple providers, including the visitors themselves, raises questions relating to the actors involved. Account needs to be taken not only of the direct provision of products and services but also of broader issues and functions relating to destination management and marketing (Fig. 6.8). Who is providing which products and services and undertaking which functions? The actors involved can be classified in numerous ways and on many dimensions but the preceding chapters suggest the following are particularly relevant questions which need to be addressed for each sector and for the destination as a whole:

- How many actors are involved: one, a few, or many? What is the degree of concentration in each sector?
- Who are they? This question can be addressed in terms of multiple variables:
  - origin: are the actors local, regional, national or international?
  - location: where are they based – within the destination or elsewhere?
  - size: are they small, medium, large or very large?
  - type: are they public, para-public or private sector; NGOs; civil society; self-providers?
  - ownership and management: are businesses independent, corporations or affiliated to a chain?
- What factors influence the composition of the actors and the degree of concentration?

Relationships

Most importantly, the likely presence of multiple providers and other destination stakeholders means we need to consider the relationships and levels of interdependence between these:

- What links exist between which sets of actors?
- Are the links horizontal, vertical or diagonal?
- Are the relationships mutually beneficial or asymmetric? What is the nature and direction of dependency? Who is dependent on whom and for what?
- Are the ties weak or strong, formal and contractual or informal and pragmatic, purposeful or accidental?
- Are the relationships stable and ongoing or temporary and occasional?
- Are the relationships transactional, regulatory or social?
- Are the ties internal to the destination (at whatever scale), or do they extend beyond it?

- How are the relationships configured? Are they dyads, sets, clusters, networks, systems or various combinations of these?
- In what ways and to what extent are independent self-providing visitors and local residents incorporated in the destination's organizational structure?
- What degree of coordination exists and who coordinates?
- In what fields of activity does interaction between actors take place: strategic, operational, governance, knowledge sharing and innovation, capacity building, conservation?
- Is the behaviour underlying these relationships best described as cooperative, competitive or some degree of coopetition?
- What factors influence the behaviour of the actors and shape the relationships between them?
- What impact do geographic and organized proximity have?

It is clearly a challenging task to take a comprehensive approach to addressing each of these questions in turn and to analyse the relationships between them but one which would begin to reveal a potentially wider range of organizational structures in more detail and to explain how and why they come about. Chapter 4 and subsequent chapters suggest many of the necessary tools are available; the challenge now is to draw these different threads together. Even if not all these questions can be addressed within the scope of a single study, setting them explicitly in this broader context would enable the findings of more partial projects to be integrated more effectively with others and contribute to a more complete understanding of destinations, how they are structured and function.

Spatial-Organizational Structure of Destinations

As Chapters 5 to 9 have shown, much of the work on the spatial and organizational structure of destinations has been carried out from two distinct perspectives and in two largely non-convergent streams of literature. Systematic consideration of both streams has begun to reveal how research in one may enrich the other. In particular, the majority of the spatial structure studies have focused on issues of demand and supply without giving much attention to organizational matters. The latter have been shown by some organizational researchers to have a geographical dimension. Much research on organizational structure on the other hand has not taken into account issues of location, distribution and spatial interaction. This lack of convergence is somewhat surprising given that researchers from both streams have drawn attention to two essentially similar sets of relationships, namely those between proximity and association put forward by Jansen-Verbeke and Ashworth (1990) (Table 1.1) and between geographic and organizational proximity as adapted by Bocquet (2008) (Table 9.1) from the work of Torre and Rallett (2005). Both frameworks make provision for linkages between (spatial) proximity and association/organizational proximity while also recognizing that in other situations destination actors can be spatially clustered without having any other association or, conversely, they may have organizational ties but be geographically separate. These key ideas and the implications arising from them have only occasionally been addressed explicitly and offer much scope for further research.

There is interest in both streams of research in the nature and implications of the locational and distributional characteristics of destinations. Much of the work in this field from a geographical perspective has been in terms of identifying patterns of location and distribution and accounting for them in terms of functional linkages, resources, access and the underlying geography of the countries, regions or localities in question. Some studies framed either in terms of industrial districts or clusters have also been concerned with identifying the functional linkages of spatial concentrations and determining the weight of agglomerative forces relative to resource endowments as determinants of the success of destinations. These latter studies have taken destination research a step further by linking spatial clustering to performance and provided some theoretical basis to the topic. This is seen in analyses of the spatial or horizontal clustering of various tourism indicators in the country-wide analyses discussed in Chapter 5 (Figs 5.4–5.8); in some of the regional studies in Chapter 6; in research on urban hotel agglomeration (Chapter 7); and on hotel agglomeration in coastal resorts in Spain (Chapter 8). While some researchers have argued their results validate the theories they have tested (Capone and Boix, 2008; Marco-Lajara *et al.*, 2016a), the findings overall regarding the impact of spatial clustering on performance are variable and the results relating to the externalities arising from the nature and extent of agglomeration are mixed. These studies are based on quantitative measures of spatial association and while they identify aggregate spatial patterns and offer some objective assessment of destination performance they do not investigate the actual nature and level of inter-firm interaction, factors which lie at the heart of destinations being conceptualized as clusters or industrial districts. A broader view of performance is required. The negative externalities of agglomeration, for example, need to be explored more widely and extended from the economic to the environmental and social.

The more qualitative organizational research has provided insights into interaction between co-located firms. The mixed results support the argument that co-location is not necessarily a synonym for cooperation. That finding is consistent with Tables 1.1 and 1.2, with what Jansen-Verbeke and Ashworth (1990) term spatial coincidence and Torre and Rallett (2005) refer to as agglomeration without interactions. However, much of the co-location research from an organizational perspective is characterized by an absence of detail on the actual spatial patterns of where the firms or other actors in question are co-located other than at the general level of a particular local or regional destination and it ignores destination morphologies. Conversely, the big data analyses of urban tourism fail to make connections with organizational matters (Chapter 7). Few studies have considered details of proximity along with levels of interaction (Selin and Beason, 1991; Bocquet, 2008; Luthe *et al.*, 2012; Longjit and Pearce, 2013; Kelman *et al.*, 2016) and much scope exists to examine these relationships in greater depth.

Other research drawing related studies more closely together involves those relationships categorized by Jansen-Verbeke and Ashworth (1990) as non-spatial associations, labelled supra-local organizations by Torre and Rallet (2005) and illustrated by Bocquet (2008) with the 'brand resort' limited to the frontiers of the firm concerned (Table 9.1). The basic notion is that some firms and/or destinations have organizational ties which extend beyond the spatial limits of the local level destination. What becomes important here in terms of a broader approach to destinations is the question of scale; that is, how do these organizational ties connect a local destination to other such destinations at a regional, national or even international scale? An obvious empirical example is the empirical research on route structures and the spatial strategies of carriers as the role of transport is to connect places and in this case to enable travellers to reach their destination (Chapter 5). Such work, however, is not yet explicitly integrated into the literature on the structure of regional or national destinations and largely neglects other modes of transport. Likewise, a well-developed body of research deals with horizontal integration in the hotel sector and the structure and strategies of hotel chains but this too is rarely linked to the spatial-organizational structure of national destinations even though there is a clear spatial element to the distribution of chain hotels (Chapter 5). The interaction of chain hotels with other hotels or firms at the local level has also received little attention (Ingram and Baum, 1997) and their role in circuit travel and the structure of gateways has been neglected. Some horizontal integration across resorts is evident in the ski industry (Chapter 9) but on the whole there is an absence of country-wide research on the commercial structure of the attractions sector and even less on supporting services despite chains being common in some of these, such as fast food restaurants, duty-free outlets or currency exchange offices.

External ties also involve most forms of intermediation, those that link providers in a destination with distribution channel partners located in markets or en route to destinations (Pearce, 2008). Such external links have been identified and incorporated in some of the destination studies (Bocquet, 2008; Erkuş-Öztürk, 2009; Fabry, 2009). Institutional frameworks for tourism are commonly hierarchical in nature, with DMOs at a local and regional level being connected both vertically and horizontally in differing degrees and varying ways (Fig. 1.1; Chapters 5 and 6). Spatial variations in the composition of tourism within and between destinations highlight the challenges which DMOs and NTOs continually face in developing and maintaining relationships between stakeholders, in marketing destinations through creating a strong and cohesive image and in destination management requiring joint action. The relationships between stakeholder networks and the networks of places revealed by the analysis of tourist travel patterns would appear to offer much potential for a more integrated approach to research on the spatial and organizational structure of destinations given that they are based on a common technique.

More generally, adopting a more integrated multi-scale approach and giving greater attention to external ties, both spatially and organizationally, would highlight the fact that destinations are not self-contained places or entities. By their very nature through the consumption of products and services *in situ* away from the tourists' place of origin, destinations depend in varying degrees on their connectedness with other places and the organizations and institutions located there. This is most apparent in the regional and national studies of tourist travel patterns but is also evident in the external organizational and institutional relationships outlined here.

The value of taking a more integrated approach to the spatial and organizational structure of destinations can be readily illustrated with respect to the increasing pressures being experienced in many places as a consequence of the substantial growth of tourism in recent years, popularized as overtourism. Much of that pressure results from the concentration of supply and demand at the various scales discussed earlier. Other pressures result from the expansion of tourism into new neighbourhoods following the rapid growth of Airbnb and similar forms of accommodation. One response to concentration is to disperse tourist flows into new or less developed areas. One response to the spread of Airbnb is to attempt to curb its expansion into residential areas. Both strategies are spatial. Neither is guaranteed of success when viewed at a larger scale. Redistributing demand may simply shift the pressures to other places, including some that are even less prepared to accommodate them. Curbing growth in new neighbourhoods may only exacerbate problems in those that are already under pressure. In some contexts concentration may also have advantages, as is seen in protected area management strategies which consolidate infrastructure and visitor activity in defined zones leaving other areas in a less disturbed natural state. Having a sound understanding of the spatial structure of destinations will assist effective policy making on these issues. They require an understanding of the factors which underlie the geography of demand, explain concentrations of supply, influence visitor movements and show how destinations are interconnected in various ways and at different scales. That, however, is only part of the picture. Understanding the organizational structure of destinations, their composition, how the different sets of actors interact in various ways as they carry out their different tasks and perform different functions, how the organizational dynamics evolve are all vital to developing and implementing effective strategies. Moreover, as this section has stressed, the spatial and the organizational are intimately linked.

Conclusions

Destinations will continue to be a central focus of tourism research and offer important and interesting challenges to researchers. As earlier chapters have identified and outlined, a wide range of different facets of destinations remain to be addressed, many of which have been drawn together in this final chapter. New issues and research questions will emerge as destinations evolve further. The trend of growing specialization and fragmentation of the literature is also likely to continue with the result that our cumulative knowledge of destinations is not going to increase at the same rate unless more is done to develop a more coherent approach to such research. There will always be a place for specialized studies which bring specific expertise to focus on particular aspects of destinations but there is also a parallel need for synthesis, to bring the resultant findings together, to fit them into the bigger picture of destinations so that research on one aspect is linked effectively to others and that the knowledge derived as a whole is greater than that from the sum of the individual parts.

The challenge of synthesis can be addressed in a number of ways. First, the need for greater integration has to be recognized more widely and explicitly. It is hoped this book has made that case clearly through the approach adopted, the issues raised, the examples cited and the different literatures drawn upon, not only from different disciplines but also from studies published in a range of languages between which there is often little cross-pollination. Second, there is the issue of specialization to be tackled. One way forward would be to undertake more comparative research on a specific topic or sub-topic in multiple destinations so as to elucidate the generality or specificity of patterns and findings. Some of the work being done on urban tourism attractions and visitor movements in different cities highlights what might be done in this regard. Comparative studies across different types of destinations – urban, coastal, alpine, rural – would appear to be an especially fruitful avenue for further research given the insights which emerge when this is done. Another way forward would be to engage in more inter-disciplinary research from the outset of a project. What new insights would emerge if researchers with skills in spatial interaction were joined by those with expertise in interorganizational relationships? Big data analyses might benefit from having input from other researchers who have a deeper understanding of what is happening on the ground and are able to bring this to a more informed interpretation of results. Third, there is still room for in-depth case studies, what might perhaps be considered old-school but which provide the scope to draw the many different components of a destination together. Given the breadth of these, inter-disciplinary teams might again yield additional insight. However, to increase their value and impact, these case studies must be set clearly in the literature and problematized in terms of wider questions and concepts. In this regard a fourth consideration is that of adopting concepts, frameworks and techniques derived in other fields and applying them in destination research. To what extent is this appropriate, do we need to be more critical before we do so or should we start out with a blank slate and develop more targeted tools and frameworks in our destination research? The work reviewed here calls for a balanced approach; we need to carefully assess what comes from other fields, not continually reinvent the wheel but also recognize destinations have their own characteristics which may open up new and exciting opportunities which require originality and innovation.

References

Achin, C. and George-Marcelpoil, E. (2013) Sorties de piste pour la performance touristique des stations de sports d'hiver. *Tourisme & Territoires*, 3, 67–92.

Acuña Ortega, M. and Brugnoli, A. (2007) Innovación en el cluster ecoturístico de Monteverde. *Revista Centroamericana de Ciencias Sociales*, IV, 1, 115–144.

Adam, I. (2013) Urban hotel development patterns in the Kumasi Metropolis, Ghana. *Tourism Planning & Development*, 10, 1, 85–98. DOI:10.1080/21568316.2012.724706

Adam, I. and Amuquandoh, F.E. (2014) Hotel characteristics and location decisions in Kumasi Metropolis, Ghana. *Tourism Geographies*, 16, 4, 653–668. DOI:10.1080/14616688.2012.762689

Adam, I. and Mensah, A.M. (2014) Perceived spatial agglomeration effects and hotel location choice. *Anatolia*, 25, 1, 49–60. DOI:10.1080/13032917.2013.822818

Akehurst, G.P. (1984) The measurement of concentration in the hospitality industry. *International Journal of Hospitality Management*, 3, 1, 25–33.

Aldape Perez, G. (2013) La configuración del espacio turístico en Cancún, México. In: Pie i Ninot, R. and Rosa Jimenez, C.J. (eds) *Turismo Líquido*. Instituto Habitat Turismo Territorio and Iniciativa Digital Politecnica, Universitat Politecnica de Catalunya, Barcelona/Universidad de Málaga, Málaga, pp. 228–245.

Aldrich, H. (1979) *Organizations and Environments*. Prentice-Hall, Englewood Cliffs, NJ.

Alegre, J., Cladera, M. and Sard, M. (2013) Tourist areas: examining the effects of location attributes on tour-operator package holiday prices. *Tourism Management*, 38, 131–141.

Aliagaoglu, A. and Ugur, A. (2008) Hotels as a model of regional life: the Erzurum sample. *European Planning Studies*, 16, 10, 1403–1422.

Álvarez-Díaz, M., d'Hombres, B. and Ghisetti, C. (2017) Modelling inter- and intra-regional tourism flows in Spain – a spatial econometric approach. *Regional Statistics*, 7, 2, 3–34. DOI:10.15196/RS070205

Amar, G. (1993) Pour une écologie urbaine des transports. *Les Annales de la Recherche Urbaine*, 59–60, 141–151. DOI:10.3406/aru.1993.1736

Ambrosie, L.M. (2015) Myths of tourism institutionalization and Cancún. *Annals of Tourism Research*, 54, 65–83.

Andergassen, R., Candela, G. and Figini, P. (2013) An economic model for tourism destinations: product sophistication and price coordination. *Tourism Management*, 37, 86–98.

Andria, J., di Tollo, G. and Pesenti, R. (2015) Detection of local tourism systems by

threshold accepting. *Computer Management Science*, 12, 559–575. DOI:10.1007/s10287-015-0238-x
Anselin, L. (1995) Local indicators of spatial association LISA. *Geographical Analysis*, 27, 2, 93–115.
Aranburu, I., Plaza, B. and Esteban, M. (2016) Sustainable cultural tourism in urban destinations: does space matter? *Sustainability*, 8, 699. DOI: 10.3390/su8080699
Arbel, A. and Pizam, A. (1977) Some determinants of urban hotel location: the tourists' inclinations. *Journal of Travel Research*, 15, 3, 1–22.
Asero, V., Gozzo, S. and Tomaselli, V. (2016) Building tourism networks through tourist mobility. *Journal of Travel Research*, 55, 6, 751–763.
Ashworth, G.J. (1989) Accommodation and the historic city. *Built Environment*, 15, 2, 92–100. Available at: http://www.jstor.org/stable/23286293 (accessed 2 July 2020).
Ashworth, G.J. (1990) The historic cities of Groningen: which is sold to whom? In: Ashworth, G. and Goodall, B. (eds) *Marketing Tourism Places*. Routledge, London, pp. 138–155.
Ashworth, G.J. and de Haan, T.Z. (1985) *The Tourist Historic City: A Model and Initial Application in Norwich, U.K.* Geographical Institute, University of Groningen, Groningen.
Ashworth, G.J. and de Haan, T.Z. (1987) *Regionalizing the Resort System: Tourist Regions on the Languedoc Coast*. University of Groningen, Groningen.
Baerenholdt, J., Haldrup, M., Larsen, J. and Urry, J. (2004) *Performing Tourist Places*. Ashgate, Aldershot, UK.
Baggio, R. (2008) Symptoms of complexity in a tourism system. *Tourism Analysis*, 13, 1, 1–20.
Baggio, R. (2014) Complex tourism systems: a visibility graph approach. *Kybernetes*, 43, 3/4, 445–461.
Baggio, R. and Cooper, C. (2010) Knowledge transfer in a tourism destination: the effects of a network structure. *The Service Industries Journal*, 30, 10, 1757–1771.
Balseinte, R. (1958) Les stations de sports d'hiver en France. *Revue de Géographie Alpine*, 46, 1, 129–180. DOI:10.3406/rga.1958.1819
Barbaza, Y. (1970) Trois types d'intervention du tourisme dans l'organisation de l'espace littoral. *Annales de Géographie*, 434, 446–469.
Barbier, B. (1968) Vars, une grande station des Alpes du sud. *Revue de Géographie Alpine*, 56, 1, 265–289.
Barbosa, M.A.C. and Zamboni, R.A. (2001) El cluster del ecoturismo en Bonito, MS, Brasil. In: Buitelaar, R. and Gómez, J.J. (eds) *Serie Seminarios y conferencias*. CEPAL, Santiago, Chile, 17, 115–142.
Barrado Timón, D.A. (2004) El concepto de destino turístico: una aproximación geográfico-territorial. *Estudios Turísticos*, 160, 45–68.
Barrett, J.A. (1958) The seaside resort towns of England and Wales. PhD thesis, University of London, London.
Barros, C.P., Botti, L., Peypoch, N., Robinot, E., Solonsandrasana, B. and Assaf, A.G. (2011) Performance of French destinations: tourism attraction perspectives. *Tourism Management*, 32, 1, 141–146.
Bauder, M. and Freytag, T. (2015) Visitor mobility in the city and the effects of travel preparation. *Tourism Geographies*, 17, 5, 682–700. DOI:10.1080/14616688.2015.1053971
Baum, J.A.C. and Haveman, H.A. (1997) Love thy neighbor? Differentiation and agglomeration in the Manhattan hotel industry, 1898–1990. *Administrative Science Quarterly*, 42, 2, 304–338.
Becattini, G. (1990) The Marshallian industrial district as a socio-economic notion. In: Pyke, F. (ed.) *Industrial Districts and Inter-Firm Co-operation in Italy*. International Institute for Labour Studies, Geneva, pp. 37–51.
Bégin, S. (2000) The geography of a tourist business: hotel distribution and urban development in Xiamen, China. *Tourism Geographies*, 2, 4, 448–471. DOI:1404-1422.10.1080/146166800750035530
Bengtsson, M. and Kock, S. (2000) 'Coopetition' in business networks to cooperate and compete simultaneously. *Industrial Marketing Management*, 29, 5, 411–426.

Benson, J.K. (1975) The interorganizational network as political economy. *Administrative Science Quarterly*, 20, 229–249.

Beritelli, P. and Laesser, C. (2011) Power dimensions and influence reputation in tourist destinations: empirical evidence from a network of actors and stakeholders. *Tourism Management*, 32, 1299–1309.

Beritelli, P., Bieger, T. and Laesser, C. (2014) The new frontiers of destination management: applying variable geometry as a function-based approach. *Journal of Travel Research*, 53, 4, 403–417.

Bessy, O. (2016) Innovations événementielles et structuration des destinations touristiques. Pour une hybridation des approches: l'exemple de l'Ultra-Trail du Mont-Blanc. *Mondes du Tourisme*. Hors-série 2016. DOI:10.4000/tourisme.1277. Available at: http://tourisme.revues.org/1277 (accessed 2 July 2020).

Bieger, T. (1998) Reengineering destination marketing organizations: the case of Switzerland. *The Tourist Review*, 53, 3, 4–17.

Blasco, D., Guia, J. and Prats, L. (2014) Tourism destination zoning in mountain regions: a consumer-based approach. *Tourism Geographies*, 16, 3, 512–528. DOI:10.1080/14616688.2013.851267

Blondy, C. (2016) Le tourisme, un facteur de développement durable des territoires insulaires tropicaux? Tourisme, aménagement, environnement et société locale à Bora Bora (Polynésie française). *Mondes du Tourisme*. Hors-série 2016. DOI:10.4000/tourisme.1283. Available at: http://tourisme.revues.org/1283 (accessed 30 September 2020).

Bocquet, R. (2008) Proximités et coordination inter-firmes: une analyse empirique de la compétitivité des stations de sports d'hiver. *Revue d'Économie Régionale & Urbaine*, 2 June, 159–178. DOI:10.397/reru.082.0159

Bodega, D., Cioccarelli, G. and Denicolai, S. (2004) New inter-organizational forms: evolution of relationship structures in mountain tourism. *Tourism Review*, 59, 3, 13–19.

Boerwinkel, H.W.J. (1995) Management of recreation and tourist behaviour at different spatial levels. In: Ashworth, G.J. and Dietvorst, A.G.J. (eds) *Tourism and Spatial Transformations – Implications for Policy and Planning*. CAB International, Wallingford, Oxon, pp. 241–263.

Bonetti, E., Petrillo, C.S. and Simoni, M. (2006) Tourism system dynamics: a multi-level destination approach. In: Lazzeretti, L. and Petrillo, C.S. (eds) *Tourism Local Systems and Networking*. Elsevier, Oxford, pp. 111–133.

Bordas Rubies, E. (2001) Improving public–private sectors cooperation in tourism: a new paradigm for destinations. *Tourism Review*, 56, 3/4, 38–41.

Bornhorst, T., Ritchie, J.R.B. and Sheehan, L. (2010) Determinants of tourism success for DMOs and destinations: an empirical examination of stakeholders' perspectives. *Tourism Management*, 31, 572–589.

Bote, V., Huéscar, A., Goy, L. and Sánchez, D. (2003) Los viajes combinados de turismo cultural del mercado europeo e interior. *Documentos de Economía del Turismo*, Documento Serie A, No. 0301. Escuela Universitaria de Turismo, Universidad de Málaga, Málaga.

Bote Gómez, V., Álvarez Cuervo, R., González García, L. and Nieto González, B. (2003) *La Demanda Turística Nacional e Internacional con Destino Ronda*. Empresa Pública Municipal 'Turismo de Ronda', Ronda.

Botti, L., Peypoch, N. and Solonandrasana, B. (2008) Time and tourism attraction. *Tourism Management*, 29, 594–596.

Botti, L., Peypoch, N., Robinot, E. and Solonadrasana, B. (2009) Tourism destination competitiveness: the French regions case. *European Journal of Tourism Research*, 2, 5–24.

Bourne, L.S. (1982) Urban spatial structure: an introductory essay on concepts and criteria. In: Bourne, L.S. (ed.) *Internal Structure of the City: Readings on Urban Form, Growth and Policy*, 2nd edn. Oxford University Press, New York and Oxford, pp. 28–45.

Bowden, J. (2003) A cross-national analysis of international tourist flows in China, *Tourism Geographies*, 5, 3, 257–279. DOI:10.1080/14616680309711

Bramwell, B. (2004) Mass tourism, diversification and sustainability in Southern

Europe's coastal regions. In: Bramwell, B. (ed.) *Coastal Mass Tourism. Diversification and Sustainable Development in Southern Europe*. Channel View, Bristol, pp. 1–31.

Brandenburger, A. and Nalebuff, B. (1996) *Co-opetition*. Doubleday, New York.

Briassoulis, H. (2017) Tourism destinations as multiplicities: the view from assemblage thinking. *International Journal of Tourism Research*, 19, 304–317.

Brida, J.G., González, M.N. and Lanzilotta, B. (2017) Análisis de los determinantes del turismo interno en Uruguay. *Revista de Estudios Regionales*, 108, 43–78.

Britton, S.G. (1982) The political economy of tourism in the Third World. *Annals of Tourism Research*, 9, 331–358.

Britton, S. (1990) The economic role and organization of island tourist industries. In: Kissling, C. (ed.) *Destination South Pacific: Perspectives on Island Tourism*. Cahiers du Tourisme, Serie B, 60. Centre des Hautes Etudes Touristiques, Aix-en-Provence, pp. 51–71.

Brown, R.M. (1935) The business of recreation. *Geographical Review*, 2, 5, 467–475.

Brown, K.G. and Geddes R. (2007) Resorts, culture and music: the Cape Breton tourism cluster. *Tourism Economics*, 13, 1, 129–141.

Bruttomesso, E. (2018) Making sense of the square: facing the touristification of public space through playful protest in Barcelona. *Tourist Studies*, 18, 4, 467–485.

Budović, A. and Ratkaj, I. (2018) Spatial behaviour of tourists in Belgrade. *Collection of Papers*, Faculty of Geography at the University of Belgrade, 66, 1, 71–87. DOI:10.5937/zrgfub1866071B

Buhalis, D. (2000a) Marketing the competitive destination of the future. *Tourism Management*, 21, 1, 97–116.

Buhalis, D. (2000b) Relationships in the distribution channel of tourism: conflicts between hoteliers and tour operators in the Mediterranean region. *International Journal of Hospitality and Tourism Administration*, 1, 1, 113–139.

Bujosa, A., Riera, A. and Pons P.J. (2015) Sun-and-beach tourism and the importance of intra-destination movements in mature destinations. *Tourism Geographies*, 17, 5, 780–794.

Bull, A.O. (1994) Pricing a motel's location. *International Journal of Contemporary Hospitality Management*, 6, 6, 10–15. DOI:10.1108/09596119410070422

Bull, A. (1995) *The Economics of Travel and Tourism*, 2nd edn. Longman, Melbourne.

Burghardt, A.F. (1971) A hypothesis about gateway cities. *Annals of the Association of American Geographers*, 61, 2, 269–285.

Burkart, A.J. and Medlik, S. (1974) *Tourism: Past, Present and Future*. Heinemann, London.

Burnet, L. (1963) *Villégiature et Tourisme sur les Côtes de France*. Librairie Hachette, Paris.

Burtenshaw, D., Bateman, M. and Ashworth, G.J. (1981) *The City in West Europe*. Wiley, Chichester.

Butler, R.W. (1980) The concept of a tourist area life cycle of evolution: implications for the management of resources. *Canadian Geographer*, 24, 5–12.

Butler, R.W. and Boyd, S.W. (eds) (2000) *Tourism and National Parks: Issues and Implications*. Wiley, Chichester.

Byrne, P. and Skinner, S. (2007) International business tourism: Destination Dublin or Destination Ireland? *Journal of Travel & Tourism Marketing*, 22, 3–4, 55–65. DOI:10.1300/J073v22n03_05

Caldeira, A.M and Kastenholz, E. (2018) Tourists' spatial behaviour in urban destinations: the effect of prior destination experience. *Journal of Vacation Marketing*, 24, 3, 247–460.

Campbell, C.K. (1967) An approach to research in recreational geography. In: *B.C. Occasional Papers*, 7. Department of Geography, University of British Columbia, Vancouver, pp. 85–90.

Campillo-Besses, X., Priestley, G. and Romagose, F. (2004) Using EMAs and local agenda 21 as tools towards sustainability: the case of a Catalan coastal resort. In: Bramwell, B. (ed.) *Coastal Mass Tourism. Diversification and Sustainable Development in Southern Europe*. Channel View, Bristol, pp. 220–248.

Camprubi, R., Guia, J. and Comas J. (2008) Destination networks and induced tourism. *Tourism Review*, 63, 2, 47–58.

Candela, G. and Figini, P. (2012) *The Economics of Tourism Destinations*. Springer, Berlin and Heidelberg.

Canina, L., Enz, C.A. and Harrison, J.S. (2005) Agglomeration effects and strategic orientations: evidence from the U.S. lodging industry. *Academy of Management Journal*, 48, 4, 565–581.

Cao, G., Shi, Q. and Liu, T. (2016) An integrated model of urban spatial structure: insights from the distribution of floor area ratio in a Chinese city. *Applied Geography*, 75, 116–126.

Capone, F. (2006) Systemic approaches for the analysis of tourism destination: towards the tourist local systems. In: Lazzeretti, L. and Petrillo, C. (eds) *Tourism Local Systems and Networking*. Elsevier, Amsterdam, pp. 7–23.

Capone, F. and Boix, R. (2008) Sources of growth and competitiveness of local tourist production systems: an application to Italy (1991–2001). *Annals of Regional Science*, 42, 209–224. DOI:10.1007/s00168-007-0133-7

Carlson, A.S. (1938) Recreation industry of New Hampshire. *Economic Geography*, 14, 255–270.

Castro Fernández, B.M., Lois González, R.C. and Lopez, L. (2016) Historic city, tourism performance and development: the balance of social behaviours in the city of Santiago de Compostela (Spain). *Tourism and Hospitality Research*, 16, 3, 282–293.

Caves, R.E. and Gosling, G.D. (1999) *Strategic Airport Planning*. Pergamon, Oxford.

Cawley, M., Marsat, J.-B. and Gillmor, D.A. (2007) Promoting integrated rural tourism: comparative perspectives on institutional networking in France and Ireland. *Tourism Geographies*, 9, 4, 405–420.

Celata, F. (2012) Suburban hotels and the atomization of tourist space in large cities: the case of Rome. *Working Papers*, Università degli Studi di Roma 'La Sapienza', Rome, 102.

Chadefaud, M. (1981) *Lourdes: Un Pèlerinage, une Ville*. Edisud, Aix-en-Provence.

Chang, T.C. (1997) From 'instant Asia' to 'multi-faceted jewel': urban imaging strategies and tourism development in Singapore. *Urban Geography*, 18, 542–562.

Chappis, L. (1965) A la recherche d'une architecture en montagne. *Économie et Prospective de la Montagne*, 6, 43.

Chhetri, A., Chhetri, P., Arrowsmith, C. and Corcoran, J. (2017) Modelling tourism and hospitality employment clusters: a spatial econometric approach. *Tourism Geographies*, 19, 3, 398–424. DOI:10.1080/14616688.2016.1253765

Chim-Miki, A.F., Medina-Brito, P. and Batista-Canino, R.M. (2019) Integrated management in tourism: the role of coopetition. *Tourism Planning & Development*. DOI:10.1080/21568316.2019.1574888

Chin, W.L., Haddock-Fraser, J. and Hampton, M.P. (2015) Destination competitiveness: evidence from Bali. *Current Issues in Tourism*. DOI:10.1080/13683500.2015.1111315

Christaller, W. (1966) *Central Places in Southern Germany*, trans. Baskin, C.W. Prentice-Hall, Englewood Cliffs, NJ.

Chu, R.K.S. and Choi, T. (2000) An importance-performance analysis of hotel selection factors in the Hong Kong hotel industry: a comparison of business and leisure travellers. *Tourism Management*, 21, 363–377.

Chung, W. and Kalnins, A. (2001) Agglomeration effects and performance: a test of the Texas lodging industry. *Strategic Management Journal*, 22, 10, 969–988.

Cilliers, P. (1998) *Complexity and Postmodernism: Understanding Complex Systems*. Routledge, London.

Cirer-Costa, J.C. (2013a) The role of geography in the success of the Balearic tourism industry. *MPRA Paper*, No. 47701. Available at: http://mpra.ub.uni-muenchen.de/47701 (accessed 2 July 2020).

Cirer-Costa, J.C. (2013b) Price formation and market segmentation in seaside accommodations. *International Journal of Hospitality Management*, 33, 446–455.

Cirer-Costa, J.C. (2014a) Majorca's tourism cluster: the creation of an industrial district, 1919–36. *Business History*, 56, 8,

1243–1261. DOI:10.1080/00076791.2013.876532

Cirer-Costa, J.C. (2014b) The explosive expansion and consolidation of the Balearic hotel sector, 1964–2010. *Revista de Historia Industrial*, 56, 3, 189–216.

Clary, D. (1977) *La Façade Littorale de Paris: le tourisme sur la côte normande*. Masson, Paris.

Clergeau, C. and Violier, P. (2013) Is 'cluster' a relevant concept for tourism? *Journal of Hospitality & Tourism*, 11, 1, 54–70.

Clifford, H. (2002) *Downhill Slide: Why the Corporate Ski Industry is Bad for Skiing, Ski Towns and the Environment*. Sierra Club Books, San Francisco CA.

Coccossis, H., Janssen, H., Kiers, M. and Nijkamp, P. (1991) *Tourism and Strategic Development*. Serie Research Memoranda, Faculteit der Economische Wetenschappen en Econometrie, Vrije Universiteit, Amsterdam.

Cohen, E. (2018) Backpacker enclaves research: achievements, critique and alternative approaches. *Tourism Recreation Research*, 43, 1, 105–116. DOI:10.1080/02508281.2017.1388572

Cohen, E. and Neal, M. (2012) A Middle Eastern Muslim tourist enclave in Bangkok. *Tourism Geographies*, 14, 570–598. DOI:10.1080/14616688.2012.647320

Costa, C., Panyik, E. and Buhalis, D. (eds) (2014) *European Tourism Planning and Organisation Systems: The EU Member States*. Channel View, Bristol.

Costa, H.A., Costa, A.C. and Miranda Júnior, N.S. (2012) Arranjos produtivos locais (APL) no turismo: estudo sobre a competitividade e o desenvolvimento local na Costa dos Corais – AL. *Revista Acadêmica do Observatório de Inovação do Turismo, VII*, 1, 1–27.

Costa, H.A, Pereira, M.F.A. and Hoffmann,V. E. (2006) Compreendendo o espaço turístico de Balneário Camboriú (SC) como insumo para o estudo da competitividade local. *Turismo – Visão e Ação*, 8, 2, 223–234.

Cox, N., Gyrd-Jones, R. and Gardiner, S. (2014) Internal brand management of destination brands: exploring the roles of destination management organisations and operators. *Journal of Destination Marketing and Management*, 3, 85–95.

Cracolici, M.F. and Nijkamp, P. (2008) The attractiveness and competitiveness of tourist destinations: a study of Southern Italian regions. *Tourism Management*, 30, 3, 336–344.

Cruz, S.C.S. and Teixeira, A.A.C. (2010) The evolution of the cluster literature: shedding light on the regional studies–regional science debate. *Regional Studies*, 44, 9, 1263–1288.

Cuadrado-Ciuraneta, S., Durà-Guimerà, A. and Salvati, L. (2017) Not only tourism: unravelling suburbanization, second-home expansion and 'rural' sprawl in Catalonia, Spain. *Urban Geography*, 38, 1, 66–89. DOI:10.1080/02723638.2015.1113806

Cumin, G. (1970) Les stations intégrées. *Urbanisme*, 116, 50–53.

Czajkowski, A. and Cunha, S. K. (2010) Organização e coordenação da rede de cooperação em aglomerados de turismo rural. *Revista Turismo Visão e Ação – Eletrônica*, 12, 1, 92–113.

d'Angella, F. and Go, F.M. (2009) Tale of two cities' collaborative tourism marketing: towards a theory of destination stakeholder assessment. *Tourism Management*, 30, 429–440.

Daskalopoulou, I. and Petrou, A. (2009) Urban tourism competitiveness: networks and the regional asset base. *Urban Studies*, 46, 4, 779–801.

Datzira-Masip, J. and Poluzzi, A. (2014) Brand architecture management: the case of four tourist destinations in Catalonia. *Journal of Destination Marketing & Management*, 3, 48–58. DOI:10.1016/j.jdmm.2013.12.006

Dauphiné, A. (1979) *Espace, Région et Système*. Economica, Paris.

Davern, M. (1997) Social networks and economic sociology: a proposed research agenda for a more complete social science. *American Journal of Economics and Sociology*, 11, 13–25.

Davidson, R. and Maitland, R. (1997) *Tourism Destinations*. Hodder and Stoughton, London.

Dean Runyan Associates, Inc. (2013) *A Regional Analysis of the California Travel Industry and Other Industry Sectors 2007–2012*. California Travel & Tourism Commission, Sacramento, CA.

Dearden, P. (2000) Tourism, national parks and resource conflicts. In: Butler, R.W. and Boyd, S.W. (eds) *Tourism and National Parks: Issues and Implications*. Wiley, Chichester, pp. 187–202.

Deasy, G.F. (1949) The tourist industry in a North Woods County. *Economic Geography*, 25, 2, 240–259.

De Cantis, S., Ferrante, M., Kahani, A. and Shoval, N. (2016) Cruise passengers' behaviour at the destination: investigation using GPS technology. *Tourism Management*, 52, 133–150.

Defert, P. (1960) Introduction à une géographie touristique et thermale de l'Europe. *Acta Geographica*, 36, 4–I 1.

Dejbakhsh, S., Arrowsmith, C. and Jackson, M. (2011) Cultural influence on spatial behaviour. *Tourism Geographies*, 13, 1, 91–111. DOI:10.1080/14616688.2010.516396

de la Mata T. and Llano-Verduras, C. (2012) Spatial pattern and domestic tourism: an econometric analysis using inter-regional monetary flows by type of journey. *Papers in Regional Science*, 91, 2, 437–470. DOI:10.1111/j.1435-5957.2011.00376.x

de la Maza, F. (2018) Tourism in indigenous territories: the impact of public policies and tourism value of indigenous culture. *Latin American and Caribbean Ethnic Studies*, 13, 1, 94–111. DOI:10.1080/17442222.2018.1416894

Denicolai, S., Cioccarelli, G. and Zucchella, A. (2010) Resource-based local development and networked core-competencies for tourism excellence. *Tourism Management*, 31, 2, 260–266.

Dennis, N. (1994) Airline hub operations in Europe. *Journal of Transport Geography* 2, 4, 219–233.

Derek, M. (2018) Spatial structure of tourism in a city after transition: the case of Warsaw, Poland. In: Müller, D.K. and Więckowski M. (eds) *Tourism in Transitions, Geographies of Tourism and Global Change*. Springer International, pp. 157–171. DOI:10.1007/978-3-319-64325-0_9

Destination NSW (2017) *Holiday Travel to Sydney Year Ended December 2016*. Destination NSW, Sydney.

Dewailly, J.-M. (1990) *Tourisme et Aménagement en Europe du Nord*. Masson, Paris.

Direction Générale des Entreprises (2019) *Guide Méthodologique: Procédure Relative aux Communes Touristiques et aux Stations Classées de Tourisme*. Direction Générale des Entreprises. Paris. Available at: https://www.entreprises.gouv.fr/files/files/directions_services/tourisme/territoires/stations-classees/guide_methodologique_version__version_16_mars_2017.pdf (accessed 5 February 2020).

Dixit, A.K. and Stiglitz, J.E. (1977) Monopolistic competition and optimum product diversity. *American Economic Review*, 67, 297–308.

Dobruszkes, F. (2006) An analysis of European low-cost airlines and their networks. *Journal of Transport Geography*, 14, 249–264.

Dolnicar, S. and Grün, B. (2017) In a galaxy far, far away ... market yourself differently. *Journal of Travel Research*, 56, 5, 593–598.

Dominguez Perez, M. (2016) Domestic tourism in Uruguay: a matrix approach. *Pasos*, 14, 4, 811–825.

Dorward, S. (2006) The evolution of village form and its relevance as a model for resort design and development. In: Clark, T.A., Gill, A. and Hartmann R. (eds) *Mountain Resort Planning and Development in an Era of Globalization*. Cognizant Communication Corporation, Elmsford, NY, pp. 253–277.

Dragusin, M., Constantin, D.L. and Petrescu, R.M. (2010) Clustering in transition economies: the case of Romanian tourism industry. *Tourism & Hospitality Management 2010 Conference Proceedings*, 287–301.

Dredge, D. (1999) Destination place planning and design. *Annals of Tourism Research*, 26, 4, 772–791.

Duhamel, P. and Knafou, R. (2007) Le tourisme dans la centralité parisienne. In:

Saint-Julien, T., le Goix, R., Duhamel, P. and Knafou, R. (eds) *La Métropole Parisienne: Centralités, Inégalités, Proximités*. Belin, Paris, pp. 39–64.

Edwards, D. and Griffin, T. (2013) Understanding tourists' spatial behaviour: GPS tracking as an aid to sustainable destination management. *Journal of Sustainable Tourism*, 21, 4, 580–595. DOI:10.1080/09669582.2013.776063

Egan, D. and Nield, K. (2000) Towards a theory of intraurban hotel location. *Urban Studies*, 37, 3, 611–621.

Eiselen, E. (1945) The tourist industry of a modern highway, US 16 in South Dakota. *Economic Geography*, 21, 221–230.

Ejarque, J. (2005) *Destinos Turísticos de Éxito: Diseño, Creación, Gestión y Marketing*. Ediciones Pirámide, Madrid.

Ekinci, Y., Sirakaya-Turk, E. and Baloglu, S. (2007) Host image and destination personality. *Tourism Analysis*, 12, 5–6, 433–446.

Elbe, J., Hallén, L. and Axelsson, B. (2009) The destination-management organisation and the integrative destination-marketing process. *International Journal of Tourism Research*, 11, 283–296. DOI:10.1002/jtr.695

Encalada, L., Rocha, J. and Ferreira, C.C. (2018) Análise exploratória da pegada digital dos turistas para a identificação de padrões espaciais e temporais em destinos urbanos. *Revista Cartográfica*, 97, 17–35.

Erkuş-Öztürk, H. (2009) The role of cluster types and firm size in designing the level of network relations: the experience of the Antalya tourism region. *Tourism Management*, 30, 589–597.

Erkuş-Öztürk, H. (2011) Emerging importance of institutional capacity for the growth of tourism clusters: the case of Antalya. *European Planning Studies*, 19, 10, 1735–1753.

Erkuş-Öztürk, H. and Eraydin, A. (2010) Environmental governance for sustainable tourism development: collaborative networks and organisation building in the Antalya tourism region. *Tourism Management*, 31, 1, 113–124.

Estevão, C. and Ferreira, J.J. (2012) Tourism cluster positioning and performance evaluation: the case of Portugal. *Tourism Economics*, 18, 4, 711–730. DOI:10.5367/te.2012.0137

Everitt, J., Massam, B.H., Chávez-Dagostino, R.M., Espinosa Sánchez, R. and Andrade Romo, E. (2008) The imprints of tourism on Puerto Vallarta, Jalisco, Mexico. *The Canadian Geographer*, 52, 1, 83–104.

Ewert, A.W. (1996) Gateways to adventure tourism: the economic impacts of mountaineering on one portal community. *Tourism Analysis*, 1, 1, 59–63.

Fablet, G. (2013) La croissance immobilière des stations de sports d'hiver en Tarentaise. *Revue de Géographie Alpine*, 101, 3. Available at: http://rga.revues.org/2188 (accessed 1 October 2016). DOI:10.4000/rga.2188

Fabry, N. (2009) Clusters de tourisme et compétivité des acteurs: une solution pour conforter le tourisme d'affaires et événementiel parisien? In: Lemasson, J.-P. and Violier, P. (eds) *Destinations et Territoires: Co-Présence à l'Œuvre*. Téoros, Montréal/Presses de l'Université de Québec, Québec, pp. 55–68.

Fabry, N. (2015) Potential collaboration of collaborative forms in tourism innovation: a focus on tourism cluster. In: Gursoy, D., Saayman, M. and Sotiriadis, M. (eds) *Collaboration in Tourism Businesses and Destinations: A Handbook*. Emerald, Bingley, UK, pp. 285–298.

Fabry, N. and Zeghni, S. (2012) Tourisme et développement local: une application aux clusters de tourisme. *Mondes en Développement*, 1, 157, 97–110. DOI: 10.3917/med.157.0097

Favre-Bonté, V., Gardet, E. and Thevenard-Puthod, C. (2016) Inter-organisational network configurations for ski areas innovations. *European Journal of Innovation Management*, 19, 1, 90–110. DOI:10.1108/EJIM-04-2014-0039

Favre-Bonté, V., Gardet, E. and Thevenard-Puthod, C. (2019) The influence of territory on innovation network design in mountain tourism resorts. *European Planning Studies*, 27, 5, 1035–1057. DOI:10.1080/09654313.2019.1588856

Fayos-Solà, E., Fuentes Morales, L. and Muñoz Mazón, A.I. (2014) The FAS Model: destination analysis and governance. In: Fayos-Solà, E., Alvarez, M.D. and Cooper, C. (eds) *Tourism as an Instrument for Development: A Theoretical and Practical Study*. Emerald, Bingley, UK, pp. 55–86.

Fernández Cuevas, M. del C., Manzanera Díaz, E. and Aurioles Martín, J. (2008) Los distritos turísticos en España. Paper presented at the X Congreso de la Asociación Andaluza de Ciencia Regional, Baeza-Jaen, 27–29 November 2008.

Ferrer Regales, M. (1988) Las nuevas tendencias de crecimiento regional, urbano y rural. *Situation*, 3, 78–116.

Field, K. (2010) Gravity is your friend but every turn is a leap of faith: design and testing a schematic map for ski resort trails. *The Cartographic Journal*, 47, 3, 222–237. DOI:10.1179/000870410X12849977317444

Fistola, R. and La Rocca, R.A. (2017) Driving functions for urban sustainability: the double-edged nature of urban tourism. *International Journal of Sustainable Development and Planning*, 12, 3, 425–434.

Flagestad, A. and Hope, C.A. (2001) Strategic success in winter sports destinations: a sustainable value creation perspective. *Tourism Management*, 22, 445–461. DOI:10.1 108/095644230610651589

Fleischer, A. (2012) A room with a view – a valuation of the Mediterranean Sea view. *Tourism Management*, 33, 598–602.

Fleming, D.K. and Hayuth, Y. (1994) Spatial characteristics of transportation hubs: centrality and intermediacy. *Journal of Transport Geography*, 2, 1, 3–18.

Flores, L.C.S. and Mendes, J.C. (2014) Perspectivas do destino turístico: repensando o sentido do conceito. *Revista Brasileira de Pesquisa em Turismo*, 8, 2, 222–237.

Flores Ruiz, D. (2014) *Manual de Gestión de Destinos Turísticos*. Tirant Humanidades, Valencia.

Forer, P. (2005) Tourist flows and dynamic geographies: applying GI Science to understanding tourist movement and impact. In: Simmons, D.G. and Fairweather, J.R. (eds) *Understanding the Tourism Host–Guest Encounter in New Zealand: Foundations for Adaptive Planning and Management*. Eos Ecology, Christchurch, pp. 23–56.

Forer, P.C. and Pearce, D.G. (1984) Spatial patterns of package tourism in New Zealand. *New Zealand Geographer*, 40, 1, 34–42.

Framke, W. (2002) The destination as a concept: a discussion of the business-related perspective versus the socio-cultural approach in tourism theory. *Scandinavian Journal of Hospitality and Tourism*, 2, 2, 92–108. DOI:10.1080/15022250216287

Franch, M., Martini, U. and Buffa, F. (2010) Roles and opinions of primary and secondary stakeholders within community-type destinations. *Tourism Review*, 65, 4, 74–86.

François, H. and Billet, S. (2010) De la station au territoire: quels contours pour les destinations touristiques? *Mondes du Tourisme*, 2, 87–98.

François, H., George-Marcelpoil, E., Fablet, G., Bray, F., Achin, C , Torre, A. and Barré, J.-B. (2012) *Atlas des Stations du Massif des Alpes*. Institut national de recherche en sciences et technologies pour l'environnement et l'agriculture (IRSTEA), Paris.

Freeman, L.C. (1978/79) Centrality in social network: I. Conceptual clarification. *Social Networks*, 1, 215–239.

Freytag, T. (2010) Déjà-vu: tourist practices of repeat visitors in the city of Paris. *Social Geography*, 5, 49–58. DOI:10.5194/sg-5-49-2010

Frisk, L. (1999) Separate worlds: attitudes and values towards tourism development and co-operation among public organisations and private enterprises in Northern Sweden. *Proceedings of Forksarforum: Local och Regional Utveckling Conference*. Östersund, Sweden.

Fuchs, M. and Weiermair, K. (2004) Destination benchmarking: an indicator-system's potential for exploring guest satisfaction. *Journal of Travel Research*, 42, 3, 212–225.

Gaido, L. (2002) Du concept de station au concept du district. *Revue de Géographie Alpine*, 4,109–114.

Gajdošík, T., Gajdošíková, Z., Maráková, V. and Flagestad, A. (2017) Destination structure revisited in view of the community and corporate model. *Tourism Management Perspectives*, 24, 54–63. DOI:10.1016/j.tmp.2017.07.009

Gallarza, M.G., Gil Saura, I. and Calderón García, H. (2002) Destination image: towards a conceptual framework. *Annals of Tourism Research*, 29, 1, 56–78.

García-Palomares, J.C., Gutiérrez, J. and Mínguez, C. (2015) Identification of tourist hot spots based on social networks: a comparative analysis of European metropolises using photo-sharing services and GIS. *Applied Geography* 63, 408–417.

Gardiner, S. and Scott, N. (2014) Successful tourism clusters: passion in paradise. *Annals of Tourism Research*, 46, 171–173.

Gaviría, M., Iribas, J.M., Sabbah, F. and Sanz, J.R. (1975) *Turismo de Playa en España*. Ediciones Turner, Madrid.

Getz, D. (1993) Planning for tourism business districts. *Annals of Tourism Research*, 20, 583–600.

Gibson, L.A., Lynch, P.A. and Morrison, A. (2005) The local destination tourism network: development issues. *Tourism and Hospitality Planning & Development*, 2, 2, 87–99.

Gilbert, E.W. (1939) The growth of inland and seaside health resorts in England. *Scottish Geographical Magazine*, 55, 16–35.

Gill, A. and Williams, P. (2006) Corporate responsibility and place: the case of Whistler, British Columbia. In: Clark, T.A., Gill, A. and Hartmann, R. (eds) *Mountain Resort Planning and Development in an Era of Globalization*. Cognizant Communication Corporation, Elmsford, NY, pp. 26–40.

Goffman, E. (1959) *The Presentation of Self in Everyday Life*. Doubleday, Garden City, NY.

Gollub, J., Hosier, A. and Woo, G. (2003) *Using Cluster-Based Economic Strategy to Minimize Tourism Leakages*. UNWTO, Madrid. Available at: http://www.e-unwto.org/doi/book/10.18111/9789284406166 (accessed 1 March 2017).

Gómez de Salazar, N.N. (2013) Génesis y evolución de la marina turística. In: Pie i Ninot, R. and Rosa Jimenez, C.J. (eds) *Turismo Líquido*. Instituto Habitat Turismo Territorio and Iniciativa Digital Politecnica, Universitat Politecnica de Catalunya, Barcelona/Universidad de Málaga, Málaga, pp. 122–141.

Gómez Diaz, J., Torres Valdez, J.C. and Miguel Velasco, A.E. (2017) El cluster turístico de un destino cultural en México: su ciclo de vida y la red de actores. *Cuadernos de Turismo*, 39, 265–285. DOI:10.6018/turismo.39.290531

Gordon, I. and Goodall, B. (2000) Localities and tourism. *Tourism Geographies*, 2, 3, 290–311.

Gospodini, A. (2001) Urban design, urban space morphology, urban tourism: an emerging new paradigm concerning their relationship. *European Planning Studies*, 9, 7, 925–934. DOI:10.1080/09654310120079841

Gray, H.P. (1970) *International Travel-International Trade*. Heath Lexington, Lexington, MA.

Greater Miami Convention & Visitors Bureau (2018) *Greater Miami and the Beaches: 2017 Visitor Industry Overview*. Greater Miami Convention & Visitors Bureau, Miami, FL.

Greer, T. and Wall, G. (1979) Recreational hinterlands: a theoretical and empirical analysis. In: Wall, G. (ed.) *Recreational Land Use in Southern Ontario*. Department of Geography Publications Series, 14. University of Waterloo, Waterloo, pp. 227–245.

Griffin, T., Hayllar, B. and Edwards, D. (2008) Places and precincts: a precinct typology. In: Hayllar, B., Griffin, T. and Edwards, D. (eds) *City Spaces–Tourist Places: Urban Tourism Precincts*. Butterworth-Heinemann, Oxford and Burlington, MA, pp. 41–61.

Grinberger, A.Y., Noam Shoval, N. and McKercher, B. (2014) Typologies of tourists' time–space consumption: a new approach using GPS data and GIS tools. *Tourism Geographies*, 16, 1, 105–123. DOI:10.1080/14616688.2013.869249

Guala Catalan, C. and Pearce, D.G. (2019) Changes to the national institutional framework for tourism in Chile: the use and impact of evaluation reports in the policy-making process. In: Andriotis, K., Stylidis, D. and Weidenfeld, A. (eds) *Tourism Policy and Planning Implementation*. Routledge, Abingdon, Oxon, pp. 115–139.

Guardia Galvez, T., Muro Romero, J. and Such Devesa, M.J. (2014) Measuring and analyzing domestic tourism: the importance of an origin and destination matrix. *Tourism Economics*, 20, 3, 451–472.

Gunn, C. (1972) *Vacationscape: Designing Tourist Regions*. University of Texas, Austin, TX.

Gunn, C. (1979) *Tourism Planning*. Crane, Russak and Co., New York.

Guthrie, H.W. (1961) Demand for tourists' goods and services in a world market. *Papers, Regional Science Association*, 7, 159–175.

Gutiérrez, J., García-Palomares, J.C., Romanillos, G. and Salas-Olmedo, M.H. (2017) Airbnb in tourism cities: comparing spatial patterns of hotels and peer-to-peer accommodation in Barcelona. *Tourism Management*, 62, 278–291. DOI: 10.1016/j.tourman.2017.05.003

Gutiérrez Ronco, S. (1980) Evolución en la localización de la hostelería madrileña. *Jornadas de Estudios sobre la Provincia de Madrid*. Deputación Provincial de Madrid, Madrid, pp. 283–288.

Gyimóthy, S. (1999) Visitors' perceptions of holiday experiences and service providers: an exploratory study. *Journal of Travel & Tourism Marketing*, 8, 2, 57–74.

Haddad, E.A, Porsse, A.A. and Rabahy, W. (2013) Domestic tourism and regional inequality in Brazil. *Tourism Economics*, 19, 1, 173–186.

Hägerstrand, T. (1973) The domain of human geography. In: Chorley R.J. (ed.) *Directions in Geography*. Methuen, London, pp. 67–87.

Han, H., Kim, S. and Otoo, F.E. (2018) Spatial movement patterns among intra-destinations using social network analysis. *Asia Pacific Journal of Tourism Research*, 23, 8, 806–822. DOI:10.1080/10941665.2018.1493519

Hanlan, J., Fuller, D. and Wilde, S. (2006) An evaluation of how market segmentation approaches aid destination marketing. *Journal of Hospitality and Leisure Marketing*, 15, 1, 5–26.

Hartmann, R. (1988) Combining field methods in geography. *Annals of Tourism Research*, 15, 1, 88–105.

Hartmann, R. (2006) Downstrean and down-valley: essential components and directions of growth and change in the sprawling resort landscapes of the Rocky Mountain West. In: Clark, T.A., Gill, A. and Hartmann, R. (eds) *Mountain Resort Planning and Development in an Era of Globalization*. Cognizant Communication Corporation, Elmsford, NY, pp. 278–293.

Hawkins, D.E. and Calnan, L.M. (2009) Comparative study of the sustainability of donor-supported tourism clusters in developing economies. *Tourism Recreation Research*, 34, 3, 269–282. DOI:10.1080/02508281.2009.11081602

Hayllar, B. and Griffin, T. (2005) The precinct experience: a phenomenological approach. *Tourism Management*, 26, 4, 517–528.

Hayllar, B., Griffin, T. and Edwards, D. (eds) (2008) *City Spaces – Tourist Places: Urban Tourism Precincts*. Butterworth-Heinemann, Oxford and Burlington, MA.

Hazebroucq, J.-M. (2009) De l'espace touristique à la 'destination touristique'. Un 'territoire' qui se projette. In: Lemasson, J.-P. and Violier, P. (eds) *Destinations et Territoires: Co-Présence à l'Œuvre*. Téoros, Montréal/Presses de l'Université de Québec, Québec, pp. 7–25.

Held, J. (1996) Regional variation and economic drivers: an application of the Hill and Brennan methodology. *Economic Development Quarterly*, 18, 4, 384–405.

Heo, C.Y., Blal, I. and Choi, M. (2019) What is happening in Paris? Airbnb, hotels, and the Parisian market: a case study. *Tourism Management*, 70, 78–88.

Hernández-Martín, R., Simancas-Cruz, R., González-Yanesc, J.A., Rodríguez-Rodríguez, Y., García-Cruz, J.I. and González-Mora, Y.M. (2016) Identifying micro-destinations and providing statistical information: a pilot study in the Canary Islands. *Current Issues in Tourism*, 19, 8, 771–790. DOI:10.1080/13683500.2014.916657

Hervas-Oliver, J.-L., Gonzalez, G., Caja, P. and Sempere-Ripoll, F. (2015) Clusters and industrial districts: where is the literature going? Identifying emerging subfields of research. *European Planning Studies*, 23, 9, 1827–1872. DOI:10.1080/09654313.2015.1021300

Hidalgo, R., Azar, P., Borsdorf, A. and Paulsen, A. (2016) Hospedándose en la ciudad global: patrones de localización de los hoteles de lujo en Santiago de Chile. *Cuadernos de Geografía – Revista Colombiana de Geografía*, 25, 2, 221–236.

Hillier, B. (1996) *Space is the Machine*. Cambridge University Press, Cambridge.

Hillier, B. (2005) The art of place and the science of space. *World Architecture*, 11, 185, 96–102.

Hills, T.L. and Lundgren, J. (1977) The impact of tourism in the Caribbean: a methodological study. *Annals of Tourism Research*, 4, 5, 248–267.

Hjalager, A.-M. (2000) Tourism destinations and the concept of industrial districts. *Tourism and Hospitality Research*, 2, 3, 92–112.

Horwath HTL (2018) *Asia Pacific Hotels & Chains Report 2018*. Available at: https://horwathhtl.com/publication/2018-asia-pacific-chains-report (accessed 13 March 2020).

Hottola, P. (1992) Tourist time–space budgets in Eilat, Israel. MSc thesis, University of Joensuu, Joensuu, Finland.

Howitt, R. (1998) Scale as relation: musical metaphors of geographical scale. *Area*, 30, 1, 49–58.

Hristov, D. and Petrova, P. (2015) Destination management plans – a new approach to managing destinations in England: evidence from Milton Keynes. *Current Issues in Tourism*, 21, 133–153. DOI:10.1080/13683500.2015.1070800

Hu, Y. and Ritchie, J.R.B. (1993) Measuring destination attractiveness: a contextual approach. *Journal of Travel Research*, 32, 2, 25–34.

Hu, Y., Gao, S., Janowicz, K., Yu, B., Li, W. and Prasad, S. (2015) Extracting and understanding urban areas of interest using geotagged photos. *Computers, Environment and Urban Systems*, 54, 240–254.

Hughes, G. (1998) Tourism and the semilogical realization of space. In: Ringer, G. (ed.) *Destinations: Cultural Landscapes of Tourism*. Routledge, London, pp. 17–32.

Huybers, T. and Bennett, J. (2003) Inter-firm cooperation at nature-based tourism destinations. *Journal of Socio-Economics*, 32, 571–587.

Hwang, Y.-H. and Fesenmaier, D.R. (2003) Multidestination pleasure travel patterns: empirical evidence from the American Travel Survey. *Journal of Travel Research*, 42, 166–171. DOI:10.1177/0047287503253936

Hwang, Y.-H. and Fesenmaier, D.R. (2005) The nature of Japanese travellers' multidestination trips in the United States. *Tourism Review International*, 9, 271–280.

Ingram, P and Baum, J.A.C. (1997) Chain affiliation and the failure of Manhattan hotels, 1898–1980. *Administrative Science Quarterly*, 42 , 68–102.

Institut de la Statistique de la Polynésie Française (2018) *Points Forts de la Polynésie Française: Bilan Tourisme 2017*. Institut de la Statistique de la Polynésie Française, Papeete, Tahiti.

Instituto de Turismo de España (2013) Movimientos turísticos de los españoles. *Familitur Informe Anual 2012*. Instituto de Turismo de España, Madrid.

Ipsos (2017) *Attractivité de Paris et sa région*. Ipsos, Paris.

Ivaniš, M. (2011) General model of small entrepreneurship development in tourism destinations in Croatia. *Tourism & Hospitality Management*, 17, 2, 231–250.

Ivanova, M. (2017) *Air Transport–Tourism Nexus: A Destination Management Perspective*. Zangador, Varna, Bulgaria.

Ivanova, M. and Ivanov, S. (2015) The nature of hotel chains: an integrative framework. *International Journal of Hospitality & Tourism Administration*, 16, 122–142.

Iwashita, C. (2003) Media construction of Britain as a destination for Japanese tourists: social constructionism and tourism. *Tourism Hospitality Research*, 4, 4, 331.

Jaakson, R. (2004) Beyond the tourist bubble? Cruiseship passengers in port. *Annals of Tourism Research*, 31, 1, 44–60.

Jackson, J. and Murphy, P. (2002) Tourism destinations as clusters: analytical experiences from the New World. *Tourism & Hospitality Research*, 4, 1, 36–52.

Jackson, J. and Murphy, P. (2006) Clusters in regional tourism: an Australian case. *Annals of Tourism Research*, 33, 4, 1018–1035. DOI: 10.1016/j.annals.2006.04.005

Jansen-Verbeke, M.C. and Ashworth, G. (1990) Environmental integration of recreation and tourism. *Annals of Tourism Research*, 17, 4, 618–622.

Jansen-Verbeke, M. and Lievois, E. (1999) Analysing heritage resources for urban tourism in European cities. In: Pearce, D.G. and Butler, R.W. (eds) *Contemporary Issues in Tourism Development*. Routledge, London, pp. 81–107.

Jimenez Zurita, F.F. (2013) Forma y tipología de los puertos deportivos del Arco Mediterráneo. In: Pie i Ninot, R. and Rosa Jimenez, C.J. (eds) *Turismo Líquido*. Instituto Habitat Turismo Territorio and Iniciativa Digital Politecnica, Barcelona/ Universidad de Málaga, Málaga, pp. 152–158.

Jones, P. and Chen, M.-M. (2011) Factors determining hotel selection: online behaviour by leisure travellers. *Tourism and Hospitality Research*, 11, 83–95. DOI:10.1057/thr.2010.20

Jones, S.B. (1933) Mining tourist towns in the Canadian Rockies. *Economic Geography*, 9, 368–378.

Johnson, C. and Vanetti, M. (2005) Locational strategies of international hotel chains. *Annals of Tourism Research*, 32, 4, 1077–1099.

Judd, D.R. (1993) Promoting tourism in US cities. *Tourism Management*, 16, 3, 583–600.

Judd, D.R. (2003) Visitors and the spatial ecology of the city. In: Hoffman, L.M., Fainstein, S.S. and Judd, D.R. (eds) *Cities and Visitors: Regulating, People, Markets and City Space*. Blackwell, Malden, MA, pp. 23–38.

Kachniewska, M. (2013) Toward the definition of a tourism cluster. *Journal of Entrepreneurship Management and Innovation*, 9, 1, 33–56.

Kadar, B. (2014) Measuring tourist activities in cities using geotagged photography. *Tourism Geographies*, 16, 1, 88–104. DOI: 10.1080/14616688.2013.868029

Kadar, B. (2015) A network-based spatial planning method for sustainable urban tourism. In: Tózsa, I. and Zátori, A. (eds) *Metropolitan Tourism Experience Development: Selected Studies from the Tourism Network Workshop of the Regional Studies Association, Budapest Hungary*. Department of Economic Geography and Futures Study, Corvinus University of Budapest, Budapest, pp. 9–31.

Kadri, B., Khomsi, M.R. and Bondarenko, M. (2011) Le concept de destination: diversité sémantique et réalité organisationnelle. *Téoros*, 30, 1, 12–24. DOI:10.7202/1012104ar

Karski, A. (1990) Urban tourism: a key to urban regeneration. *The Planner*, 76, 13, 15–17.

Kelman, I., Luthe, T., Wyss, R., Tørnblad, S.H., Evers,Y., Curran, M.M. *et al.* (2016) Social network analysis and qualitative interviews for assessing geographic characteristics of tourism business networks. *PLoS ONE*, 11, 6, e0156028. DOI:10.1371/journal.pone.0156028

Kennedy, V. and Augustyn, M.M. (2014) Stakeholder power and engagement in an English seaside context: implications for destination leadership. *Tourism Review*, 69, 3, 187–201. DOI:10.1108/TR-06-2013-0030

Keul, A. and Kühberger, A. (1997) Tracking the Salzburg tourist. *Annals of Tourism Research*, 24, 4, 1008–1012.

Khiati, A. (2018) *Atlas du Tourisme en France*. Direction Générale des Entreprises, Paris.

Koo, T.T.R., Lau, P.-L. and Dwyer, L. (2017) The geographic dispersal of visitors: insights from the power law. *Journal of Travel Research*, 56, 1, 108–121.

Kotus, J., Rzeszewski, M. and Ewertowski, W. (2015) Tourists in the spatial structures of a big Polish city: development of an uncontrolled patchwork or concentric spheres? *Tourism Management*, 50, 98–110.

Krakover, S. and Wang, Y. (2008) Spatial dimensions of the Orlando destination region. *Tourism Analysis*, 13, 245–258.

Kureha, M. (2014) Changes in Japanese ski resorts with the development of inbound tourism. A case study of Niseko-Hirafu district, Hokkaido. *Asia Pacific World*, 5, 2, 32–43. DOI:10.3167/apw.2014.050204. Available at: http://www.apu.ac.jp/iaaps/modules/publication/apw/vol5/no2/content0004.html (accessed 2 July 2020).

Kylänen, M. and Rusko, R. (2011) Unintentional coopetition in the service industries: the case of Pyhä-Luosto tourism destination in the Finnish Lapland. *European Management Journal*, 29, 193–205.

Lane, B. (2009) Rural tourism: an overview. In: Jamal, T. and Robinson, M. (eds) *The SAGE Handbook of Tourism Studies*. Sage, London, pp. 355–371.

Lane, B. and Kastenholz E. (2015) Rural tourism: the evolution of practice and research approaches – towards a new generation concept? *Journal of Sustainable Tourism*, 23, 8–9, 1133–1156. DOI: 10.1080/09669582.2015.1083997

Lanquar, R. (1994) *L'Économie du Tourisme*, 4th edn. Presses Universitaires de France, Paris.

Latinopoulos, D. (2018) Using a spatial hedonic analysis to evaluate the effect of sea view on hotel prices. *Tourism Management* 65, 87–99.

Laws, E. (1995) *Tourism Destination Management: Issues and Policies*. Routledge, London.

Lazerson, M. and Lorenzoni, G. (1999) Resisting organizational inertia: the evolution of industrial districts. *Journal of Management and Governance*, 3, 361–377.

Leask, A. (2010) Progress in visitor attraction research: towards more effective management. *Tourism Management*, 31, 155–166.

Lee, S.K. and Jang, S.C. (2011) Room rates of U.S. airport hotels: examining the dual effects of proximities. *Journal of Travel Research*, 50, 2, 186–197.

Lee, S.-H., Choi, J.-Y., Yoo, S.-H. and Oh,Y.-G. (2013) Evaluating spatial centrality for integrated tourism management in rural areas using GIS and network analysis. *Tourism Management*, 34, 14–24.

Leiper, N. (1979) The framework of tourism: towards a definition of tourism, tourist, and the tourist industry. *Annals of Tourism Research*, 6, 4, 390–407.

Leiper, N. (1990) Tourist attraction systems. *Annals of Tourism Research*, 17, 367–384.

Leiper, N. (2000) Systems theory. In: Jafari, J. (ed.) *Encyclopedia of Tourism*. Routledge, London, pp. 507–571.

Lemasson, J.-P. and Violier P. (2009) *Destinations et Territoires: Co-Présence à l'Œuvre*. Téoros, Montréal/Presses de l'Université de Québec, Québec.

Le Ngoc T., Pearce, D.G. and Smith, K.A. (2018) Distribution channels and ownership of upmarket hotels in a transition economy. *International Journal of Hospitality Management*, 68, 50–58.

Lepan, L. (2013) L'espace touristique de la grande ville: une approche par les pratiques et les mobilités touristiques. Le cas de la destination Paris. PhD thesis, Université d'Angers, Angers.

Levine, S. and White, P.E. (1961) Exchange as a conceptual framework for the study of interorganizational relations. *Administrative Science Quarterly*, 5, 583–601.

Lew, A.A. (1987) A framework of tourist attraction research. *Annals of Tourism Research*, 14, 553–575.

Lew, A.A. and McKercher, B. (2002) Trip destinations, gateways and itineraries: the example of Hong Kong. *Tourism Management*, 23, 6, 609–621.

Lew, A. and McKercher, B. (2006) Modeling tourist movements, a local destination analysis. *Annals of Tourism Research*, 33, 2, 403–423.

Lewis, R.C. (1985) The basis of hotel selection. *Cornell Hotel and Restaurant Administration Quarterly*, 25, 91, 54–69.

Li, M., Fang, L., Huang, X. and Goh, C. (2015) A spatial-temporal analysis of hotels in urban tourism destination. *International Journal of Hospitality Management*, 45, 34–43.

Lieshout, R., Malighetti, P., Redondi, R. and Burghouwt, G. (2016) The competitive landscape of air transport in Europe. *Journal of Transport Geography*, 50, 68–82.

Litteljohn, D. (1997) Internationalization in hotels: current aspects and developments. *International Journal of Contemporary Hospitality Management*, 9, 5/6, 187–192.

Liu, F., Zhang, J., Zhang, J., Chen, D., Liu, Z. and Lu, S. (2012) Roles and functions of tourism destinations in tourism region of South Anhui: a tourist flow network perspective. *Chinese Geographical Science*, 22, 6, 755–764.

Liu, J. and Wall, G. (2009) Resort morphology research: history and future perspectives. *Asia Pacific Journal of Tourism Research*, 14, 4, 339–350.

Lockyer, T. (2005) Understanding the dynamics of the hotel accommodation purchase decision. *International Journal of Contemporary Hospitality Management*, 17, 6, 481–492.

Lohmann, G. and Pearce, D.G. (2010) Conceptualizing and operationalizing tourism nodal functions. *Journal of Transport Geography*, 18, 266–275. DOI:10.1016/j.jtrangeo.2009.05.003

Longjit, C. and Pearce, D.G. (2013) Managing a mature coastal destination: Pattaya, Thailand. *Journal of Destination Marketing & Management*, 2, 3, 165–175. DOI:10.1016/j.jdmm.2013.05.002

López Bonilla, J.M. and López Bonilla, L.M. (2007) Diferencias territoriales en la planificación y la gestión del destino turístico. *Cuadernos de Turismo*, 19, 71–90.

López-López, A., Cukier, J. and Sánchez-Crispín, A. (2006) Segregation of tourist space in Los Cabos, Mexico. *Tourism Geographies*, 8, 4, 359–379. DOI: 10.1080/14616680600922054

López Palomeque, F. (2015) Barcelona, de ciudad con turismo a ciudad turística. Notas sobre un proceso complejo e inacabado. *Documents d'Anàlisi Geogràfica*, 61, 3, 483–506. DOI:10.5565/rev/dag.296

Lopez-Toro, A.A., Diaz-Munoz, R. and Perez-Moreno, S. (2010) An assessment of the quality of a tourist destination: the case of Nerja, Spain. *Total Quality Management and Business Excellence*, 21, 3, 269–289.

Lue, C.C., Crompton, J.L. and Fesenmaier, D.R. (1993) Conceptualization of multi-destination pleasure trip decisions. *Annals of Tourism Research*, 20, 289–301.

Lundgren, J.O.J. (1982) The tourist frontier of Nouveau Quebec: functions and regional linkages. *Tourist Review*, 37, 2, 17–25.

Luo, H. and Yang, Y. (2016) Intra-metropolitan location choices for star-rated and non-rated budget hotels: the role of agglomeration economies. *International Journal of Hospitality Management*, 59, 72–83.

Luthe, T., Wyss, R. and Schuckert, M. (2012) Network governance and regional resilience to climate change: empirical evidence from mountain tourism communities in the Swiss Gotthard region. *Regional Environmental Change*, 12, 839–854. DOI:10.1007/s10113-012-0294-5

Lynch, K. (1960) *The Image of the City*. MIT Press, Cambridge, MA.

Lynch, P. and Morrison, A. (2007) The role of networks. In: Michael, E. (ed.) *Micro-Clusters and Networks*. Elsevier, Oxford and Amsterdam, pp. 43–67.

Ma, M., Weng, J. and Yu, L. (2015) Market size, scale economies, and tourism market structure: a case of historic water town tourism in China. *Tourism Management*, 49, 119–137.

MacCannell, D. (1973) Staged authenticity: arrangements of social space in tourist settings. *American Journal of Sociology*, 79, 589–603.

MacCannell, D. (1976) *The Tourist: A New Theory of the Leisure Class*. Macmillan, London.

Machiavelli, A. (2001) Tourist destinations as integrated systems. *Tourism Review*, 56, 3/4, 6–11.

MacLeod, N. (2017) The role of trails in the creation of tourist space. *Journal of Heritage Tourism*, 12, 5, 423–430. DOI:10.1080/1743873X.2016.1242590

Maitland, R. (2013) Backstage behaviour in the global city: tourists and the search for the 'real London'. *Procedia – Social and Behavioral Sciences*, 105, 12–19.

Majewska, J. (2015) Inter-regional agglomeration effects in tourism in Poland. *Tourism Geographies*, 17, 3, 408–436. DOI: 10.1080/14616688.2014.997279

Manente, M. and Cerato, M. (2000a) Destination management: the conceptual framework. In: Manente, M. and Cerato, M. (eds) *From Destination to Destination Marketing and Management: Designing and Repositioning Tourism Products*. CISET International Centre of Studies on the Tourist Economy, Padua, pp. 15–27.

Manente, M. and Cerato, M. (2000b) Understanding the destination as a system. In: Manente, M. and Cerato, M. (eds) *From Destination to Destination Marketing and Management: Designing and Repositioning Tourism Products*. CISET International Centre of Studies on the Tourist Economy, Padua, pp. 131–146.

Manente, M. and Minghetti, V. (2005) Destination management organizations and actors. In: Buhalis, D. and Costa, C. (eds) *Tourism Business Frontiers: Consumers, Products and Industry*. Routledge, ProQuest Ebook Central, London and New York, pp. 228–237.

Marcelpoil, E. and François, H. (2008) Le processus d'articulation des proximités dans les territoires touristiques: l'exemple des stations de montagne. *Revue d'Économie Régionale et Urbaine*, 2, 179–191.

Marco-Lajara, B., Úbeda-García, M., Sabater-Sempere, V. and García-Lillo, F. (2014) Territory impact in the performance of Spanish vacation hotels. *Tourism Economics*, 20, 779–796. DOI:10.5367/te.2013.0301

Marco-Lajara, B., Claver-Cortés, E., Úbeda-García, M. and Zaragoza-Sáez, P.C. (2016a) A dynamic analysis of the agglomeration and performance relationship. *Journal of Business Research*, 69, 1874–1879.

Marco-Lajara, B., Zaragoza-Sáez, P.C., Claver-Cortés, E. and Úbeda-García, M. (2016b) Knowledge sources, agglomeration, and hotel performance. *Journal of Business Research*, 69, 4856–4861.

Marco-Lajara, B., Zaragoza-Sáez, P.C., Claver-Cortés, E., Úbeda-García, M. and García-Lillo, F. (2017) Tourist districts and internationalization of hotel firms. *Tourism Management*, 61, 451–464.

Marez Lopez, I.E. (2013) Proyecto de ordenación del Languedoc-Roussillon, Francia. In: Pie i Ninot, R. and Rosa Jimenez, C.J. (eds) *Turismo Líquido*. Instituto Habitat Turismo Territorio and Iniciativa Digital Politecnica, Universitat Politecnica de Catalunya, Barcelona/Universidad de Málaga, Málaga, pp. 206–226.

MarketLine (2018) *MarketLine Industry Profile: Hotels and Motels in Spain*. December 2018. MarketLine, London.

Markusen, A. (1996) Sticky places in slippery space: a typology of industrial districts. *Economic Geography*, 72, 3, 293–313.

Marshall A. (1920) *Principles of Economics*. Macmillan, London.

Martin, R. and Sunley, P. (2003) Deconstructing clusters: chaotic concept or policy panacea? *Journal of Economic Geography*, 3, 5–35.

Martínez Pérez, Á., Elche Hortelano, D. and García Villaverde, P.M. (2010) Delimitación de los distritos turísticos: problemática y orientaciones. *Documento de trabajo 2010*. Seminario Permanente de Ciencias Sociales, Facultad de Ciencias Sociales de Cuenca, Spain, SPCS 16.

Matley, I.M. (1976) The geography of international tourism. *Resource Paper*, No. 76-1. Association of American Geographers, Washington, DC.

Maulet, G. (2006) A framework to identify a localised tourism system. In: Lazzeretti, L. and Petrillo, C.S. (eds) *Tourism Local Systems and Networking*. Elsevier, Amsterdam, pp. 43–55.

Mazanec, J., Wober, K. and Zins, A.H. (2007) Tourism destination competitiveness: from definition to explanation? *Journal of Travel Research*, 46, 1, 86–95.

Mazaro, R.M. and Varzin, G. (2008) Modelos de competitividad para destinos turísticos en el marco de la sostenibilidad. *RAC – Revista de Administração Contemporânea*, 12, 3, 789–809.

McCleary, K.W., Weaver, P.A. and Hutchinson, J.C. (1993) Hotel selection factors as they relate to business travel situations. *Journal of Travel Research*, 32, 2, 42–48.

McCleary, K.W., Choi, B.M. and Weaver, P.A. (1998) A comparison of hotel selection criteria between U.S. and Korean business travelers. *Journal of Hospitality & Tourism Research*, 22, 1, 25–38.

McIntyre, G. (1993) *Sustainable Tourism Development: Guide for Local Planners*. World Tourism Organization, Madrid.

McKercher, B. and Lau,G. (2008) Movement patterns of tourists within a destination. *Tourism Geographies*, 10, 3, 355–374. DOI:10.1080/14616680802236352

McKercher, B., Shoval, N., Ng, E. and Birenboim, A. (2012) First and repeat visitor behaviour: GPS tracking and GIS analysis in Hong Kong. *Tourism Geographies*, 14, 1, 147–161. DOI:10.1080/14616688.2011.598542

Melian-Gonzalez, A. and Garcia-Falcon, J.M. (2003) Competitive potential of tourism in destinations. *Annals of Tourism Research*, 30, 3, 720–740.

Meng, F., Zou, T., Li, H., Ren, Y. and Zhang, P. (2016) International tourism hub: function assessment and application. *Tourism Economics*, 22, 6, 1225–1244. DOI:10.1177/1354816616670504

Meriläinen, K. and Lemmetyinen, A. (2011) Destination network management: a conceptual analysis. *Tourism Review*, 66, 3, 25–31.

Merinero Rodríguez, R. (2008) Micro-cluster turísticos: el papel del capital social en el desarrollo económico local. *Revista de Estudios Empresariales*, Segunda época, 2, 67–92.

Merinero Rodríguez, R. and Pulido Fernández, J.I. (2009) Desarrollo turístico y dinámica relacional. Metodología de análisis para la gestión activa de destinos turísticos. *Cuadernos de Turismo*, 23, 173–193.

Merinero-Rodríguez, R., Pulido-Fernández, J.I. and Navarro-Yáñez, C. (2014) Propuesta metodológica para la caracterización territorial del turismo a escala local. *Boletín de la Asociación de Geógrafos Españoles*, 66, 291–321. DOI:10.21138/bage.1791

Meyer-Arendt, K.J. (1990) Recreational districts in Gulf of Mexico seaside resort. *Journal of Cultural Geography*, 11, 1, 39–55.

Michael, E.J. (2003) Tourism micro-clusters. *Tourism Economics*, 9, 2, 133–145.

Michael, E.J. (2007) Development and cluster theory. In: Michael, E. (ed.) *Micro-Clusters and Networks*. Elsevier, Oxford and Amsterdam, pp. 21–31.

Miller, M.L., Auyong, J., Smenge, M., Offringa, B., Myles, P.B. and Lück, M. (2013) A field focuses on destinations: papers from the 7th International Coastal and Marine Tourism Congress. *Tourism in Marine Environments*, 9, 3, 117–128.

Miller, M.L., Lieske, S.N., Walsh S.J. and Carter, R.W. (2018) Understanding the interaction between a protected destination system and conservation tourism through remote sensing. In: Walsh, S.J. (ed.) *Remote Sensing for Societal Benefits*. Elsevier, Oxford, pp. 123–143.

Miller, M.M. and Gibson, L.J. (2006) Cluster-based development in the tourism industry: putting practice into theory. *Applied Research in Economic Development*, 2, 2, 47–64.

Miller, M.M., Henthorne, T.L. and George, B.P. (2008) The competitiveness of the Cuban tourism industry in the twenty-first century: a strategic re-evaluation. *Journal of Travel Research*, 46, 268–278.

Mings, R.C. and McHugh, K.E. (1992) The spatial configuration of travel to Yellowstone National Park. *Journal of Travel Research*, 30, 4, 38–46.

Miranda Júnior, N.S, Costa, H.A. and Hoffmann, V.E. (2016) Dificuldades para a cooperação entre hotéis aglomerados territorialmente: um estudo da hotelaria em Brasília DF. *Turismo em Análise*, 27, 1, 155–177. DOI:10.11606/issn.1984-4867.v27i1p153-177

Miyakuni, K. and Vander Stoep, G.A. (2006) Linking linkage concepts from diverse fields to build a community-based tourism planning framework: the case of Shuri, Japan. *Tourism Geographies*, 8, 3, 286–309.

Molloy, J.C., Ployhart, R.E. and Wright, P.M. (2011) The myth of 'the' micro–macro divide: bridging system-level and disciplinary divides. *Journal of Management*, 37, 2, 581–609.

Montero, C. and Parra, C. (2001) El cluster del ecoturismo en San Pedro. In: Buitelaar, R. and Gómez, J.J. (eds) *Serie Seminarios y conferencias*. CEPAL, Santiago, Chile, 17, 93–114.

Mottiar, Z. and Ryan, T. (2006) The role of SMEs in tourism development: an industrial district approach applied to Killarney, Ireland. In: Thomas, R. and Augustyn, M. (eds) *Tourism in the New Europe*. Taylor & Francis, Abingdon, Oxon, pp. 63–79.

Mules, T., (2005) Economic impacts of national park tourism on gateway communities: the case of Kosciuszko National Park. *Tourism Economics*, 11, 2, 247–259.

Murphy, P.E. (1980) Tourism management using land use planning and landscape design: the Victoria experience. *Canadian Geographer*, 24, 1, 60–71.

Murphy, P.E. and Keller, C.P. (1990) Destination travel patterns: an examination and modelling of tourist patterns on Vancouver Island, British Columbia. *Leisure Sciences*, 12, 1, 49–65. DOI:10.1080/01490409009513089

Murphy, P., Pritchard, M.P. and Smith, B. (2000) The destination product and its impact on traveller perceptions. *Tourism Management*, 21, 1, 43–52.

Murray, G. (2007) Constructing paradise: the impacts of big tourism in the Mexican coastal zone. *Coastal Management*, 35, 2, 339–355.

Mwesiumo, D. and Halpern, N. (2019) A review of empirical research on interorganizational relations in tourism. *Current Issues in Tourism*, 22, 4, 428–455. DOI: 10.1080/13683500.2017.1390554

National Travel & Tourism Office (2015) *Overseas Visitation Estimates for U.S. States, Cities, and Census Regions: 2014*. Available at: https://travel.trade.gov/outreachpages/download_data_table/2014_States_and_Cities.pdf (accessed 16 October 2018).

Newsome, D., Moore, S.A. and Dowling, R.K. (2013) *Natural Area Tourism: Ecology, Impacts and Management*, 2nd edn. Channel View, Bristol.

Nguyen, L.T.P. and Pearce, D.G. (2015) Joint destination marketing in the South Central Coast region of Vietnam. *Journal of Destination Marketing & Management*, 4, 88–97. DOI:10.1016/j.jdmm.2015.03.001 2212–571X/& 2015

Nicolau, J.L. and Más, F.J. (2006) The influence of distance and prices on the choice of tourist destinations: the moderating role of motivations. *Tourism Management*, 27, 5, 982–996.

Njite, D. and Schaffer, J. (2017) Revisiting attributes: how important is green in the consumer selection of hotel rooms? *International Journal of Hospitality & Tourism Administration*, 18, 2, 219–244. DOI:10.1080/15256480.2016.1263168

Nordin, S. and Svensson, B. (2007) Innovative destination governance: the Swedish ski resort of Åre. *Entrepreneurship and Innovation*, 8, 1, 53–66.

Normann, R. and Ramirez, R. (1993) From value chain to value constellation: designing interactive strategy. *Harvard Business Review*, July–August, 65–77.

Observatoire Économique du Tourisme Parisien (2014) *Étude 2014. Le Shopping, un Pilier du Tourisme à Paris*. Observatoire Économique du Tourisme Parisien, Paris.

O'Connor, K. (1995) Airport development in Southeast Asia. *Journal of Transport Geography*, 3, 4, 269–279.

O'Kelly, M.E. (1998) A geographer's analysis of hub-and-spoke networks. *Journal of Transport Geography*, 6, 3, 171–186.

O'Kelly, M.E. and Miller, H.J. (1994) The hub network design problem: a review and synthesis. *Journal of Transport Geography*, 2, 1, 31–40.

Oppermann, M. (1995) A model of travel itineraries. *Journal of Travel Research*, 33, 4, 57–61.

Paasi, A. (2004) Place and region: looking through the prism of scale. *Progress in Human Geography*, 28, 4, 536–546.

Paddison, R. (1983) *The Fragmented State: The Political Geography of Power*. Blackwell, Oxford.

Pan, C.-M. (2005) Market structure and profitability in the international tourist hotel industry. *Tourism Management*, 26, 845–850.

Paniagua, A. (2012) The rural as a site of recreation: evidence and contradictions in Spain from a geographical perspective. *Journal of Tourism and Cultural Change*, 10, 3, 264–275. DOI:10.1080/14766825.2012.707358

Papatheodorou, A. (2004) Exploring the evolution of tourism resorts. *Annals of Tourism Research*, 31, 1, 219–237.

Parrilla-González, J.A. and Pulido-Fernández, J.I. (2017) Dinamismo económico del turismo a escala local: una aproximación a través del análisis de clases latentes. *Cuadernos de Turismo*, 40, 521–545. DOI:10.6018/turismo.40.310091

Parroco, A.M., Vaccina, F., De Cantis, S. and Ferrante, M. (2012) Multi-destination trips and tourism statistics: empirical evidences in Sicily. *Economics: The Open-Access, Open-Assessment E-Journal*, 6, 2012–44, 1–27. DOI:10.5018/economics-ejournal.ja.2012-44

Partalidou, M. and Koutsou, K. (2012) Locally and socially embedded tourism clusters in rural Greece. *Tourismos: an International Multidisciplinary Journal of Tourism*, 7, 1, 99–116.

Pasquinelli, C. (2017) Tourism connectivity and spatial complexity: a widening bi-dimensional arena of urban tourism research. In: Bellini, N. and Pasquinelli, C. (eds) *Tourism in the City*. Springer, Cham, Switzerland, pp. 29–50.

Pavlovich, K. (2003) The evolution and transformation of a tourism destination network: the Waitomo Caves, New Zealand. *Tourism Management*, 24, 2, 203–216. DOI:10.1016/S0261-5177(02)00056-0

Pearce, D.G. (1978) Form and function in French resorts. *Annals of Tourism Research*, 5, 1, 142–56. DOI:10.1016/0160-7383(78)90008-7

Pearce, D.G. (1981) L'espace touristique de la grande ville: éléments de synthèse et application à Christchurch (Nouvelle-Zélande). *L'Espace Géographique*, 10, 3, 207–213.

Pearce, D.G. (1984) Planning for tourism in Belize. *Geographical Review*, 74, 3, 291–303.

Pearce, D.G. (1987a) Mediterranean charters: a comparative geographic perspective. *Tourism Management*, 8, 4, 291–305.

Pearce, D. G. (1987b) Motel location and choice in Christchurch. *New Zealand Geographer*, 43, 1, 10–17.

Pearce, D.G. (1988) Tourist time-budgets. *Annals of Tourism Research*, 15, 1, 106–121.

Pearce, D.G. (1992) *Tourist Organizations*. Longman, Harlow, Essex/Wiley, New York.

Pearce, D.G. (1993a) Comparative studies in tourism research. In: Pearce, D.G. and Butler, R.W. (eds) *Tourism Research: Critiques and Challenges*. Routledge, London, pp. 20–35.

Pearce, D.G. (1993b) Domestic tourist travel patterns in New Zealand. *GeoJournal*, 29, 3, 225–232.

Pearce, D.G. (1995a) *Tourism Today: A Geographical Analysis*, 2nd edn. Longman, Harlow, Essex/Wiley, New York.

Pearce, D.G. (1995b) Japanese tourists in Europe. *Geographical Review of Japan*, Ser. B, 68, 1, 63–74.

Pearce, D.G. (1996a) Domestic tourist travel in Sweden: a regional analysis. *Geografiska Annaler*, 78B, 2, 71–84.

Pearce, D.G. (1996b) Tourist organizations in Sweden. *Tourism Management*, 17, 7, 413–424.

Pearce, D.G. (1996c) Federalism and the organization of tourism in Belgium. *European Urban and Regional Studies*, 3, 3, 189–204.

Pearce, D.G. (1997a) Analysing the demand for urban tourism: issues and examples from Paris. *Tourism Analysis*, 1, 1, 5–18.

Pearce, D.G. (1997b) Tourism and the autonomous communities in Spain. *Annals of Tourism Research*, 21, 1, 156–177.

Pearce, D.G. (1998a) Tourist districts in Paris: structure and functions. *Tourism Management*, 19, 1, 49–65. DOI:10.1016/S0261-5177(97)00095-2

Pearce, D.G. (1998b) Tourism development in Paris: public intervention. *Annals of Tourism Research*, 25, 2, 457–476.

Pearce, D.G. (1999) Tourism in Paris: studies at the microscale. *Annals of Tourism Research*, 26, 1, 77–97.

Pearce, D.G. (2001) An integrative framework for urban tourism. *Annals of Tourism Research*, 28, 4, 926–946.

Pearce, D.G. (2008) A needs-functions model of tourism distribution. *Annals of Tourism Research*, 35, 1, 148–168.

Pearce, D.G. (2010) From development to destination management: international perspectives on major coastal resorts. *2010 Saemangeum Tourism Development Symposium*, Byunsan Daemyung Resort, Korea, pp. 112–123.

Pearce D.G. (2012) *Frameworks for Tourism Research*. CABI, Wallingford, Oxon.

Pearce, D.G. (2014) Toward an integrative conceptual framework of destinations. *Journal of Travel Research*, 53, 2, 141–153.

Pearce, D.G. (2015a) Destination management in New Zealand: structures and functions. *Journal of Destination Marketing & Management*, 4, 1–15. DOI:10.1016/j.jdmm.2014.12.001

Pearce, D.G. (2015b) Urban management, destination management and urban destination management: a comparative review with issues and examples from New Zealand. *International Journal of Tourism Cities*, 1, 1, 1–17.

Pearce, D.G. (2016a) Destination management: plans and practitioners' perspectives in New Zealand. *Tourism Planning & Development*, 13, 52–71.

Pearce D.G. (2016b) Interdependent destination management functions. *Tourism Recreation Research*, 41, 37–48. DOI: 10.1080/02508281.2015.1119433

Pearce, D.G. and Elliott, J.M.C. (1983) The trip index. *Journal of Travel Research*, 22, 1, 6–9.

Pearce, D.G. and Kirk, R.M. (1986) Carrying capacities for coastal tourism. *Industry and Environment*, 9, 1, 3–7.

Pearce, D.G. and Priestley, G.K. (1998) Tourism in Spain: a spatial analysis and synthesis. *Tourism Analysis*, 2, 3/4, 185–205.

Pearce, D.G. and Schänzel, H. (2013) Destination management: the tourists' perspective. *Journal of Destination Marketing & Management*, 2, 3, 137–145.

Pearce, D.G. and Schänzel, H. (2015) Destinations: tourists' perspectives from New Zealand. *International Journal of Tourism Research*, 17, 4–12.

Pearce, D.G. and Schott, C. (2005) Tourism distribution channels: the visitors' perspective. *Journal of Travel Research*, 44, 1, 50–63.

Pearce, D.G. and Tan, R. (2006) The distribution mix for tourism attractions in Rotorua, New Zealand. *Journal of Travel Research*, 44, 2, 250–258.

Pearce, D.G., Tan, R. and Schott, C. (2004) Tourism distribution channels in Wellington, New Zealand. *International Journal of Tourism Research*, 6, 6, 397–410.

Pearce, D.G., Tan, R. and Schott, C. (2007) Distribution channels in international markets: a comparative analysis of the distribution of New Zealand Tourism in Australia, Great Britain and the USA. *Current Issues in Tourism*, 10, 1, 33–60.

Pearce, D.G., Reid, L. and Schott, C. (2009) Travel arrangements and the distribution behaviour of New Zealand outbound travelers. *Journal of Travel & Tourism Marketing*, 26, 1, 80–97. DOI:10.1080/10548400802656850

Pearce, D.G., Guala, C., Veloso, K., Llano, S., Negrete, J. *et al.* (2017) Destination management in Chile: objectives, actions and actors. *International Journal of Tourism Research*, 19, 50–67. DOI:10.1002/jtr.2083

Pearce, P.L. (1991) Analysing tourist attractions. *Journal of Tourism Studies*, 2, 46–55.

Pechlaner, H. and Fuchs, M. (2002) Towards new skill requirements for destination organizations: an exploratory study. *Tourism Analysis*, 7, 1, 43–53.

Pechlaner, H., Raich, F. and Zehrer, A. (2007) The Alps: challenges and potentials of a brand management. *Tourism Analysis*, 12, 5/6, 359–369.

Pechlaner H., Herdorfer, P. and Tödter, N. (2008) European destination management: challenges for product and brand management. In: Conrady, R. and Buck, M. (eds) *Trends and Issues in Global Tourism*. Springer, Berlin, pp. 109–120.

Pechlaner, H., Volgger, M. and Herntrei, M. (2012) Destination management organizations as interface between destination governance and corporate governance. *Tourism Review*, 67, 2, 34–44.

Pejanović, R., Demirović, D., Glavaš-Trbić, D., Maksimović, G. and Tomaš-Simin, M. (2017) Clusters as a factor of competitiveness of rural tourism destinations in the Danube region of the Republic of Serbia. *Tourism Economics*, 23, 2, 475–482.

Penrose, E.G. (1959) *The Theory of the Growth of the Firm*, 3rd edn. Oxford University Press, New York.

Perles-Ribes, J.F., Rodríguez-Sánchez, I. and Ramón-Rodríguez, A.B. (2017) Is a cluster a necessary condition for success? The case of Benidorm. *Current Issues in Tourism*, 20, 15, 1575–1603. DOI:10.1080/13683500.2015.1043247

Petrić, L. and Mrnjavac, Ž. (2003) Tourism destination as a locally embedded system: analogy between tourism destination and industrial district. *Tourism*, 51, 4, 403–415.

Petrou, A., Pantziou, E.F., Dimara, E. and Skuras, D. (2007) Resources and activities complementarities: the role of business networks in the provision of integrated rural tourism. *Tourism Geographies*, 9, 4, 421–440. DOI: 10.1080/14616680701647634

Pettersson, R. and Zillinger, M. (2011) Time and space in event behaviour: tracking visitors by GPS. *Tourism Geographies*, 13, 1, 1–20. DOI:10.1080/14616688.2010.529932

Pham, V.T.T. and Pearce, D.G. (2019) Destination and tourism business brand consistency in Binh Thuan, Vietnam. *Journal of Vacation Marketing*, 25, 1, 37–50.

Pie i Ninot, R. (2013) Las arquitecturas del turismo: las piezas mínimas. In: Pie i Ninot, R. and Rosa Jimenez, C.J. (eds) *Turismo Líquido*. Instituto Habitat Turismo Territorio and Iniciativa Digital Politecnica, Universitat Politecnica de Catalunya, Barcelona/Universidad de Málaga, Málaga, pp. 15–39.

Pike, S. (2002) Destination image analysis: a review of 142 papers from 1973 to 2000. *Tourism Management*, 23, 5, 541–549.

Pike, S. and Kotsi, F. (2016) Stopover destination image – using the repertory test to identify salient attributes. *Tourism Management Perspectives*, 18, 68–73.

Pike, S. and Page, S.J. (2014) Destination marketing organizations and destination marketing: a narrative analysis of the literature. *Tourism Management*, 41, 202–227.

Piriou, J. (2016) La dimension régionale du système touristique: une lecture réticulaire par les pratiques spatiales des touristes au sein de l'espace des 'Châteaux de la Loire'. *Mondes du Tourisme*. Hors-série 2016. DOI:10.4000/tourisme.1276. Available at: http://tourisme.revues.org/1276 (accessed 2 July 2020).

Plog, S.C. (1974) Why destinations rise and fall in popularity. *Cornell Hotel and Restaurant Administration Quarterly*, 14, 4, 55–58.

Ponsford, I., Williams, P. and Gill, A. (2006) Destination structuration: evolving corporate–community relationships. In: Clark, T., Gill, A. and Hartmann, R. (eds) *New Frontiers in Global Tourism – Trends and Competitive Challenges: 37th Annual Conference Proceedings*. Travel and Tourism Research Association, Dublin, pp. 558–568.

Porter, M.E. (1985) *Competitive Advantage*. The Free Press, New York.

Porter, M.E. (1990) *The Competitive Advantage of Nations*. The Free Press, New York.

Porter, M.E. (1991) Towards a dynamic theory of strategy. *Strategic Management Journal*, 12, 95–117.

Porter, M.E. (1998) Clusters and the new economics of competition. *Harvard Business Review*, 11, 77–90.

Porter, M.E. (2000) Location, competition, and economic development: local clusters in a global economy. *Economic Development Quarterly*, 14, 1, 15–34.

Prats, L., Guia, J. and Molina, F.-X. (2008) How tourism destinations evolve: the notion of tourism local innovation system. *Tourism and Hospitality Research*, 8, 3, 178–191.

Préau, P. (1968) Essai d'une typologie des stations de sports d'hiver dans les Alpes du Nord. *Revue de Géographie Alpine*, 56, 1, 127–140. DOI:10.3406/rga.1968.3571

Préau, P. (1970) Principe d'analyse des sites en montagne. *Urbanisme*, 116, 21–25.

Presenza, A. and Cipollina, M. (2010) Analysing tourism stakeholders networks. *Tourism Review*, 65, 4, 17–30.

Presenza, A., Sheehan, L. and Ritchie, J.R.B. (2005) Towards a model of the roles and activities of destination management organizations. *Journal of Hospitality, Tourism and Leisure Science*, 3, 1–16.

Przezbórska, L. (2003) Relationships between rural tourism and agrarian restructuring in a transitional economy: the case of Poland. In: Hall, D., Roberts, L. and Mitchell, M. (eds) *New Directions in Rural Tourism*. Routledge, London and New York, pp. 205–222.

Pulido-Fernández, J.I. and Merinero-Rodríguez, R. (2018) Destinations' relational dynamic and tourism development. *Journal of Destination Marketing & Management*, 7, 140–152. DOI:10.1016/j.jdmm.2016.09.008

Raun, J., Ahas, R. and Tiru, M. (2016) Measuring tourism destinations using mobile tracking data. *Tourism Management*, 57, 202–212.

Reynolds-Feighan, A. (2010) Characterisation of airline networks: a North American and European comparison. *Journal of Air Transport Management*, 16, 109–120.

Reynolds-Feighan, A. (2018) US feeder airlines: industry structure, networks and performance. *Transportation Research, A*, 117, 142–157.

Ribaudo, G. and Domeniconi, S. (2013–2014) The penetration of international hotel chains in Italy: evidences from an updated census. *Review of Business*, 34, 1, 5–19.

Richards, G. (2002) Tourism attraction systems: exploring cultural behavior. *Annals of Tourism Research*, 29, 4, 1048–1064.

Rieucau, J. (2000a) La Grande-Motte, ville permanente, ville saisonnière. *Annales de Géographie*, 109, 616, 631–654.

Rieucau, J. (2000b) La Grande-Motte: une ville-station touristique. *Norois*, 187, 3, 341–352. DOI:10.3406/noroi.2000.7029

Rigall-I-Torrent, R., Fluvià, M., Ballester, R., Saló, A., Ariza, E. and Espinet, J.-M. (2011) The effects of beach characteristics and location with respect to hotel prices. *Tourism Management*, 32, 1150–1158

Ringer, G. (1998) *Destinations: Cultural Landscapes of Tourism*. Routledge, London.

Rivers, M.J., Toh, R.S. and Alaoiu, M. (1991) Frequent-stayer programs: the demographic , behavioral, and attitudinal characteristics of hotel steady sleepers. *Journal of Travel Research*, 30, 2, 41–45.

Robinson, H. (1976) *A Geography of Tourism*. MacDonald and Evans, London.

Rocks, S., Gilmore, G. and Carson, S. (2005) Developing strategic marketing through the use of marketing networks. *Journal of Strategic Marketing*, 13, 81–92.

Rodrigue, J.-P., Comtois, C. and Slack, B. (2006) *The Geography of Transport Systems*. Routledge, Abingdon, Oxon and New York.

Rogers, A., Castree, N. and Kitchin, R (2013) *A Dictionary of Human Geography*. Oxford University Press, Oxford. DOI: 10.1093/acref/9780199599868.001.0001

Rogerson, J.M. (2012) The changing location of hotels in South Africa's coastal cities, 1990–2010. *Urban Forum*, 23, 73–91. DOI:10.1007/s12132-012-9143-3

Rognant, L. (1990) *Un Géo-Système Touristique National, Essai Systématique*. Centre des Hautes Études Touristiques, Aix-en-Provence.

Romeiro, P. and Costa, C. (2010) The potential of management networks in the innovation and competitiveness of rural tourism: a case study on the Valle del Jerte (Spain). *Current Issues in Tourism*, 13, 1, 75–91.

Rothman, H. (1998) *Devil's Bargain: Tourism in the Twentieth Century American West*. University Press of Kansas, Lawrence, KS.

Rubin, R.M. and Joy, J.N. (2005) Where are the airlines headed? Implications of airline industry structure and change for consumers. *Journal of Consumer Affairs*, 39, 1, 215–228.

Ruekert, R.W. and Walker, O.C. (1987) Marketing's interaction with other functional units: a conceptual framework and empirical evidence. *Journal of Marketing*, 51, 1–19.

Rusko, R.T., Kylänen, M. and Saari, R. (2009) Supply chain in tourism destinations: the case of Levi resort in Finnish Lapland. *International Journal of Tourism Research*, 11, 71–87.

Ryglová, K. (2008) Destination management. *Agricultural Economics*, 54, 9, 440–448.

Saarinen, J. (1998) The social construction of tourist destinations: the process of transformation of the Saariselka tourism region in Finnish Lapland. In: Ringer, G. (ed.) *Destinations: Cultural Landscapes of Tourism*. Routledge, London, pp. 154–173.

Saarinen, J. (2004) Destinations in change: the transformation process of tourist destinations. *Tourist Studies*, 4, 2, 161–179.

Saarinen, J. (2014) Transforming destinations: a discursive approach to tourist destinations and development. In: Viken, A. and Granås, B. (eds) *Tourism Destination Development: Turns and Tactics*. Routledge, Abingdon, Oxon, pp. 47–60.

Sainaghi, R. (2006) From contents to processes: versus a dynamic destination management model (DDMM). *Tourism Management*, 27, 5, 1053–1063.

Salas-Olmedo, M.H., Moya-Gómez, B., García-Palomares, J.C. and Gutiérrez, J. (2018) Tourists' digital footprint in cities: comparing big data sources. *Tourism Management*, 66, 13–25.

Sánchez Martín, J.M., Sánchez Rivero, M. and Rengifo Gallego, J.I. (2017) Análisis del equilibrio entre el potencial turístico y la oferta de alojamientos en turismo rural mediante técnicas de estadística espacial – una aplicación a la provincia de Cáceres (España). *Cuadernos de Turismo*, 39, 547–576. DOI:10.6018/turismo.39.290701

Sánchez Rivero, M., Sánchez Martín, J.M. and Rengifo Gallego, J.I. (2016) Methodological approach for assessing the potential of a rural tourism destination: an application in the province of Cáceres (Spain). *Current Issues in Tourism*, 19, 11, 1084–1102. DOI:10.1080/13683500.2014.978745

Santos, G.E.O., Ramos, V. and Rey-Maquieira, J. (2012) Determinants of multi-destination tourism trips in Brazil. *Tourism Economics*, 18, 6, 1331–1349. DOI:10.5367/te.2012.0170

Saraniemi, S. and Kylänen, M. (2011) Problematizing the concept of tourism destination: an analysis of different theoretical approaches. *Journal of Travel Research*, 50, 2, 133–143.

Sarrión-Gavilán, M.D., Benítez-Márquez, M.D. and Mora-Rangel, E.O. (2015) Spatial distribution of tourism supply in Andalusia. *Tourism Management Perspectives*, 15, 29–45.

Sartori, A., Mottironi, C. and Antonioli Corigliano, M. (2012) Tourist destination brand equity and internal stakeholders: an empirical research. *Journal of Vacation Marketing*, 18, 4, 327–340.

Schachar, A. and Shoval, N. (1999) Tourism in Jerusalem: a place to pray. In: Judd, D.R. and Fainstein, S.S. (eds) *The Tourist City*. Yale University Press, New Haven, CT, pp. 198–211.

Schmidt, C.J. (1979) The guided tour: insulated adventure. *Urban Life*, 7, 441–467.

Schroeder, R.V., Zimmermann, F.W. and Formiga, N. (2016) Growing complexity in rural areas: networking through tourism and recreation. *American Journal of Rural Development*, 4, 1, 15–23. DOI:10.12691/ajrd-4-1-3.

Scott, J. (2000) *Social Network Analysis: A Handbook*. Sage, London.

Scott, N., Cooper, C. and Baggio, R. (2008a) *Network Analysis and Tourism: From Theory to Practice*. Channel View, Bristol.

Scott, N., Cooper, C. and Baggio, R. (2008b) Destination networks: four Australian cases. *Annals of Tourism Research*, 35, 1, 169–186.

Segarra-Oña, M., Miret-Pastor, L.G., Peiro-Signes, A. and Verma, R. (2012) The effects of localization on economic performance: analysis of Spanish tourism clusters. *European Planning Studies*, 20, 8, 1319–1334. DOI:10.1080/09654313.2012.680586

Selin, S. and Beason, K. (1991) Interorganizational relations in tourism. *Annals of Tourism Research*, 18, 639–652.

Selke, A.C. (1936) Geographic aspects of the German tourist trade. *Economic Geography*, 12, 206–216.

Senftleben, W. (1973) Some aspects of the Indian hill stations: a contribution towards a geography of tourist traffic. *Philippines Geographical Journal*, 17, 1, 21–29.

Sfandla, C. and Björk, P. (2013) Tourism experience network: co-creation of experiences in interactive processes. *International Journal of Tourism Research*, 15, 495–506.

Sharpley, R. (2004a) Tourism, modernisation and development on the island of Cyprus: challenges and policy responses. In: Bramwell, B. (ed.) *Coastal Mass Tourism.*

Diversification and Sustainable Development in Southern Europe. Channel View, Bristol, pp. 321–340.

Sharpley, R. (2004b) Tourism and the countryside. In: Lew, A.A., Hall, M. and Williams, A.M. (eds) *A Companion to Tourism*. Blackwell, Malden, MA, pp. 374–386.

Shaw, S.-H. (1993) Hub structures of major US passenger airlines. *Journal of Transport Geography*, 1, 1, 47–48.

Sheppard, E. and Gregory, D. (2009) Spatial structure. In: Gregory, D., Johnston, R.J., Pratt, G., Watts, M. and Whatmore, S. (eds) *The Dictionary of Human Geography*, 5th edn. Wiley-Blackwell, Chichester.

Shih, H.-Y. (2006) Network characteristics of drive tourism destinations: an application of network analysis in tourism. *Tourism Management*, 27, 1029–1039. DOI:10.1016/j.tourman.2005.08.002

Shoval, N. (2006) The geography of hotels in cities: an empirical validation of a forgotten model. *Tourism Geographies*, 8, 1, 56–75. DOI:10.1080/14616680500392499

Shoval, N. and Raveh, A. (2004) Categorization of tourist attractions and the modelling of tourist cities: based on the co-plot method of multivariate analysis. *Tourism Management*, 25, 741–740.

Shoval, N., McKercher, B., Ng, E. and Birenboim, A. (2011) Hotel location and tourist activity in cities. *Annals of Tourism Research*, 38, 4, 1594–1612.

Silva Lira, I. (2002) Desarrollo local y alternativas de desarrollo productivo: el impulso de un *cluster* eco-turístico en la región de Aysén. Serie *Gestión Pública*, No. 24. CEPAL, Instituto Latinoamericano y del Caribe de Planificación Económica y Social-ILPES, Santiago, Chile. Available at: https://repositorio.cepal.org/bitstream/handle/11362/7276/S0210859_es.pdf (accessed 30 September 2020).

Simon, G. (2010) Pratiques touristiques dans la métropole parisienne: une analyse des mouvements intra-urbains. PhD thesis. Université Paris-Est, Paris.

Smith, S.L.J. (1994) The tourism product. *Annals of Tourism Research*, 21, 3, 582–595.

Soares, J.C., Ivars Baidal, J.A. and Gândara, J.M. (2016) Dinámica evolutiva y path dependence en los destinos turísticos litorales. El papel y la percepción de los agentes locales desde una perspectiva comparada. *Estudios y Perspectivas en Turismo*, 25, 164–185.

Sohrabi, B., Vanani, I.R., Tahmasebipur, K. and Fazli, S. (2012) An exploratory analysis of hotel selection factors: a comprehensive survey of Tehran hotels. *International Journal of Hospitality Management*, 31, 96–106.

Sowman, P. and Pearce, D.G. (2000) Tourism, national parks and visitor management. In: Butler, R.W. and Boyd, S.W. (eds) *Tourism and National Parks: Issues and Implications*. Wiley, Chichester, pp. 223–243.

Squire, S.J. (1998) Rewriting languages of geography and tourism: cultural discourses of destinations, gender and tourism. In: Ringer, G. (ed.) *Destinations: Cultural Landscapes of Tourism*. Routledge, London, pp. 80–100.

Standley, S. (1971) *The Impact of the Vail Ski Resort: An Input–Output Analysis*. Graduate School of Business Administration, University of Colorado, Boulder, CO.

Stange, J. and Brown, D. (2011) *Tourism Destination Management: Achieving Sustainable and Competitive Results*. Sustainable Tourism: International Cooperation for Development. Online Tool Kit and Resource Series. International Institute for Tourism Studies, Washington, DC. Available at: https://www.usaid.gov/sites/default/files/documents/2151/DMOworkbook_130318.pdf (accessed 2 July 2020).

Stansfield, C.A. (1969) Recreational land use patterns within an American seaside resort. *Tourist Review*, 24, 128–136.

Stansfield, C.A. (1971) The geography of resorts: problems and potentials. *Professional Geographer*, 23, 164–166.

Stansfield, C.A. (1978) Atlantic City and the resort cycle: background to the legalization of gambling. *Annals of Tourism Research*, 5, 2, 238–251.

Stansfield, C.A. and Rickert, J.E. (1970) The recreational business district. *Journal of Leisure Research*, 2, 4, 213–25.

Stors, N. and Baltes, S. (2018) Constructing urban tourism space digitally: a study of Airbnb listings in two Berlin neighborhoods. *Proceedings of the ACM on Human–Computer Interaction*, 2, CSCW, Article 166. DOI:0.1145/3274435

Subsecretaria de Turismo (2015) *Plan Nacional de Desarrollo Turístico Sustentable*. Subsecretaria de Turismo, Santiago, Chile. Available at: http://www.subturismo.gob.cl/documentos/plan-nacional-de-desarrollo-turistico-sustentable (accessed 2 May 2019).

Subsecretaria de Turismo (2018) *Informe de Intensidad Turística y Definición de Destinos Turísticos*. Subsecretaria de Turismo, Santiago, Chile. Available at: http://www.subturismo.gob.cl/wp-content/uploads/2015/09/Informe-de-Intensidad-Turística-y-Definición-de-Destinos-Turísticos-2018-1.pdf (accessed 2 May 2019).

Such Devesa, M.J. and Mendieta Peñalver, L.F. (2013) Size, efficiency and productivity in the Spanish hotel industry – independent properties versus chain-affiliated hotels. *Tourism Economics*, 19, 4, 801–809. DOI:10.5367/te.2013.0320

Sundbo, J., Orfila-Sintes, F. and Sørensen, F. (2007) The innovative behaviour of tourism firms – comparative studies of Denmark and Spain. *Research Policy*, 36, 1, 88–106.

Swisscontact (2014) *Conceptos básicos para la gestión de destinos turísticos*. Programa de Cooperación al Desarrollo Económico. Secretaría de Estado para Asuntos Económicos SECO, Lima, Perú.

Symanski, R. and Newman, J.L. (1973) Formal, functional, and nodal regions: three fallacies. *The Professional Geographer*, 25, 4, 350–352. DOI:10.1111/j.0033-0124.1973.00350.x

Szarka, J. (1990) Networking and small firms. *International Small Business Journal*, 8, 10–21.

Tait, A. (2010) Mountain ski maps of North America: a preliminary survey and analysis of style. *Cartographic Perspectives*, 67, 5–17.

Terhorst, P. and Erkuş-Öztürk, H. (2011) Scaling, territoriality, and networks of a tourism place, Anatolia. *An International Journal of Tourism and Hospitality Research*, 22, 2,168–183. DOI:10.1080/13032917.2011.597932

Thomazi, S. (2006) *Cluster de Turismo: Introdução ao Estudo de Arranjo Produtivo Local*. Aleph, São Paulo, Brasil.

Timur, S. and Getz, D. (2008) A network perspective on managing stakeholders for sustainable urban tourism. *International Journal of Contemporary Hospitality Management*, 20, 4, 445–461. DOI: 10.1108/09596110810873543

Tinsley, R. and Lynch, P.A. (2008) Differentiation and tourism destination development: small business success in a close-knit community. *Tourism and Hospitality Research*, 8, 161–177. DOI:10.1057/thr.2008.26

Torre, A. and Rallet, A. (2005) Proximity and localization. *Regional Studies*, 39, 1, 47–59. DOI:10.1080/0034340052000320842

Torres, R. (2002) Cancun's tourism development from a Fordist spectrum of analysis. *Tourist Studies*, 2, 1, 87–116.

Torres, R.M. and Momsen, J.D. (2005a) Planned tourism development in Quintana Roo, Mexico: engine for regional development or prescription for inequitable development. *Current Issues in Tourism*, 8, 4, 259–285.

Torres, R.M. and Momsen, J.D. (2005b) Gringolandia: the construction of a new tourist space in Mexico. *Annals of the Association of American Geographers*, 95, 2, 314–335.

Trippl, M., Grillitsch, M., Isaksen, A. and Sinozic, T. (2015) Perspectives on cluster evolution: critical review and future research issues. *European Planning Studies*, 23, 10, 2028–2044. DOI:10.1080/09654313.2014.999450

Truchet, S., Piguet, V., Aubert, F. and Callois, J.-M. (2016) Spatial influence of attractions on tourism development. *Tourism Geographies*, 18, 5, 539–560. DOI:10.1080/14616688.2016.1221985

Tsai, H., Yeung, S. and Yim, P.H.L. (2011) Hotel selection criteria used by mainland Chinese and foreign individual travelers to Hong Kong. *International Journal of*

Hospitality and Tourism Administration, 12, 252–267.

Tsang, E.W.K. and Yip, P.S.L. (2009) Competition, agglomeration, and performance of Beijing hotels. *The Service Industries Journal*, 29, 2, 155–171.

Twyoniak, S. (1998) Le modèle des ressources et des compétences: un nouveau paradigme pour le management stratégique? In: Laroche, H. and Nioche, J.-P. (eds) *Repenser la Stratégie*. Vuibert, Paris, pp. 166–204.

Ullman, E.L. (1940) A theory of location for cities, reprinted in Mayer, H.M. and Kohn, C.F. (eds) *Readings in Urban Geography*. University of Chicago Press, Chicago, IL.

Urtasun, A. and Gutiérrez, I. (2006) Hotel location in tourism cities: Madrid 1936–1998. *Annals of Tourism Research*, 33, 2, 382–402.

Uysal, M. and McDonald, C.D. (1989) Visitor segmentation by trip index. *Journal of Travel Research*, 27, 38–42.

van der Borg, J. (2008) Place marketing, governance and tourism development. Or how to design the perfect regional tourist board? *Working Paper*, 4, Department of Economics, Ca'Foscari University of Venice, Venice.

van der Zee, E. and Vanneste, D. (2015) Tourism networks unravelled: a review of the literature on network in tourism management studies. *Tourism Management Perspectives*, 15, 46–56.

Vanhove, N. (2005) *The Economics of Tourism Destinations*. Elsevier Butterworth-Heinemann, Oxford.

Varisco, C. (2004) El cluster turístico de Miramar. *Aportes y Transferencias*, 8, 2, 61–88.

Vera Rebollo, J.F. (1987) *Turismo y Urbanización en el Litoral Alicantino*. Instituto de Estudios Juan Gil-Albert, Alicante.

Viken, A. (2014) Destination discourses and the growth paradigm. In: Viken, A. and Granås, B. (eds) *Tourism Destination Development: Turns and Tactics*. Routledge, Abingdon, Oxon, pp. 21–46.

Viken, A. and Aarsaether, N. (2013) Transforming an iconic attraction into a diversified destination: the case of North Cape tourism. *Scandinavian Journal of Hospitality and Tourism*, 13, 1, 38–54. DOI:10.1080/15022250.2013.771994

Ville de La Grande Motte (2017) *La Grande Motte: Projet d'Aménagement et de Développement Durables*. Ville de La Grande Motte, La Grande Motte, Département de l'Hérault, France.

Violier, P. (2009) Introduction: destinations et territoires touristiques: co-présence à l'œuvre. In: Lemasson, J.-P. and Violier, P. (eds) *Destinations et Territoires: Co-Présence à l'Œuvre*. Téoros, Montréal/Presses de l'Université de Québec, Québec, pp. 1–3.

VisitBritain (2013) London and beyond. *Foresight*, 117 (July). Available at: https://www.visitbritain.org/sites/default/files/vb-corporate/markets/2013-7%20%E2%80%98Beyond%20London%E2%80%99%20Research_8.pdf (accessed 3 July 2020).

VisitBritain Research (2017) First-time and repeat visits to Britain. *Foresight*, 149. Available at: https://trade.visitbritain.com/wp-content/uploads/2017/07/foresight_149_-_frequency_of_inbound_tourism_to_britain_feb2017.pdf (accessed 2 July 2020).

Visit Florida (2016) *2015 Florida Visitor Study*. Visit Florida, Tallahassee, FL.

Vlès, V. (2010) Du moderne au pastiche: questionnement sur l'urbanisme des stations de ski et d'alpinisme. *Mondes du Tourisme*, 1, 39–48.

Vlès, V. (2012) Stations de ski en crise et construction territoriale en Catalogne française: la gestion intercommunale de la ressource touristique, un débat confisqué. *Revue de Géographie Alpine*, 100, 2.

von Friedrichs Grängsjö, Y. (2003) Destination networking: co-opetition in peripheral surroundings. *International Journal of Physical Distribution & Logistics Management*, 33, 5 427–448.

von Friedrichs Grängsjö, Y. and Gummesson, E. (2006) Hotel networks and social capital in destination marketing. *International Journal of Service Industry Management*, 17, 1, 58–75. DOI:10.1108/09564230610651589

Vu, H.Q., Li, G., Law, R. and Ye, B.H. (2015) Exploring the travel behaviors of inbound tourists to Hong Kong using geotagged photos. *Tourism Management*, 46, 222–232.

Wall, G. (1975) Form and function in British seaside resorts. *Society and Leisure*, 4, 217–226.

Wall, G. (1997) Tourism attractions: points, lines, and areas. *Annals of Tourism Research*, 24, 1, 240–243

Wang, D., Wang, L., Chen, T., Lu, L., Niu, Y. and Lew, A.A. (2016) HSR mechanisms and effects on the spatial structure of regional tourism in China. *Journal of Geographical Science*, 26, 12, 1725–1753. DOI:10.1007/s11442-016-1355-3

Wang, Y., Hutchinson, J., Okumus, F. and Naipaul, S. (2013) Collaborative marketing in a regional destination: evidence from Central Florida. *International Journal of Tourism Research*, 15, 3, 285–297.

Weaver, D. and Oppermann, M. (2000) *Tourism Management*. Wiley, Brisbane.

Weick, K. (2001) *Making Sense of the Organization*. Blackwell, Oxford.

Westermann, G., Pohle, B. and Sehl, I. (2005) Strategic management of tourism regions applying balanced scorecards. In: Alejiak, W. and Winiarski, R. (eds) *Tourism in Scientific Research*. Academy of Physical Education in Krakow, Krakow/University of Information Technology and Management in Rzeszow, Rzeszow, pp. 191–200.

Whitehand, J.W.R. (1992) *The Making of the Urban Landscape*. Blackwell, Oxford.

Wolfe, R.I. (1951) Summer cottagers in Ontario. *Economic Geography*, 27, 1, 10–32.

Wolfe, R.I. (1952) Wasaga Beach: the divorce from the geographic environment. *Canadian Geographer*, 1, 2, 57–65.

Woodside, G. and King, R.I. (2001) An updated model of travel and tourism purchase-consumption systems. *Journal of Travel and Tourism Marketing*, 10, 3–27.

Wozniak, M. (2002) Les stations de ski: quelles représentations des clientèles pour quel cadre bâti? *Revue de Géographie Alpine*, 90, 4, 17–31. DOI:10.3406/rga.2002.3098

Wu, C.-L. and Carson, D. (2008) Spatial and temporal tourist dispersal analysis in multiple destination travel. *Journal of Travel Research*, 46, 3, 311–317.

Xi, J., Wang X., Kong, Q. and Zhang, N. (2015) Spatial morphology evolution of rural settlements induced by tourism: a comparative study of three villages in Yesanpo tourism area, China. *Journal of Geographical Science*, 25, 4, 497–511. DOI:10.1007/s11442-015-1182-y

Xie, P.F., Chandra, V. and Gu, K. (2013) Morphological changes of coastal tourism: a case study of Denarau Island, Fiji. *Tourism Management Perspectives*, 5, 75–83.

Xu, S., Leung, Y.-F. and Barbieri, C. (2016) Characterizing themed touring routes: a geospatial and tourism evaluation of wine trails. *Tourism Planning & Development*, 13, 2, 168–184.

Yang, Y., Wong, K.K.F. and Wang, T. (2012) How do hotels choose their location? Evidence from hotels in Beijing. *International Journal of Hospitality Management*, 31, 3, 675–685.

Yang, Y., Mao, Z. and Tang, J. (2018) Understanding guest satisfaction with urban hotel location. *Journal of Travel Research*, 57, 2, 243–259.

Yavas, U. and Babakus, E. (2003) What do guests look for in a hotel? A multi-attribute approach. *Services Marketing Quarterly*, 25, 2, 1–9.

Yeoh, B.S.A. and Huang, S. (1996) The conservation–redevelopment dilemma in Singapore: the case of the Kampong Glam historic district. *Cities*, 13, 411–422.

Yokeno, N. (1968) La localisation de l'industrie touristique: application de l'analyse de Thunen-Weber. *Cahiers du Tourisme*, CHET, Aix-en-Provence, C–9.

Zaman, M., Botti, L. and Thanh, T. (2016) Weight of criteria in hotel selection: an empirical illustration based on TripAdvisor criteria. *European Journal of Tourism Research*, 13, 132–138.

Zehrer, A., Pechlaner, H. and Hölzl, B. (2005) The development of a destination management system (DMS) in South Tyrol. *Anatolia*, 16, 2, 147–161.

Zhang, H.Q., Qu, H. and Tang, V.M.Y. (2004) A case study of Hong Kong residents' outbound leisure travel. *Tourism Management*, 25, 2, 267–273.

Zhou, X., Xu, C. and Kimmons, B. (2015) Detecting tourism destinations using scalable geospatial analysis based on cloud computing platform. *Computers, Environment and Urban Systems*, 54, 144–153.

Zillinger, M. (2007) Tourist routes: a time-geographical approach on German car-tourists in Sweden. *Tourism Geographies*, 9, 1, 64–83. DOI:10.1080/14616680601092915

Zurick, D.N. (1992) Adventure travel and sustainable tourism in the peripheral economy of Nepal. *Annals of the Association of American Geographers*, 82, 4, 608–628.

Author Index

Page numbers in **bold** type refer to figures and tables. For reasons of space, only the first-named of multiple-author publications is included.

Subject Index

Page numbers in **bold** type refer to figures and tables. Under country entries, nation-wide topics are listed first, followed by regions and within those regions any specific places.

This book is published by **CABI**, an international not-for-profit organisation that improves people's lives worldwide by providing information and applying scientific expertise to solve problems in agriculture and the environment.

CABI is also a global publisher producing key scientific publications, including world renowned databases, as well as compendia, books, ebooks and full text electronic resources. We publish content in a wide range of subject areas including: agriculture and crop science / animal and veterinary sciences / ecology and conservation / environmental science / horticulture and plant sciences / human health, food science and nutrition / international development / leisure and tourism.

The profits from CABI's publishing activities enable us to work with farming communities around the world, supporting them as they battle with poor soil, invasive species and pests and diseases, to improve their livelihoods and help provide food for an ever growing population.

CABI is an international intergovernmental organisation, and we gratefully acknowledge the core financial support from our member countries (and lead agencies) including:

UKaid from the British people

Ministry of Agriculture People's Republic of China

Australian Government Australian Centre for International Agricultural Research

Agriculture and Agri-Food Canada

Ministry of Foreign Affairs of the Netherlands

Schweizerische Eidgenossenschaft Confédération suisse Confederazione Svizzera Confederaziun svizra Swiss Agency for Development and Cooperation SDC

Discover more

To read more about CABI's work, please visit: **www.cabi.org**

Browse our books at: **www.cabi.org/bookshop**,
or explore our online products at: **www.cabi.org/publishing-products**

Interested in writing for CABI? Find our author guidelines here:
www.cabi.org/publishing-products/information-for-authors/